Practical Handbook of
INDUSTRIAL TRAFFIC MANAGEMENT

Sixth Edition

Practical Handbook of
INDUSTRIAL TRAFFIC MANAGEMENT

Leon Wm. Morse

Previous editions by

Charles H. Wager
Richard C. Colton
Edmond S. Ward

Published by THE TRAFFIC SERVICE CORPORATION
Washington • New York • Chicago • Philadelphia
Boston • Atlanta • Palo Alto

© The Traffic Service Corporation, 1980
1435 G Street, N.W., Washington, D.C. 20005

All rights reserved. No part of the book may be reproduced without permission from the publisher, except for brief passages included in a review appearing in a newspaper or magazine.

Sixth Edition, Completely Revised and Reset

Library of Congress Catalog Card Number 77-9240

ISBN 0-877408-020-7

Printed in the United States of America

Produced by Stephen R. Hunter

To
my wife Goldie
for her patience and
understanding and love

. . . and to acknowledge the patience and encouragement of Jeff and Sue, Saul and Bob and Sandy, Josh, Aaron, Ben, Lisa and Eric, Morse's all.

But not to forget the spirit and willingness of many people who become reality when you seek the aids needed by one who undertakes this type of activity. I suspect that I may have forgotten someone so you will, I hope, understand if I've failed to name you . . . . I suspect my notes are lost somewhere.

Three old time friends who gave me moral support must be mentioned—in alphabetical order—Bernie James, Will Kominsky and Joe Queenan.

Assistance in various forms, some of which is evident in the book, was given by Jim Kelso of Flying Tigers, John F. Dubbleman of Intercontinental Transport (ICT) BV; Al Sobelman of Maersk Line Agency; John Pruitt of Eastern Lift Truck of Maple Shade, N.J. and Judith E. Bently of Allis-Chalmers; and Michael E. Curley of the Southern Railway System.

Lou Creekmur, Director of Customer and Community Relations of Ryder Truck Rental Inc.; Dr. John Doggett of the American Warehouseman's Association and J.R. Candvan of the Eastern Weighing and Inspection Bureau as well as the people at the Contract Carriers Conference and the Private Carrier Conference.

To Warren Blanding, author, lecturer, educator and entrepreneur, my particular thanks for introducing me to this project and my appreciation to Dick Coleman for his patience and to Steve Hunter for his help in producing this book and to Ken Marshall for his editorial guidance and finally to Doris Erb for her secretarial work.

I cannot disregard the ever so many who throughout the many years of my participation in traffic management and physical distribution have taught me.

Leon Wm. Morse

TABLE OF CONTENTS

Chapter   *Page*

Managing Transportation 1
    Functions 4
    Principles 5

II
Transportation Regulation 9
    Interstate Commerce Commission 21
    Revised I.C. Act 23
    State Commissions 27
    Federal Maritime Commission 28
    Civil Aeronautics Board 33
    Department of Transportation 33

III
Freight Classification 37
    Purpose 37
    R.R. Classification 39
    Motor Classification 42
    Water Classification 46
    Freight Forwarder Classification 46
    Using Classification 48
    Classification Rules 55

IV
Shipping Documents 67
    Bill of Lading 69
    Carrier's Duty 70
    Terms of Bill of Lading 72
    Preparing the Bill of Lading 74
    Ocean Bill of Lading 78
    Government Bill of Lading 78
    Air Waybill 79

R.R. Waybill 79
Freight Bills 79

V
Freight Rates 83
   Kinds of Rates 84
   Class Rates 84
   New England Scale 89
   Exception Rates 90
   Commodity Rates 91
   R.R. Regional Groups 93
   Motor Carrier Groups 94
   Accessorial Charges 95
   Freight Forwarder Rates 95
   Motor Carrier Rates 97
   How Published 98
   How Established 101
   How To Check Rates 107
   Auditing Freight Bills 109
   Sales Terms 111

VI
Freight Routing 117
   Right To Route 118
   Fundamentals 118
   By Rail 121
   Piggyback 127
   Motor 132
   Household Goods 137
   Freight Forwarder 139
   Parcel Post 140
   Water 142
   Domestic Air 145
   U.P.S. 148
   Pipe Line 149
   Consolidation 150

VII
Industrial Packaging & Material Handling 153
   Packaging. 153
   Loss and Damage 156
   Shipping Package 157
   Regulations 160
   Classification Rules 161
   Packaging Specifications. 162
   National Safe Transit Program. 168
   Export 170
   Material Handling for Distribution 172
   Loading 175
   Containerization 179

VIII
Claims 183
   Bill of Lading Terms 184
   Terms of Shipment 185
   Carrier v. Warehouseman's Liability 186
   Released Valuation 186
   Overcharge Claims 187
   Loss or Damage Claims 190
   Preparing Claims. 192
   Concealed Damage 196
   Transportation Arbitration Board 197
   Role of the I.C.C. 198
   Record Keeping 199

IX
Distribution & Warehousing 203
   Distribution 203
   Kinds of Warehouses. 204
   Role of Public Warehouse. 206
   Private Warehouse 211
   Location. 212

X
Contract/Private Carriage 219
- Private 219
- Contract 238
- The Contract 240

XI
Expediting and Tracing 245
- Carloads 248
- Truckloads 250
- Piggyback 251
- LTL 252
- Freight Forwarder 253
- Air Freight 253
- Maritime 254
- Small Packages 255

XII
Transportation Special Services 261
- Demurrage 261
- Average Agreement 266
- Records 269
- Detention 271
- Storage 272
- Transit Privileges 273
- Reconsignment & Diversion 275
- Sidetrack Agreement 276
- Weight Agreement 279

XIII
EDP and Transportation 285
- Questions 285
- Uses 286
- Languages 287
- STCC 288
- Freight Bill Payments 295

XIV
Managing The Traffic Department 297
- Functions 300
- Organization 301
- Operations 313
- Equipping 321
- Filing Tariffs 323
- Cooperative Functions 325
- Public Relations 331

XV
Government Traffic 337
- Military 339
- Overseas 342
- Air 342
- Documentation 343

XVI
Import-Export Traffic 347
- Ocean Carriers 349
- Ocean Freight Rates 351
- Insurance and Claims 359
- Documents 367
- Paying for Exports 376
- Drafts 379
- Foreign Freight Forwarders & Custom Brokers 387
- Import & Custom Duties 388
- Financing 392
- Shipping by Air 396
- Packaging 397

XVII
Shipping Hazardous Material 405
- The Law 406
- Compliance Program 410
- Exporting and Importing 413
- By Air 414

XVIII
Physical Distribution Management 417
    Definitions 418
    Organization 420
    The Traffic Manager in Physical Distribution 422
    Functions of Physical Distribution (Traffic Management, Inventory Control, Warehousing, Materials Handling, Packaging, Customer Service, and Site Location) 423

Appendix 1
Foreign Trade Glossary 429

Appendix 2
International Chamber of Commerce Documentary Credits 435

Appendix 3
American Foreign Trade Definitions 455

Appendix 4
Uniform Straight Bill of Lading 469

Appendix 5
Railroad Bill of Lading 481

Appendix 6
Ocean Bill of Lading 489

Appendix 7
A Selection of Traffic Publications 500

Appendix 8
Transportation Safety Act of 1974 509

Appendix 9
Hazardous Materials Definitions 521

Index 525

ILLUSTRATIONS

	Page
Government Executive Branch Organizational Chart	10
U.S. Department of Transportation Organization	11
Transportation Responsibilities in Executive Branch	12-13
U.S. Department of Transportation Responsibilities	14-15
Congressional Committees	16-17
U.S Regulatory Agencies	18-19
Interstate Commerce Committee	22
Federal Maritime Committee Organizational Chart	29
Civil Aeronautics Board Organizational Chart	32
Index to Articles U.F.C.	44
Bill of Lading Descriptions, U.F.C.	45
Released Valuations in N.M.F.C.	48
Uniform Straight Bill of Lading	71
Straight Bill of Lading—Short Form	75
Stop-Off Bill of Lading	76
Carrier's Signature	80
1900-1971 Class 100 Rail Rates	85
Excerpt Page of National Rate Base Tariff	86
Excerpt Class Tariff-Application of Rate Bases	87
Excerpt Class Tariff-Rate Base Numbers	88
New England Motor Rate Bureau Class Rate Excerpts	89
Reed-Bulwinkle Veto Message	99
Chart of Kinds of R.R. Rates	102
Procedure Chart on Rate Proposal	105
Auto Rack Railroad Car	124
Big Boy Box Car	125
Loading & Unloading Piggyback Trailers	128
Over The Road Tractor-Trailer	132
Double Bottoms On The Road	133
Container Ship Loading	134
Household Goods Road Unit	136
Inland Waterways Barges	143
Three Deck Ro-Ro Barge	144
D.C. 8 Freighter Loading	146

Air Line Container	147
Specifications Fibreboard Boxes	156
Certificate of Boxmaker	158
Box and Package Certificates	159
Package Rules	162
Container Characteristics	163
Handling Hazards	164
Rule 41 Equation	167
Fork Lift as a Tractor	172
Fork Lift with Push-Pull Attachment	173
Fork Lift with Telescoping Mast	174
Air Line Container	177
Pallet Patterns	178
Pallet Construction	178
Container Ship Loading	180
Container on Roller Deck in Airplane	181
Claim Form—Overcharge	189
Indemnity Agreement	190
Standard Loss & Damage Claim Form	194
Industrial Loss and Damage Claim Form	195
Claim Record	200
Non-Negotiable Warehouse Receipt	209
Warehouse Service Specifications	210
Short Form Warehouse Bill of Lading	211
Truck Lease	222-228
Truck Lease-Schedule A	229-230
Private Carriage Cost Worksheet	234
Private Carriage Cost Analysis	235
Demurrage Calendar	263
Demurrage Rates	264
Page of Demurrage Tariff	267
Page of Demurrage Tariff	268
Weight Agreement	280
Bureau Identification	281
U.F.C. Page with STCC Numbers	289
Railroad Car Location System	292
Railroad System Tied Into Shipper System	293

Traffic Organization in Large Corporation 306
Distribution Department in a Large Corporation 309
Traffic Organization in a Small Corporation 310
Industrial Traffic Report Form 327
Government Bill of Lading................................ 345
Ocean Freight Conference Title Page 352
Ocean Freight Conference Tariff Page 353
Ocean Freight Conference Analysis Form................... 357
Ocean Dock Receipt 369
Ocean Bill of Lading...................................... 371
Ocean Arrival Notice 373
Certificate of Origin 374
Shipper's Export Declaration 375
Bank Draft.. 380
Letter of Credit Application............................... 382
Irrevocable Letter of Credit 383
Bank Notice to Shipper 385
Shippers Letter of Instructions............................ 389
Shippers Letter of Instructions............................ 390
Container Ship 402
Ro-Ro Vessel Loading.................................... 403
Hazardous Material Table 408-409
Hazardous Material Placards 416
Physical Distribution Functions Chart 420
Physical Distribution Chart of Staff Functions.............. 421
Physical Distribution Chart of Line Positions............... 421
Physical Distribution Combined Line & Staff Chart 421

Practical Handbook of

INDUSTRIAL TRAFFIC MANAGEMENT

1

MANAGING TRANSPORTATION

Very little of what is produced in this country is consumed where it is grown, mined or manufactured. To have value, it must first be transported to other, wider markets. As products move in commerce from the basic raw material to the ultimate consumer, across town or half way around the world, decisions must be taken that involve transportation. Taking those decisions constitutes the profession of traffic management.

Traffic management absorbs about one third of the production costs of most commodities; and of the total costs of physical distribution—transportation, warehousing, packaging, materials handling, and inventory control—transportation accounts for about half.

Whether a product moves by rail, truck, water, air or pipeline may be ultimately dictated by its nature. But in the many instances where a choice exists, it is the traffic manager, exercising knowledge, management skills, and understanding of the customer's service needs, who must make the decision.

The range of options is large: What mode of transportation? Private, contract, common or charter carriage? What type of packaging will best protect this product in transit? Where can we best locate our warehouses or distribution centers? Will we use stop-off privileges? What are the most economical transportation quantities for shipment of this product? Under what terms of sale will it be offered? What kind of equipment is needed to transport it? The complexities increase when the traffic manager's responsibilities are broadened to take in the scheduling of inbound raw materials to co-

INDUSTRIAL TRAFFIC MANAGEMENT

ordinate with manufacturing "runs," and other new concepts, new technology, new business and the new costs of all this activity. Even "politics" must be taken into account for its effect on traffic management and transportation.

Indeed, the complexity of the transportation industry and of the many rules and regulations surrounding the purchase and use of transportation is such that getting the most for a company's transportation dollar involves much more than "getting the best rate." Indeed, the many radical changes in the transportation industry in recent years, such as new price levels, new concepts of setting them, increased world-wide trade, the application of electronic data processing techniques, the push toward lower inventories and exciting new transportation services and shipping practices, have made it almost essential for companies of any size to retain competent traffic management, if only to keep abreast of these changes and to measure and control their effect on buying and marketing practices. *An industrial traffic department, given sufficient authority and latitude with which to carry out its ideas, can guarantee the ultimate in value for every dollar paid out for transportation and increase the profit picture of the corporation.*

An important specific objective of every purchaser of transportation is to satisfy *service* requirements at the lowest possible cost. That is to say, no traffic manager worth the title would arrange to ship a vital and needed electronic part by a comparatively slow water service, merely because of low rates; nor would he get "caught short" and order several hundreds of tons of industrial sand shipped by a premium transportation service. The added cost of crating and packing shipments for low-cost transportation might be greater than the added cost of shipping the same products with little or no crating via premium transportation. Similarly, there are conditions where part of a shipment might be sent via premium transportation in order to keep a production line going, while the rest might be sent by slower and cheaper methods and still arrive in time. Then there is the potential of high-speed, premium transportation for reducing over-all costs by lessening warehouse and inventory requirements and thus freeing capital that might otherwise have been tied up in both warehouse space and large inventories.

MANAGING TRANSPORTATION

While there can be no hard and fast rules about managing transportation that will apply for every company, there is no doubt that the function requires an individual who is both trained and skilled and is granted the authority to formulate a transportation policy that is suited to his company's requirements and *is binding upon the company's other departments.*

For example, it should not be the production department's decision as to what *form* of transportation will overcome a supplier's failure to ship on the date specified; the factory expediter should not dictate the form of expenditure of extra transportation dollars to make up for delays in production schedules; nor should the sales department necessarily have the last word as to the *routing* to be used to preserve its customer relations. The tendency of other departments out of touch with day-to-day domestic and world-wide transportation services and facilities is frequently to demand the most expensive service without regard for the possibility that less costly alternatives could achieve the desired ends. The responsibility for those decisions should devolve on an organization that is equipped to consider all factors and make its choice accordingly. It is an essential function of a traffic department.

Today's traffic manager is a far cry from the glorified shipping clerk who so often misused the title in years past. The complexity of the field of *physical distribution* requires persons of many talents.

The traffic administrator should operate on a level high enough to allow him a voice among those who set his company's over-all policies and to permit his decisions to be effective. He must be well versed in transportation law and the requirements of the various governmental regulatory bodies, not only to keep his company from violation of regulations, but to protect fully its rights under the law. More and more he is expected to handle overseas shipments entailing complicated routing, documentation, etc. And because the details of his job may not be too well known to his management, he must be a *self-starter*, with resourcefulness, a creative imagination, a nose for research, and the ability to *sell* his ideas and services. He must have the type of personality that will enable him to deal effectively with both subordinates and superiors and at the same time be a credit to his company from a public relations standpoint. As will be developed in subsequent chapters, he must be equipped to supervise and perform the following clerical and managerial traffic functions:

Clerical functions

- Audit of freight bills
- Claims
- Classification
- Demurrage
- Diversion and reconsignment
- Expediting and tracing
- Household goods moving
- Passenger reservations
- Rate and routing guides
- Rate quotations
- Routing
- Stop-off cars and pooling

Managerial functions

- Classification adjustments
- Collaboration with packing and materials handling engineers
- Consolidations
- Contracts
- Credit arrangements
- Demurrage, siding and weight agreements
- Distribution and warehouse studies
- Economical distribution
- Insurance
- Intraplant and interplant company truck operations
- Inventory control
- Location of new plants
- Packaging and materials handling studies
- Private motor carriage operations
- Public relations with carriers
- Rate adjustments
- Relations with other company departments such as accounting, legal, manufacturing, purchasing, receiving, sales, shipping, warehousing
- Tonnage distribution policy
- Transit privileges

Not every company can support a traffic department of the size and scope generally indicated in this book. But even if a company is too small to justify even one trained traffic professional there is no need for it to be deprived of professional traffic services. From one end of the country to the other there are traffic consultant firms or individual traffic consultants. They will assume any or all of these functions on a fee basis. Chambers of commerce and manufacturers' associations usually provide certain transportation services and counsel. *The point is that skilled counsel in the many aspects of traffic management is a good investment!*

Well operated traffic departments can not only effect substantial savings by efficient handling of incoming and outgoing shipments but can also be instrumental in widening company markets, opening up new sources of supply, and building customer good-will, thus lowering costs and increasing sales. Too many industrial executives lack a clear understanding of what can be expected of a traffic department. As a result, they deprive themselves of experienced counsel on a subject that can have a profound effect on company profits.

The Principles of Traffic Management

No student of transportation, nor any individual in the industrial traffic department, should approach the procurement of transportation without a sound knowledge of the basic principles of the common carrier transportation system.

The first of these is that carriers may not discriminate. They are obligated to offer the same services to small shippers as to large shippers, and the same rates, although the equivalents of volume discounts are permissible in much the same way they are in any business. Another important aspect is that carriers may not establish and discontinue services at will; in other words, a carrier may not discontinue unprofitable operations in order to concentrate on profitable ones. Although this principle has been questioned by advocates of "deregulation" and "increased competition" in transportation, it has, for the most part, been held that in the granting of rights to a carrier, the volume of business in its profitable operations will be balanced to compensate for possible losses in its unprofitable services, and that therefore the same scales of rates must apply for shippers in remote and out-of-the-way places as for those in the large urban centers. In other words, the right of a carrier to engage in interstate transportation carries with it certain obligations which have as their net result that large and small shippers, no matter where they may be located, shall receive equal treatment and equal rates.

INDUSTRIAL TRAFFIC MANAGEMENT

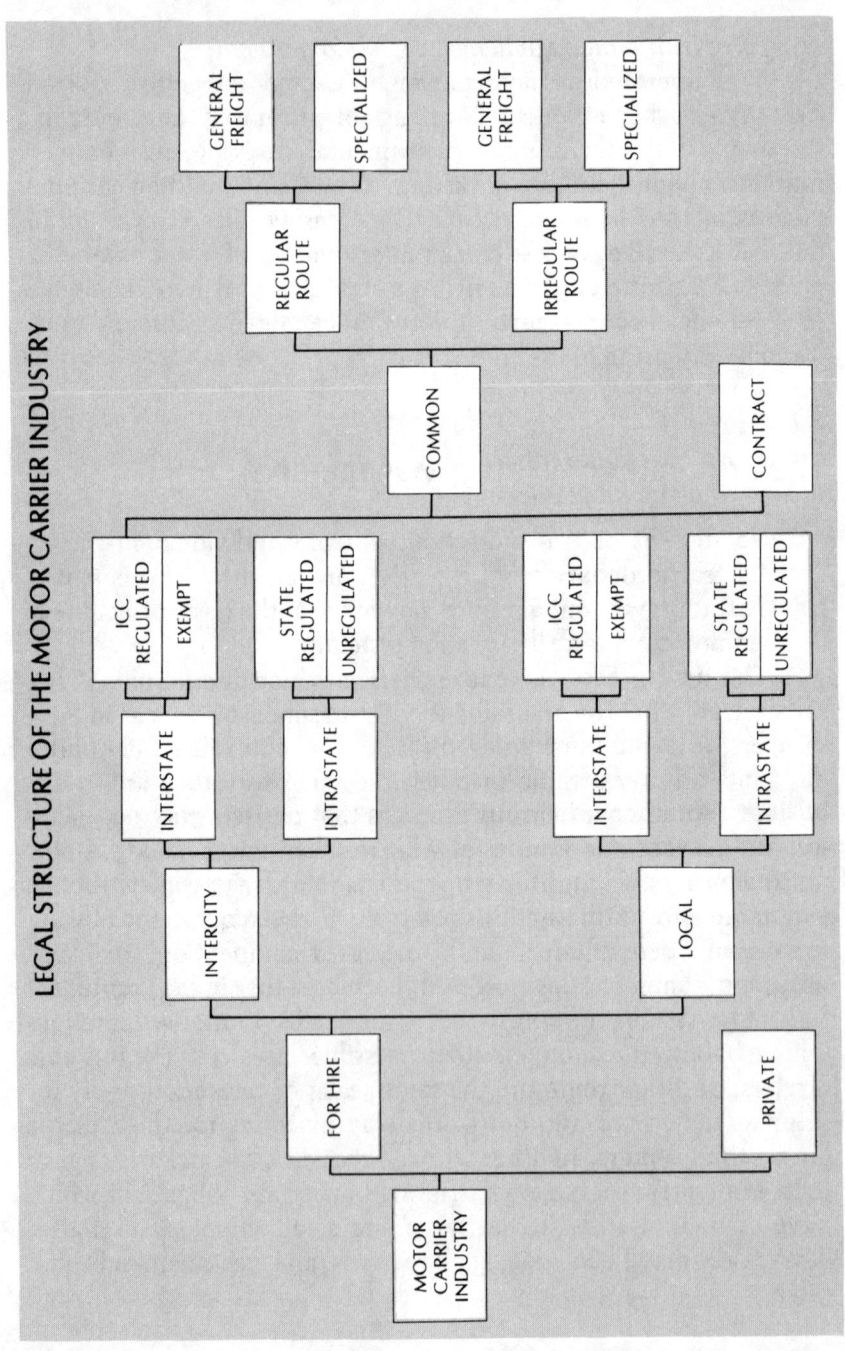

6

MANAGING TRANSPORTATION

A second important concept is that of *inherent advantages.* This is no more than saying that each form of common carrier transportation has its distinct advantages for shippers in terms of speed, capacity, flexibility or cost. At the two extremes of the scale would be the air carriers, with tremendous speed but limited load-carrying capacity and relatively high rates, on the one hand, and the water carriers, with their tremendous capacity, much lower rates and longer transit intervals, on the other. The increasing use of pipelines for transportation of liquids and semi-solids adds another dimension to the competitive picture.

The transportation complex is changing constantly, and it is essential to be well informed. For example, consider private motor carriage. It has made tremendous inroads into the common carrier world primarily for service reasons and therefore economic reasons. It has made it necessary for the railroads and common motor carriers to depart from the age-old rate making bases and operating methods that have lost much of the traffic to private carriage. Common carriers must compete, and they now quote all-commodity and specific commodity rates to meet the competition of private carriage. Developing or pioneering rural areas really are aided by private carriage, since there often is not sufficient cargo volume to support regularly scheduled common carrier service.

Contentions are heard that private carriage takes the cream of the cargo, and this is not to be denied completely. Too often, however, losses of important tonnage by common carriers occur for reasons known to the carriers to be resting at their own door steps. Such reasons are similar in many ways to the reasons for loss by an industrial company of particular pieces of its business.

A threat to profitable operations of the nation's common carrier system exists, and has existed for some time, in the transportation of freight by motor carriers operating in ways that are questionable or are not authorized under the law. It has been estimated by the Interstate Commerce Commission that millions of dollars in freight revenue are annually diverted from the common carrier system by such operators. Their activities persist and appear to have their origins in: (1) The complexity of the governing laws and attendant confusion as to what is or is not permissible and (2) pressures to reduce transportation costs, which lead to unauthorized steps in the attempts to accomplish the reductions. Dangers of fines and other penalties are, of course, ever present in these matters. Basically, however, if the common carrier system is to realize its full potential as a dynamic force in the economy, it must have the full cooperation of every shipper and receiver of freight. They must differentiate between the legal and the unauthorized and steadfastly refuse to be drawn into situations that are not proper or are not recognized under the law.

Various leaders in transportation have expressed concern about the future of the common carrier transportation system. It has stood us in good stead. Many features of our transportation policy were formulated before trucks and airplanes were considered carriers. Modernizing the legal structure and its anachronisms is needed but will take time. The process is not an easy one. In the years ahead we shall undoubtedly see more attempts to change the common carrier system in the interests of our national economy and defense, and in a continuing effort to avoid government ownership of transportation facilities.

The transportation and transportation-supply industries employ millions of persons in the United States and pay billions each year for wages, equipment, and supplies as well as for local, state and federal taxes. The additional billions of private dollars invested in transportation plant and equipment, plus the know-how and superlative capabilities of those participating have given the United States a transportation system that is unequalled anywhere in the world. While the common carrier system is by no means perfect, it is generally in the best interest of shippers to take advantage of its many potentialities for maintaining the high standards of living and business that are possible under free enterprise!

2

TRANSPORTATION REGULATION

Regulation of industry in the United States, in the sense the word is understood today, had its origin in 1887 when Congress passed the Act to Regulate Commerce and established the Interstate Commerce Commission to perform that function.

The economics and technology of transportation have changed over the years. Recognizing that a transportation system made up of strong carriers in each of the various modes is indispensable to the nation's economy and defense, Congress, aided by recommendations of various parties in interest, from time to time amends the laws and enacts new ones designed to maintain that strength in the face of changing conditions.

The Civil Aeronautics Act of 1938 established the Civil Aeronautics Authority (which later became the Civil Aeronautics Board—CAB) and the Merchant Marine Act of 1936 gives regulatory authority to the Federal Maritime Commission (FMC).

The Department of Transportation (DOT), headed by a Secretary of cabinet status, came into existence in 1967. Some of the functions of the Congressionally-established agencies—such as those of the ICC for safety—were transferred to the DOT when it was created. The DOT, however, is an "arm" of the executive branch of government, while the ICC, FMC, and CAB are "arms of Congress."

The DOT, the only transportation agency headed by a cabinet-level official, has a considerably larger sphere of responsibilities than economic regulation, while the ICC, FMC and CAB

INDUSTRIAL TRAFFIC MANAGEMENT

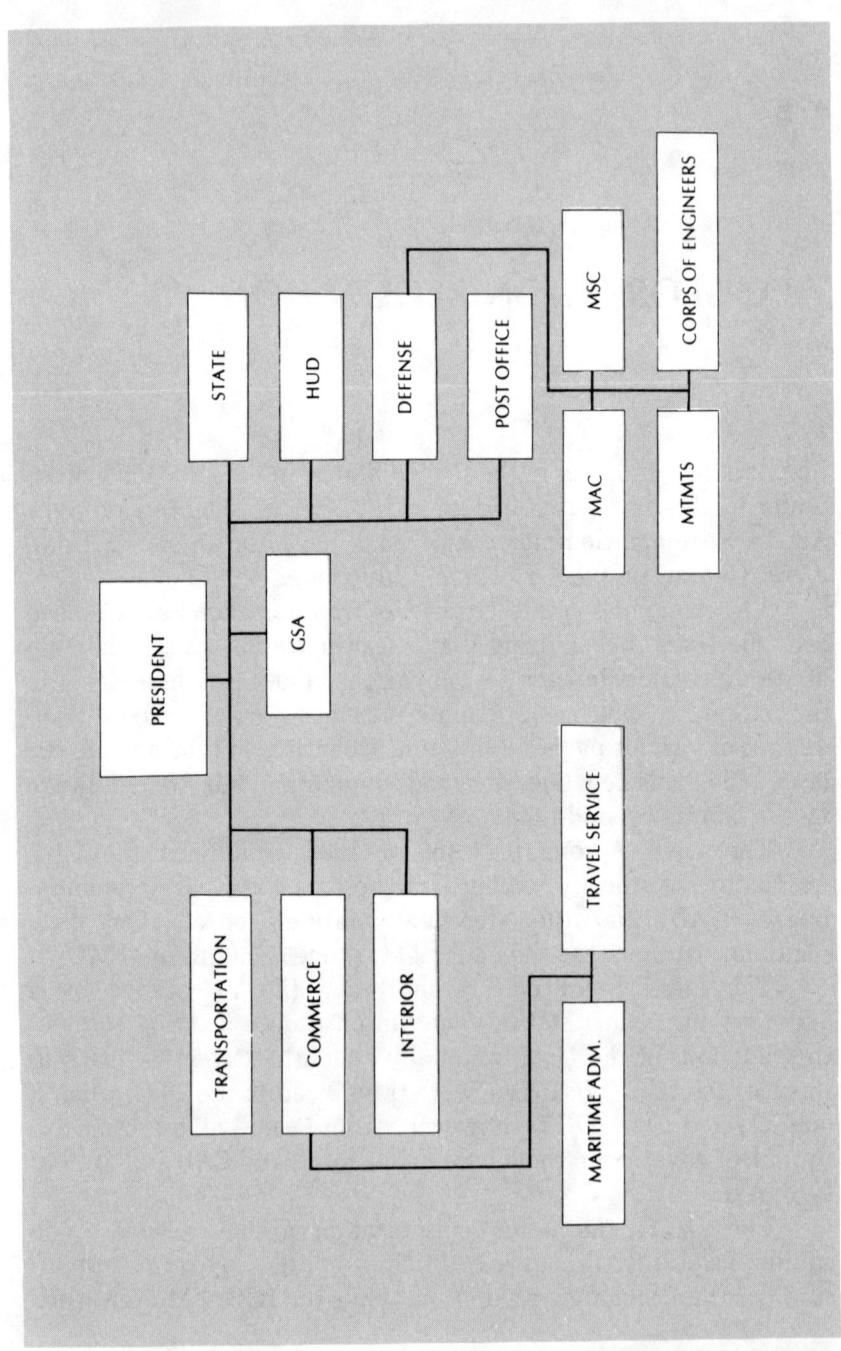

TRANSPORTATION REGULATION

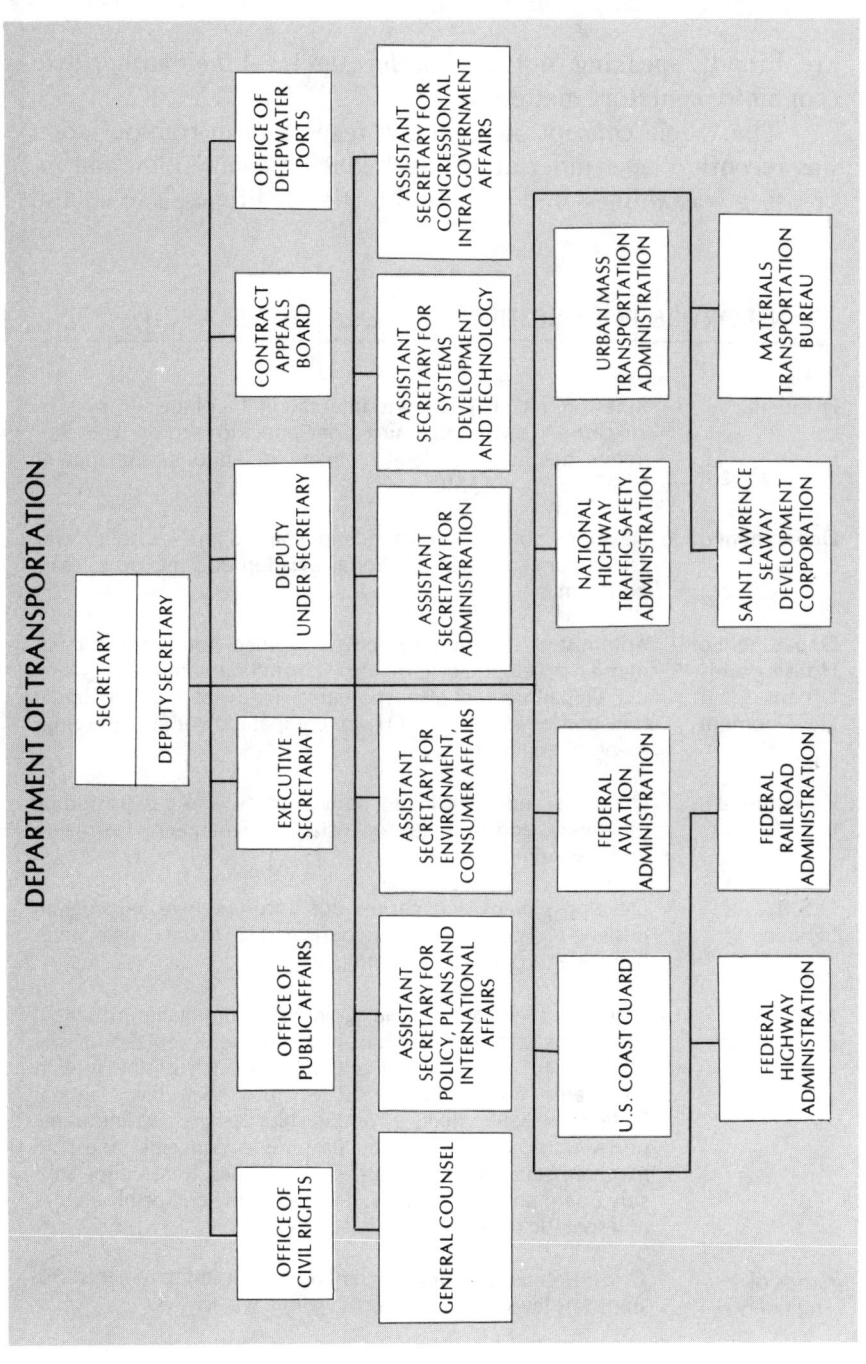

INDUSTRIAL TRAFFIC MANAGEMENT

are, broadly speaking, restricted in the exercise of their authority to economic regulatory matters.

The whole concept of economic regulation in transportation has recently come under attack as "reducing competition and increasing wastefulness and pricing distortions." Proposals to abolish

TRANSPORT RESPONSIBILITIES IN THE EXECUTIVE BRANCH

President	Rules on matters relating to international air transport by U.S. air carriers and foreign air carrier operations to the U.S. Appoints members of federal agencies and appoints chairman of I.C.C., CAB and FMC.
Department of State	Develops policy recommendations and approves policy programs concerning international aviation and maritime transportation.
Department of Housing and Urban Development	Administers a variety of federally aided housing and community development programs, consulting with and advising the Department of Transportation in order that the urban transport programs of DOT are compatible with the development programs of HUD.
Department of Interior	Develops and coordinates policy for oil and gas (including pipelines), and provides for a standby Emergency Petroleum and Gas Administration.
U.S. Travel Service	Develops, plans and carries out a comprehensive program designed to stimulate and encourage travel to the United States by residents of foreign countries.
Maritime Administration	Promotes merchant marine; grants ship mortgage insurance; determines ship requirements, ocean services, routes and lines essential for development and maintenance of the foreign commerce of the United States; maintains the National Defense Reserve Fleet; develops ship designs, marine transportation systems, advanced propulsion concepts, and ship mechanization and management techniques. Its Maritime Subsidy Board awards subsidies, determines the degree of services and specific routes of subsidized operators.
Corps of Engineers	Constructs and maintains river and harbor improvements. Administers laws for protecting navigable waterways.

TRANSPORTATION REGULATION

the principal transportation regulatory agencies have been put forward by the Administration and received sympathetic hearing and action in the Congress. One agency, the CAB, is being dismantled and will be eliminated, unless Congress has second thoughts. Similar measures to abolish or greatly reduce the powers of the ICC

General Services Administration	Develops and operates transportation programs within the Federal Government; provides and procures transportation services; develops and implements procedures for improving motor equipment management, operation and rehabilitation programs of the Federal Government including assigning, regulating or performing the operation of interagency motor pools and motor transport systems.
Military Sealift Command	Provides ocean transportation for personnel and cargo of the Department of Defense and, as directed, for other agencies and departments of the United States. Also operates ships in support of scientific projects and other programs of the Federal Government.
Military Airlift Command	Provides air transportation for personnel and cargo for all the military services on a worldwide basis; in addition furnishes weather, rescue and photographic and charting services for the Air Force.
Military Traffic Management and Terminal Service	Directs military traffic management, land transportation, and common-user ocean terminal service within the United States, and for worldwide traffic management of the DOD household goods moving and storage program. Provides for the procurement and use of freight and passenger transportation service from commercial for-hire transportation companies operating between points in the continental U.S., except for long-term contract air-lift service.
Post Office Department	Establishes and administers policies, programs, regulations, and procedures governing the procurement and utilization and transportation of mail and mail equipment, including policies governing the distribution, routing and dispatch of mail both foreign and domestic.

From 8th edition of *Transport Facts & Trends*, published by Transportation Association of America.

INDUSTRIAL TRAFFIC MANAGEMENT

DEPARTMENT OF TRANSPORTATION

Secretary of Transportation	Under direction of the President, exercises and provides leadership in transportation matters, develops national transportation policies and programs, including compliance with safety laws pertaining to all modes of transport.
National Transportation Safety Board	Determines and reports causes, facts and circumstances relating to transportation accidents, reviews on appeal the revocation, suspension or denial of any certificate or license issued by the Department, and exercises all functions relating to aircraft accident investigations.
General Counsel	Legal services, including the legal aspects and drafting of legislation.
Asst. Secy for Safety and Consumer Affairs	Safety coordination; regulation of movement of hazardous materials; pipeline safety regulations; and representation of the public viewpoint, in the Department, with regard to transportation matters.
Asst. Secy. for Policy and International Affairs	Economic and systems analysis; policy review; transport data; international transport facilitation; and, technical assistance.
Asst. Secy. for Environment and Urban Systems	Coordinate policies, programs, and resources of DOT transport program with public and private efforts to solve environmental problems having an impact on transportation.
Asst. Secy. for Systems Development and Technology	Scientific and technologic research and development relating to the speed, safety and economy of transportation; noise abatement; and, transportation information planning.
Asst. Secy. for Administration	Organization, budgeting, staffing, personnel management, logistics and procurement policy, management systems and other administrative support for the Department.
U.S. Coast Guard	Provides navigational aids to inland and offshore water and trans-oceanic air commerce; enforces federal maritime safety, including approval of plans for vessel construction and repair. Administers Great Lakes Pilotage Act of 1960. Has responsibility for water vessel anchorages, drawbridge operation, and locations and clearances of bridges over navigable water. (Previously under the Corps of Engineers).

TRANSPORTATION REGULATION

Federal Aviation Administration	Promotes civil aviation generally, including research and promulgation and enforcement of safety regulations. Develops and operates the airways, including facilities. Administers the federal airport program.
Federal Highway Administration	Responsible for implementation of the Federal-Aid Highway Program; National Traffic and Motor Vehicle Safety Act of 1966; and the Highway Safety Act of 1966. Responsibility for reasonableness of tolls on bridges over navigable waters (previously under the Corps of Engineers). Administers federal highway construction, research planning, safety programs, and Federal-Aid highway funds (formerly under Bureau of Public Roads).
Federal Railroad Administration	Responsible for the operation of the Alaska Railroad; administration of the High-Speed Ground Transportation Program; implementation of railroad laws; and advises the Secretary on matters pertaining to national railroad policy developments.
St. Lawrence Seaway Dev. Corp.	Administers operation and maintenance of the U.S. portion of the St. Lawrence Seaway, including toll rates.
Urban Mass Transportation Administration	Responsible for developing comprehensive coordinated mass transport systems for metropolitan and urban areas, including R & D and demonstration projects; aid for technical studies, planning, engineering, and designing; financial aid and grants to public bodies for modernization, equipment, and training of personnel.
National Highway Traffic Safety Administration	Formulation and promulgation of programs for use by the States in driver performance; development of uniform standards for keeping accident records and investigation of accident causes; vehicle registration and inspection and the safety aspects of highway design and maintenance. Planning, development and enforcement of federal motor vehicle safety standards relating to the manufacturing of motor vehicles.
Bureau of Motor Carrier Safety	Administers and enforces motor carrier safety regulations (formerly under the ICC) and the regulations governing the transportation of hazardous materials.

From 8th edition of *Transport Facts & Trends,* published by Transportation Association of America.

INDUSTRIAL TRAFFIC MANAGEMENT

have been introduced. In the meantime, however, the ICC and the FMC continue as the quasi-judicial, quasi-legislative regulators of freight transportation by land and water.

The charts that appear here describe briefly the regulatory functions of the various federal agencies, the areas of federal assistance, and the departments working in particular areas. They

STANDING CONGRESSIONAL COMMITTEES HAVING JURISDICTION OVER

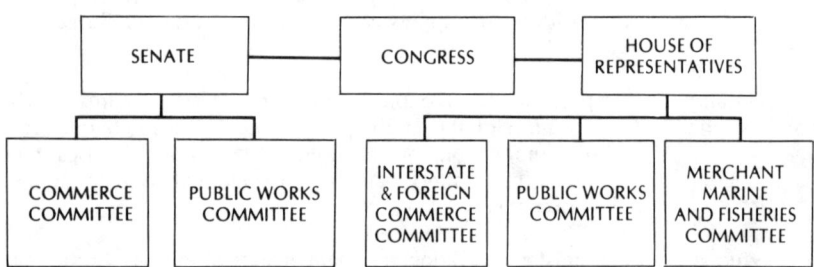

The present standing committee structure of Congress includes several committees which have both direct and indirect jurisdiction over policies affecting the transportation industry. The organizational chart above indicates the positions of these committees within the structure of the Congress. The table of functions points up the major areas of the transportation industry covered by the various committees, as well as the similarity of their respective jurisdiction.

Legislation affection transportation also comes within the jurisdiction of several other Congressional committees, such as:

Appropriations	Actual appropriation of funds.
Government Operations	Transportation operations of Federal agencies.
Judiciary	Rules and procedures for regulatory agencies.
Post Office & Civil Service	Parcel post service.
House Ways & Means	Financial matters, with House
Senate Finance	Committee originating all tax bills.
House Education & Labor	Transport labor generally, in-
Senate Labor & Public Welfare	cluding mediation or arbitration of disputes. (Note House I & FC Committee's jurisdiction over rail labor.)

16

TRANSPORTATION REGULATION

provide thumbnail sketches of agency powers and responsibilities. For convenience they are broken out to the executive and to the legislative branches.

The charts on pages 12 and 13 depict the transport responsibilities in the executive branch. The Post Office, Military Service Management, Corps of Engineers, U.S. Travel Service and others

TRANSPORTATION

SENATE COMMITTEE	HOUSE COMMITTEE
Commerce	**Interstate & Foreign Commerce**
Regulation of interstate railroads, buses, trucks, pipe lines, freight forwarders, domestic water carriers and domestic and international air carriers. Inland waterways. Promotion of civil aviation, including subsidies and airport construction.	Same basic jurisdiction, plus railroad labor and retirement and unemployment. Handles ICC regulated water transportation and domestic inland water transportation in general.
International water carriers generally, including registering and licensing of vessels and small boats; navigation and the laws relating thereto; measures relating to subsidies; and the inspection and safety of vessels. Approve Maritime Administrative programs.	**Merchant Marine & Fisheries**
	Same basic jurisdiction, plus unregulated domestic ocean-going water transportation. Approve Maritime Administration programs.
Public Works	**Public Works**
Projects for the benefit of water navigation and rivers and harbors. Measures relating to the construction or maintenance of highways.	Same basic jurisdiction.

From 8th edition of *Transport Facts & Trends,* published by Transportation Association of America.

17

INDUSTRIAL TRAFFIC MANAGEMENT

with their functions are included here, and those functions are quite substantial.

The chart on pages 14 and 15 show the organization and responsibilities of the Department of Transportation. Of particular

FEDERAL TRANSPORTATION REGULATORY AGENCIES

The Federal transportation regulatory agencies are arms of the legislative branch of the Government. They are not courts, but they do have recourse to the courts in order to enforce their orders. They exercise quasi-judicial powers as well as quasi-legislative powers. Their members are appointed by the President with Senate approval at salaries of $38,000 with chairmen receiving an additional $2,000. Not more than a majority of one can be from any one political party.

INTERSTATE COMMERCE COMMISSION

The ICC was created in 1887 by the Act to Regulate Commerce. It currently consists of nine members who serve terms of seven years. Its Chairman is appointed by the President, and its Vice Chairman is elected annually by the members. The following table indicates the types of domestic interstate carriers over which the Commission has economic jurisdiction, as well as its other major functions.

Modes Regulated	Major Functions
Railroads (1887), Express Companies, Sleeping Car Companies (1906)	Regulates, in varying degrees by mode of transport, surface carrier operations, including rates, routes, operating rights, abandonment and mergers; conducts investigations and awards damages where applicable and administers railroad bankruptcy. Prescribes uniform system of accounts and records, evaluates property owned or used by carriers subject to the Act; authorizes issuance of securities or assumption of obligations by carriers by railroad and certain common or contract carriers by motor vehicle.
Oil Pipe Lines (1906)*—Common carriers only	
Motor Carriers (1935)	
Private carriers and carriers of agricultural commodities exclusively are exempt, as are motor vehicles used by farm co-ops.	
Water Carriers (1940), Water carriers operating coastwise, inter-coastal, and on inland waters of the U.S.	
Private carriers and carriers of liquid bulk or three or less dry bulk commodities in a single vessel are exempt.	
Freight Forwarders (1942)	Develops preparedness programs covering rail, motor, and inland waterways utilization.
Non-profit shippers' associations exempt.	
*Gas Pipelines regulated by Federal Power Commission.	The Act was codified by Congress in 1978.

TRANSPORTATION REGULATION

note are the responsibilities for safety, systems development and technology, and urban mass transportation. The latter carries with it substantial burdens. Further discussions of the Department appear in later pages with particular emphasis on its regulatory ac-

---

CIVIL AERONAUTICS BOARD
The CAB, as it exists today, is an outgrowth of the Civil Aeronautics Act of 1938, Presidential Reorganization Plans of 1940, and the Federal Aviation Act of 1958. There are five Board members, each serving terms of six years. The Chairman and Vice Chairman are appointed annually by the President.

Regulates	Major Functions
U.S. Domestic and international air carriers. Foreign air carrier operations to, from, and within the U.S.	Regulates carrier operations, including rates, routes, operating rights, and mergers; determines and grants subsidies. Assists in the development of international air transport, and grants, subject to Presidential approval, foreign operating certificates to U.S. carriers and U.S. operating permits to foreign carriers.

FEDERAL MARITIME COMMISSION
The present FMC was established by Presidential Reorganization Plan 7 of 1961, although most of its regulatory powers are similar to those granted its predecessor agencies by the Shipping Act of 1916 and subsequent statutes. The Commission consists of five members appointed to four-year terms by the President with Senate approval. The President designates the Chairman. The Vice Chairman is elected annually by the members.

Regulates	Major Functions
American-flag vessels operating in the foreign commerce of the U.S. and common carriers by water operating in domestic trade to points beyond the continental U.S.	Regulates services, practices, and agreements of water common carriers in international trade. Regulates rates and practices of water common carriers operating in domestic trade to points beyond continental U.S.

*Note entry and route-designation functions of Maritime Subsidy Board on chart of executive branch.

From 8th edition of *Transport Facts & Trends*, published by Transportation Association of America.

tivities in the safety field and its systems development and related work.

Moving from the executive branch to the legislative, the charts show the standing congressional committees having jurisdiction over transportation. Of these, the Senate has two and the House three. The Commerce Committee of the Senate along with the Interstate and Foreign Commerce and the Merchant Marine and Fisheries Committees of the House have been particularly important. The Post Office, Government Operations, House Ways and Means, Senate Finance, and other which are shown in the lower section have also been involved in substantial ways in industrial transportation.

The federal transportation regulatory agencies that are arms of the legislative branch are shown on the following pages. These are the ICC, CAB, and FMC. Individual organization charts are shown for them later and additional information on their regulatory functions and activities are reviewed. The ICC has been the body with which most industrial traffic people have been involved through the years and it continues to be probably the most important arbiter of freight rates and practices for the substantial majority of transportation and traffic people.

As a part of moving to review in more detail governmental regulative activities it should be borne in mind that the Shipping Act of 1916, the Reed-Bulwinkle Act and other laws give carriers the right to set prices, rules and regulations collectively under anti-trust exemption. The regulatory philosophy considers this warranted in that it helps preserve the carriers' economic well-being and enables them to avoid price cutting and other practices which might weaken the system. But this justification has been increasingly attacked in recent years and proposals have been put before Congress to remove the rail and motor carriers' anti-trust immunity.

At the same time, there are laws that protect the users of transportation from abuses which might arise from carriers' actions, collective or otherwise. References to the various provisions of these laws appear throughout this book. High on the list of every traffic manager's responsibilities is a thorough understanding of his company's rights and limitations under the laws affecting trans-

TRANSPORTATION REGULATION

portation. A current copy of the Interstate Commerce Act should be available for his ready reference.

For the rest of this chapter we shall explore the regulatory roles of the government bodies in the administration of the law, as well as clarify the part played by the Interstate Commerce Commission, Federal Maritime Commission and the Civil Aeronautics Board in the establishment of charges for transportation.

The Interstate Commerce Commission

The establishment of the ICC in 1887 dates it as the oldest of the administrative agencies. Its board of commissioners is more than twice the size of comparable groups for the CAB and FMC. Their seven year terms also are longer.

The "Act to Regulate Commerce," under which the ICC was established, has been amended many times since its original passage and is now known as the "Interstate Commerce Act." The four major parts of the Act parallel each other in their protection of the shipping public.

Part I, covering railroads and pipe lines

Part II, added 1935, covering motor carriers

Part III, added 1940, covering water carriers

Part IV, added 1942, covering freight forwarders

The ICC may delegate certain duties and functions to individual commissioners or boards consisting of not less than three eligible employees, but the entire Commission acts on matters of national transportation importance.

The ICC's three divisions (see chart on page 22) function as "courts" of appeal for petitions for reconsideration or rehearing of decisions made by commissioners or boards of employees.

In broad terms, and within prescribed legal limits, the ICC's regulatory authority encompasses both economics and service. In the economic area, the commission settles controversies over rates and charges between competing and like modes of transportation, and is involved in accounting procedures, reparations, destructive competition, use, control and supply of railroad equipment, rebating and other matters. In the service area the commission is in-

INDUSTRIAL TRAFFIC MANAGEMENT

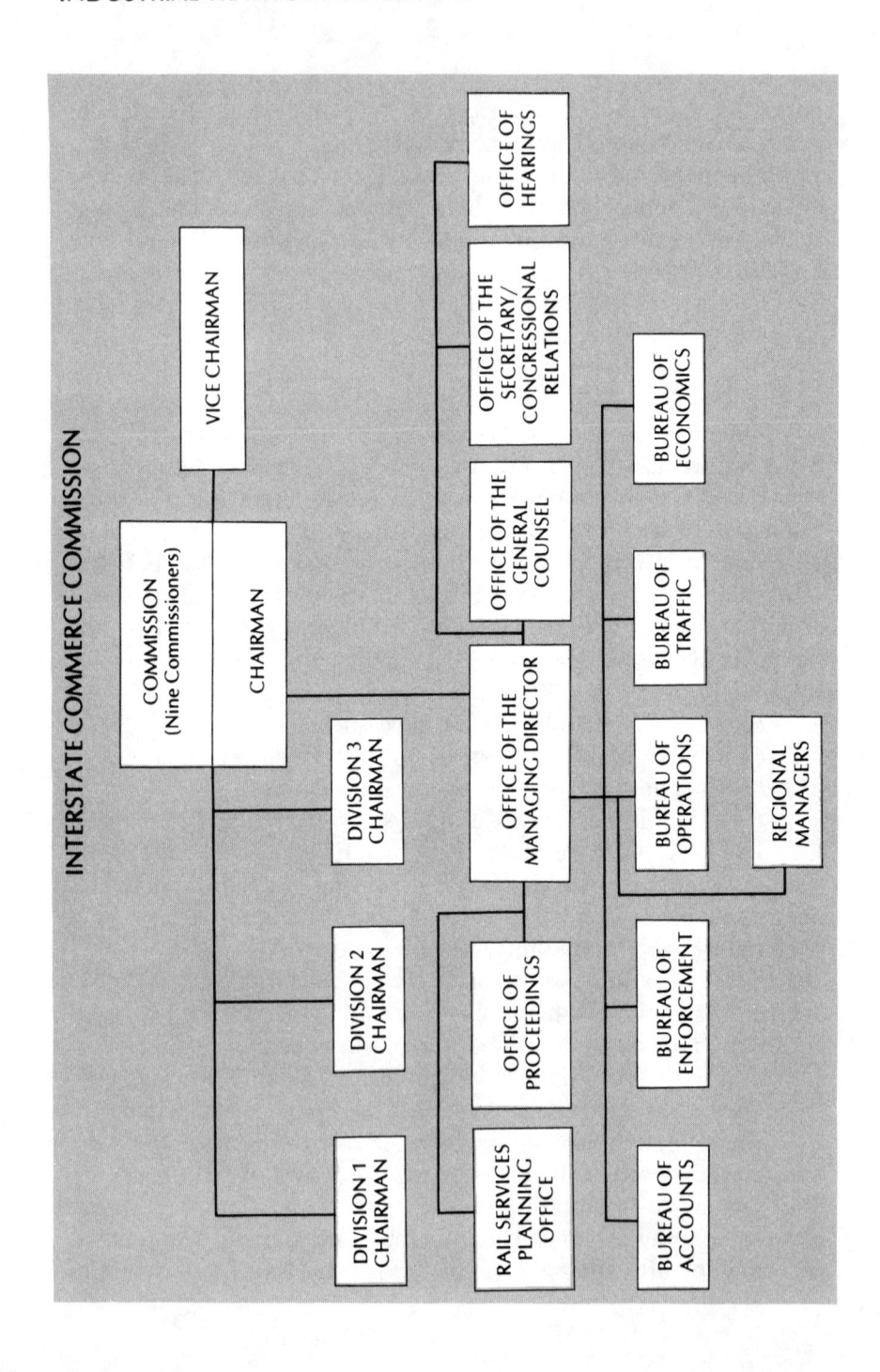

INTERSTATE COMMERCE COMMISSION

- COMMISSION (Nine Commissioners)
  - VICE CHAIRMAN
  - CHAIRMAN
    - DIVISION 1 CHAIRMAN
    - DIVISION 2 CHAIRMAN
    - DIVISION 3 CHAIRMAN
    - OFFICE OF THE MANAGING DIRECTOR
      - OFFICE OF HEARINGS
      - OFFICE OF THE SECRETARY/CONGRESSIONAL RELATIONS
      - OFFICE OF THE GENERAL COUNSEL
      - OFFICE OF PROCEEDINGS
      - RAIL SERVICES PLANNING OFFICE
      - BUREAU OF ECONOMICS
      - BUREAU OF TRAFFIC
      - BUREAU OF OPERATIONS
        - REGIONAL MANAGERS
      - BUREAU OF ENFORCEMENT
      - BUREAU OF ACCOUNTS

volved in granting operating authority, approval to construct and abandon railroads, mergers, changes in operating authority and punitive measures in enforcement matters.

Although public hearings on matters before the commission may be held at any point throughout the country, final decisions are made at its Washington, D.C. headquarters in all proceedings. These cases include rulings upon rate changes, applications to engage in for-hire transport, carrier mergers, adversary proceedings on complaint actions, and punitive measures taken in enforcement actions.

The Regional Rail Reorganization Act of 1973 created a Rail Services Planning Office (RSPO) to ensure that the public interest is represented in the restructuring and revitalization of railroads in the Northeast and Midwest. The office was given permanent status by the Railroad Revitalization and Regulatory Reform Act of 1976 (90 Stat. 31; 45 U.S.C. 801 note). In addition to its other responsibilities, RSPO provides planning support for the commission.

The ICC maintains field offices in major cities to audit carrier accounts, monitor the utilization of railroad freight cars in order to avoid severe shortages, investigate violations of the Interstate Commerce Act and related laws, and provide assistance to the public in its use of regulated carriers.

Revised Interstate Commerce Act

In 1978, Congress approved Public Law 95-473 which revised the 1887 Act and all its amendments. The revised act restates the original act, but in a comprehensive form. In effect, it is part of a program to eliminate terms that are difficult for the nonprofessional to understand and to remove obsolete and superceded statutes.

This new Act, passed by the 95th Congress is subtitle IV of title 49 of the United States Code, "Transportation." There are five parts to this code as follows:

Subtitle I   Department of Transportation
Subtitle II   Transportation Programs

INDUSTRIAL TRAFFIC MANAGEMENT

Subtitle III Air Transportation
Subtitle IV   Revised Interstate Commerce Act
Subtitle V   Miscellaneous Provisions

with subtitle IV being the first of the five to receive Congressional approval.

The new act produces a new national transportation policy of Congress and declares it to be:

• To recognize and preserve the inherent advantage of each mode of transportation;

• To promote safe, adequate, economical and efficient transportation;

• To encourage sound economic conditions in transportation, including sound economic conditions among carriers;

• To encourage the establishment and maintenance of reasonable rates for transportation without unreasonable discrimination or unfair or destructive competitive practises;

• To cooperate with each State and the officials of each State on transportation matters; and

• To encourage fair wages and working conditions in the transportation industry.

In the revised act, all four parts of the 1887 Act to Regulate Commerce and its amendments are brought together under Chapter 105 which gives the Interstate Commerce Commission continued responsibility and jurisdiction over railroads, rail-water connections, express, pipeline, motor carrier, water carrier and freight forwarders. The codification of the original act is then carried through ten chapters of law but does not change in any respect the regulatory jurisdiction of the ICC.

Every traffic manager or manager of physical distribution should have a copy of both the old and the new acts, and be acquainted with them and their effects on their employers, on their profession and on every transportation action taken.

The new act, as did the old, still requires common carriers to furnish transportation upon reasonable request and to establish and maintain just and reasonable rates, charges, classifications and rules. Carriers are still prohibited from causing one shipper or shipper group to pay more than any other for transporting the same commodity from the same point to the same point and in the same quantity. In other words, unjust discrimination is forbidden in any form whether that discrimination affects a single shipper or a community of shippers or a community itself.

The original act as amended had a provision under Section 4, Part 1 (also known as the long and short haul section) prohibiting rail and water carriers from charging more for a short haul than for a long haul over the same route. This could be circumvented by a successful petition to the ICC. Section 4 was kept in 95-473 as section 10726, still affecting only rail and water carriers.

The old act, as previously noted, had four parts to identify the modes of transportation affected by ICC jurisdiction. The new act classifies the carriers as follows (section 10706 (C) (c) (5)):

- Rail, express and sleeping car carriers
- Pipeline carriers
- Motor carriers
- Water carriers
- Freight forwarders

Congress in its Statement of Purpose to the new act said:

"The purpose of the bill is to restate in comprehensive form, without substantial change, the Interstate Commerce Act and related laws . . . .In the restatement, simple language has been substituted for awkward and obsolete terms . . . ."

Anyone claiming to have been damaged by acts of commission or omission on the part of carriers subject to parts of the Act may make a complaint to the ICC or bring suit in a federal district court, but cannot use both remedies. ICC decisions are administrative, and the test of their validity ultimately resides with the courts. Only a minute percentage of the thousands of formal cases decided each year are placed under judicial scrutiny and fewer still are overturned.

The ICC is permitted by the Act to investigate rates and charges on its own motion and to determine, after full public hearing, what rates and charges will be lawful for the future. However, the overwhelming majority of the changes in rates that take place daily are set by the carriers themselves, usually after negotiation with affected shippers. In such cases, the only part the Commission plays in the establishment of the rates is to receive the tariffs for filing (thirty days before the effective date unless ICC order permits "Sixth Section less-than-statutory notice") and to make sure that the tariffs conform to its tariff publication rules.

The divisions are each composed of three commissioners and are each assigned a definite share of the Commission's work load. Their areas of responsibility are:

Division one—Operating Rights.
Division Two—Rates, Tariffs and Valuation.
Division Three—Finance and Service.

Each division is assisted by one or more of five bureaus, which handle most of the preparation and detail work and dispose of matters that do not require the personal attention of the commissioners. The names of these bureaus and a brief summary of their functions follow:

Bureau of Traffic. Supervises the filing of tariffs, suspension of their provisions pending investigation of lawfulness and renders informal opinions concerning the application of tariff provisions.

Bureau of Economics. Performs economic, statistical and analytical work necessary to the commission's duty to foster the economic health of the country's transportation system.

Bureau of Accounts. Performs the accounting, cost finding and valuation functions necessary to bring about accurate and uniform disclosure of carriers' financial data.

Bureau of Investigation and Engorcement. Investigates violations, prosecutes in court and assists the Department of Justice in prosecuting civil and criminal proceedings arising under the Act.

Bureau of Operations. Administers rules governing lease and interchange of vehicles and inspects records of motor carriers to inform them of violations in tariffs, rebates, accounts, etc.; also administers regulation of water carriers and freight forwarders under the provisions of Parts III and IV of the Act.

In addition, there is a field organization set up on a regional basis, each office having a manager and a regional staff adequate for the needs of each of the bureaus. There are also field offices in cities throughout the country charged with responding to requests for aid and investigating complaints from shippers, carriers and the general public.

State Commissions

Many state regulatory statutes covering transportation within their borders are patterned after the interstate commerce act, follow in many ways ICC procedures, and often recognize ICC decisions as precedents. As to these, the industrial traffic manager can operate on the basis that ICC guidelines will control, as a general rule, but should know, in particular, the laws of the state in which their company has intrastate operations.

The ICC pattern by no means prevails in all states, however. Some states have no rate regulations at all. Others have laws vesting in their commissions powers considerably more extensive than those of the ICC. In the latter category are some that require approval by their commissions of each change in a rate or tariff provision prior to its publication.

Other variations occur in the territorial zones or areas which are established as being exempt from motor carrier rate regulation and operating rights requirements. These, figuratively, can vary "all over the map." The lists of products (agricultural versus others, etc.) which may be hauled by truck on an exempt basis also vary substantially. In some states motor carriers can haul without operating rights on a much broader basis than others. One state statute which comes to mind, and there are undoubtedly others, calls for freight rate regulation in a commercial zone designated as exempt by the ICC.

Questions regarding the applicability of intra- versus interstate regulations at times arise. Ordinarily the decision rests on whether the movement is within a state or whether it crosses a state line. Interstate applicability is clear where state lines are crossed, but the problem arises when there is no such crossing. An example would be, say, a combination box car and truck move where material comes from out of state by rail and moves on within the state by truck (immediately on arrival, in its original form, and in accordance with a continuing and original intent of the shipper to so deal with it from the out-of-state origin to the destination). Under applicable standards the truck part of this transportation would be considered interstate.

The intent of the shipper is accorded considerable importance in these matters and is to be given substantial weight in deciding what to do about them and the variances in them which occur. In many instances, of course, the same freight rates apply, there are no incompatible regulations, and so on. These have an end result of making the inter- versus intrastate problem a moot one. This is not always the case, however, and this feature of the traffic of many industries bears constant watching.

Where intrastate tonnage volumes are substantial, or other related conditions exist, it can be as important for a traffic manager to know the state rules and regulations as it is to know those applicable to interstate transportation. References to the state statutes or to pamphlets outlining them, which generally can be obtained at the state capitols, can be helpful in this regard. Familiarity with them is valuable and worthwhile.

As has been indicated before there is a tendency in some states to keep rail rate levels so low that intrastate traffic becomes a burden on interstate traffic and it is necessary for the ICC to correct them through proceedings under Section 13 of the Interstate Commerce Act.

The Federal Maritime Commission

The FMC should not be confused with the Maritime Administration of the Department of Commerce. The latter is responsible for the promotion of the merchant marine. The former deals with water carriers, but its functions are primarily regulatory in nature.

The Federal Maritime Commission regulates the waterborne foreign and domestic offshore commerce of the United States, and assures that United States international trade is open to all nations

TRANSPORTATION REGULATION

on fair and equitable terms. It also guards against unauthorized monopoly. This is accomplished through maintaining surveillance over steamship conferences and common carriers by water; assuring that only the rates on file with the commission are charged; approving agreements between persons subject to the Shipping Act; guaranteeing equal treatment to shippers and carriers by terminal operators, freight forwarders, and other persons subject to the shipping statutes; and ensuring that adequate levels of financial respon-

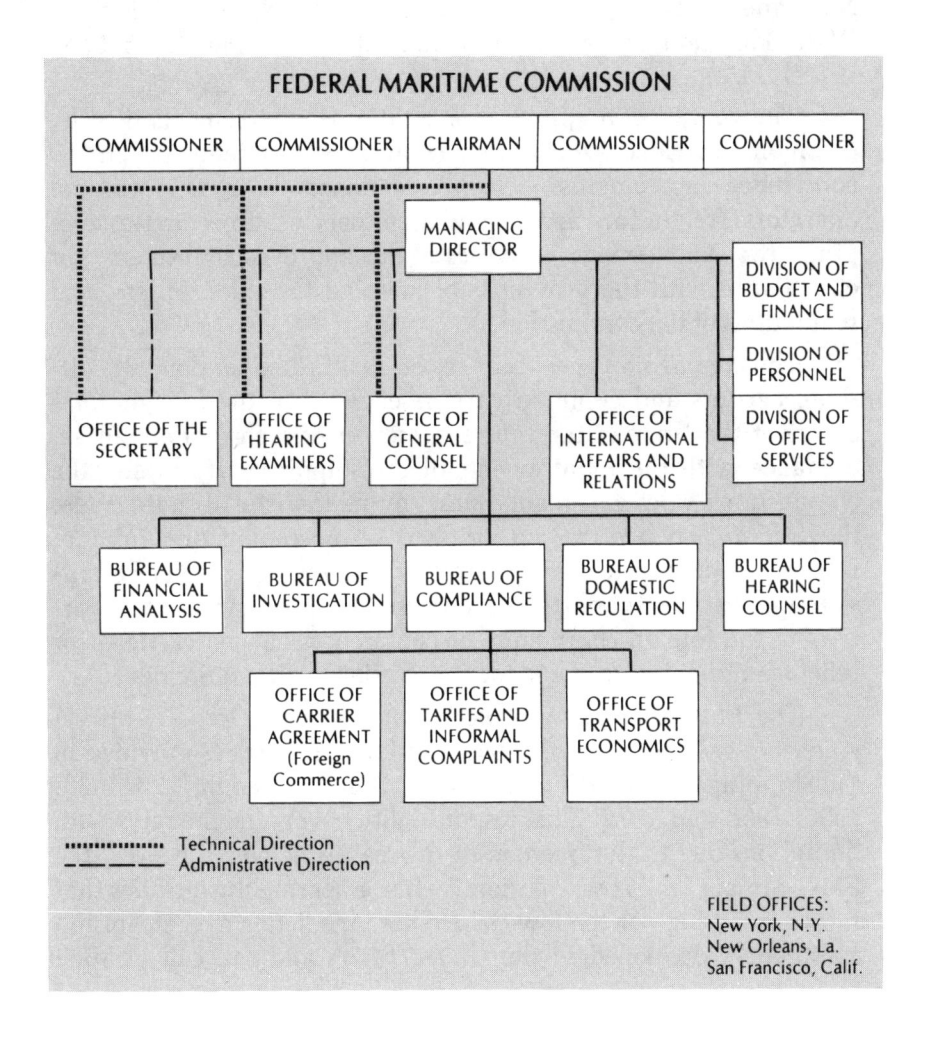

FEDERAL MARITIME COMMISSION

Organizational chart:

- Commissioners (x4) and Chairman
- Managing Director
- Division of Budget and Finance
- Division of Personnel
- Division of Office Services
- Office of the Secretary
- Office of Hearing Examiners
- Office of General Counsel
- Office of International Affairs and Relations
- Bureau of Financial Analysis
- Bureau of Investigation
- Bureau of Compliance
- Bureau of Domestic Regulation
- Bureau of Hearing Counsel
- Office of Carrier Agreement (Foreign Commerce)
- Office of Tariffs and Informal Complaints
- Office of Transport Economics

·········· Technical Direction
– – – – – Administrative Direction

FIELD OFFICES:
New York, N.Y.
New Orleans, La.
San Francisco, Calif.

29

sibility are maintained for indemnification of passengers or oil spill cleanup.

The Federal Maritime Commission was established by Reorganization Plan 7, effective August 12, 1961. It is an independent agency which administers the functions and discharges the regulatory authorities under the following statutes; Shipping Act, 1916; Merchant Marine Act, 1920; Intercoastal Shipping Act, 1933; Merchant Marine Act, 1936; and certain provisions of the act of November 6, 1966 (80 Stat. 1356; 46 U.S.C.362) and the Federal Water Pollution Control Act Amendments of 1972 (86 Stat. 816; 33 U.S.C. 1151).

The commission approves or disapproves agreements filed by common carriers, including conference agreements, inter-conference agreements between common carriers, terminal operators, freight forwarders, and other persons subject to the shipping laws, and reviews activities under approved agreements for compliance with the provisions of law and the rules, orders, and regulations of the commission.

The commission accepts or rejects tariff filings of domestic offshore carriers and common carriers engaged in the foreign commerce of the United States, or conferences of such carriers, in accordance with the requirements of the shipping statutes and the commission's rules and regulations. In the domestic offshore trade, the commission has the authority to set maximum or minimum rates or suspend rates. It approves or disapproves Special Permission applications submitted by domestic offshore carriers and carriers in the foreign commerce, or conferences of such carriers, for relief from the statutory and/or commission tariff requirements.

The FMC thus has jurisdiction over shipping procedures via water between the United States and foreign shores as provided in the Shipping Act of 1916 and the various acts passed in 1920, 1933, 1936, 1966 and 1972. This responsibility covers freight rates with limitations due to the freedom-of-the-seas principle. However, the FMC can act on its own motion if after a hearing it concludes that rates are too high or too low. It also has jurisdiction over steamship conferences domiciled in the United States and can call on such

TRANSPORTATION REGULATION

conferences domiciled in other countries for data that it might need for its own investigations.

Steamship conferences and individual steamship lines must file copies of all tariffs and notices of all rates quoted with the FMC at least 30 days before a rate increase or a new rate is effective (dual-rate contract holders are given additional days notice of increases) though a rate reduction may be made on one day's notice, or on filing.

The commission may not suspend rates without a formal hearing, but has the power to do so after such a hearing. Proceedings for suspension may be instituted by the commission on its own motion or as a result of formal complaints. Hearings are conducted in accordance with the commission's Rules of Practice and Procedure.

The Commission is also charged with the duty of seeing that water carriers do not engage in practices which are forbidden by the Act. These include the use of rebates, fighting ships, retaliation against shippers by refusing space because a shipper has patronized another carrier or has filed a complaint with the FMC. Also involved are the making of unfair or unjustly discriminatory contracts with shippers, or unjust discriminations in the allocation of cargo space and accommodations and the settlement of claims.

Because the foreign trade of our country is also the foreign trade of other countries, and neither the United States nor the foreign country of origin or destination has complete control, the Federal Maritime Commission encourages self-regulation with government supervision to prevent violations of the Act. To a large extent such self-regulation is accomplished through the organization of steamship conferences, which today exist on virtually every important trade route in the world.

Under amendments to the Shipping Act of 1916 in the 1960's, steamship conferences domiciled in the U.S.A. must regulate themselves by maintaining a *neutral body* for each conference. A neutral body is a disinterested or outside individual or company (frequently an auditing firm) that investigates all alleged and suspected violations of conference agreements or operational rules. Each member line must post a cash bond to guarantee adherence to the conference rates, practices, etc.

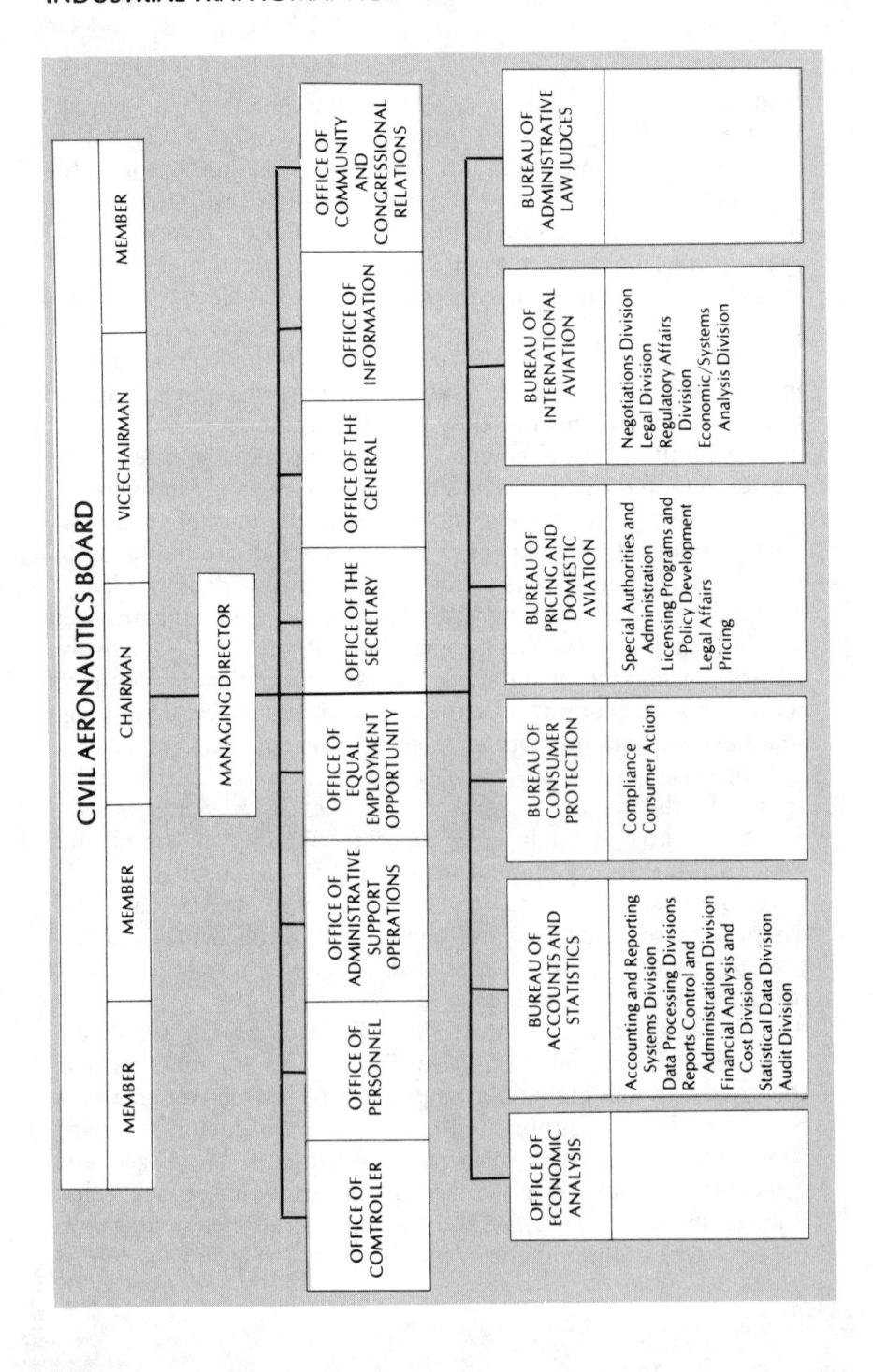

INDUSTRIAL TRAFFIC MANAGEMENT

CIVIL AERONAUTICS BOARD

| MEMBER | MEMBER | CHAIRMAN | VICECHAIRMAN | MEMBER |

MANAGING DIRECTOR

- Office of Comptroller
- Office of Personnel
- Office of Administrative Support Operations
- Office of Equal Employment Opportunity
- Office of the Secretary
- Office of the General
- Office of Information
- Office of Community and Congressional Relations

- Office of Economic Analysis

- **Bureau of Accounts and Statistics**
  - Accounting and Reporting Systems Division
  - Data Processing Divisions
  - Reports Control and Administration Division
  - Financial Analysis and Cost Division
  - Statistical Data Division
  - Audit Division

- **Bureau of Consumer Protection**
  - Compliance
  - Consumer Action

- **Bureau of Pricing and Domestic Aviation**
  - Special Authorities and Administration
  - Licensing Programs and Policy Development
  - Legal Affairs
  - Pricing

- **Bureau of International Aviation**
  - Negotiations Division
  - Legal Division
  - Regulatory Affairs Division
  - Economic/Systems Analysis Division

- **Bureau of Administrative Law Judges**

TRANSPORTATION REGULATION

The Federal Maritime Commission has indicated that it believes that conferences, when properly functioning, promote stability of rates and uniformity of treatment of all cargo interests and obviously also tend to improve the calibre of steamship service.

The Civil Aeronautics Board

The regulatory activities of the CAB are of unquestionable interest to industrial traffic managers and their staffs. The jurisdiction of the agency covers both passenger and freight activities, as does the ICC. But the agency is the first to go through a "deregulatory" process effective with the Congressional action in Public Law 95-504, the Air Transportation Regulatory Reform Act of 1978 and title IV of Public Law 95-163, the Federal Aviation Act Amendments of 1977.

The laws call for the reduction of CAB authority over domestic air line routes and its complete elimination of authority on December 31, 1981. Further, the CAB would cease to exist in 1985 barring any further Congressional action. Any remaining functions will be turned over to the Department of Justice, Transportation and State and to the Postal Service.

The law has released existing cargo carriers from route restrictions so that carriers have been expanding their routes to include many new cities where ground facilities are available or are being expanded for the new services. It has also brought into existence new smaller "air taxi" freight carriers moving freight into major airport cities for transportation by the major airlines to final domestic and overseas destinations.

Department of Transportation

Earlier it was mentioned that DOT's regulation of safety matters and some of its systems work would be discussed in greater detail. Its safety work is discussed in later chapters where pertinent as are special features of the activities of other regulatory agencies.

A non-regulatory activity of DOT of particular interest to industry has been the work it has been doing on the simplification of international trade documents. The situation regarding these has been receiving concentrated attention in recent years, and DOT has provided worthwhile help in the important progress that has been made in easing unneeded burdens there. Help also has been provided by DOT in the coordination of data processing in transportation, which has outstanding shipper-carrier possibilities.

A number of states have established departments of transportation in recent years. There are currently about 15 of them. Jurisdictions and powers vary, with mass transportation being high on the list of their responsibilities. As with state commissions functioning along lines related to the ICC, the state DOTs can be important to industrial traffic men and may be more so in the future.

General

Along with groups previously mentioned, there are numerous other aspects of the transportation regulatory picture. Illustrations could well include hours and wages of employees, weights and lengths of highway vehicles, carrier fuel taxes, and so on. Several are discussed and referred to later in the text, the chapter on hazardous materials being particularly pertinent.

Top dealings with these aspects generally require more than reading the regulations and statutes. Problem areas are to be reconciled, updated data applied, and other features covered with transportation being administered accordingly.

Illustrative of a problem area is that under the federal Motor Carrier Act a 60 hour driver work week is recognized whereas under the Fair Labor Standards Act it is 40 hours. The latter requires payment of wages at time and a half when 40 hours are exceeded. Numerous problems have arisen in this connection. Generally speaking, where interstate commerce is involved and ICC regulations apply, the 60 hour specification is recognized. Galbraith v. Gulf Oil (C.A. 1969) 413 Fed. 2 941.

To be mentioned also is that price controls applicable to carrier charges were with us in the early 1970's but have lapsed although they exist internationally in other countries. Additionally, somewhat special regulatory situations follow.

Recent activity concerning industrial safety stem particularly from the Occupational Safety and Health Act of 1970 (84 Stat. 1590) which is administered by the Department of Labor. Its provisions, and the standards and rules promulgated thereunder, are quite broad. These are framed, generally speaking, to pertain to entire plants, manufacturing works, refineries, and so on. They may also include, however, loading and unloading facilities together with other operations which involve traffic. The Act requires attention because of this.

Environment regulations stem from the National Environment Policy Act of 1969 (42 U.S.C. 4321) and related statutes. This, again, is legislation which is quite broad in scope; its administration falls to the Council of Environmental Quality, the Environmental Protection Agency, and others. Among various things, it calls upon other federal agencies to cooperate on and to consider (and make findings on) environmental aspects in matters before them. The ICC does this in line with an implementation ruling it has promulgated (Implementation—Natl. Environmental Policy Act 1969, 340 ICC 431). Where indicated, the Commission specifically rules whether or not "the quality of the human environment will be significantly affected" as a part of its decisions. Developments in this field of regulation are emerging as to the extent they bear on transportation.

Rights and liabilities of industrial companies under anti-dumping laws applicable to foreign trade are receiving more attention because the activity in this field has increased substantially. The general situation is such that this activity will be further increased. While only a limited number of transportation departments are now involved in these matters, more may be later and possibilities in this regard should be kept in mind. Laws on dumping have been in effect for a number of years and are discussed in more detail in the Customs section of the International Trade chapter. Substantial sums of money, closely allied to transportation functions, can be involved in these cases.

3

FREIGHT CLASSIFICATION

It is a fundamental requirement of every industrial traffic manager that he give adequate recognition to the importance of correctly classifying his freight. The National Motor Freight Classification has summed up the reason for this as follows:

> "To insure the assessment of correct freight charges and avoid infractions of Federal and State laws, shippers should acquaint themselves with the descriptions of articles in the tariffs under which they ship. Commodity word descriptions must be used in shipping orders and bills of lading and must conform to those in the applicable tariff, including packing specifications where different rates are provided on the same article according to the manner in which it is prepared for shipment."

Failure to use correct classification nomenclature in describing shipments on bills of lading not only will result in carriers assessing incorrect rates, but may subject the shipper to suspicion of false billing. False billing is a punishable offense by the terms of the Interstate Commerce Act. But even if no violation occurs and carriers charge high rates to protect themselves, the wasteful freight charges are absorbed by the shipper, and the customer bearing the excess freight charges may look elsewhere for his supplies. Subsequent adjustment with carriers is always a difficult process.

Purpose and Background

The need for a system of classification dates back to the time when the railroads first held themselves out to the general public as transporters of goods, for it soon became evident that to make charges *per box, per barrel, per bushel,* etc., was inadequate and that it was necessary to base freight charges on units of weight. In doing so, they had to distinguish between a commodity of which 5,000 pounds would fill a freight car and another commodity that would fill the car with 20,000 pounds or more. Similarly, they felt that they were entitled to greater compensation for transporting a shipment of expensive and easily damaged furniture, for example, than for transporting a shipment of the same weight of the much lower priced and virtually indestructible rough lumber from which furniture might be made.

To publish specific rates on each of the thousands of different articles found in transportation and to do it in a way as to make them applicable between all the different railroad stations in the United States would have been impractical, if not impossible. Consequently, a system of class rates was developed, with a limited number of graduations, or classes. By means of the classification the various articles were placed into related categories and assigned class ratings which were determined by the transportation characteristics of each article, such as weight per cubic foot, value per pound, susceptibility to damage and pilferage, competitive relationship to other articles of similar nature and use, regularity and volume of movement, etc. Each of these categories was given two class ratings, one for carload shipments coupled with minimum carload weights and a higher one for use on shipments of less-than-carload quantities.

Originally, each railroad published its own classifications, one for traffic moving locally on its own line and another for traffic handled jointly with other connecting lines. There were classifications that differed according to the direction of movement, and some railroads had separate classifications for their different divisions. One railroad had nine different classifications in effect at the same time, each with its own rules, packing requirements, minimum weights and ratings.

Under these circumstances, it is not difficult to visualize the chaotic conditions that existed in 1887, when the Interstate Commerce Commission came into being, charged with the responsibility

FREIGHT CLASSIFICATION

of putting the nation's transportation house in order within the framework of the Interstate Commerce Act. As early as 1888, the ICC had before it a resolution calling for it to order the railroads to publish a single classification for nationwide application, but the difficulties of reconciling the widely divergent views on the subject then held in various sections of the country caused the resolution to be tabled and carried on the Commission's agenda from year to year without action. Some progress did take place, however, for the railroads dropped their individual classifications and participated jointly in four publications which were issued on territorial lines, as follows:

Official Classification applied in that portion of the country situated north of the Ohio and Potomac rivers and east of the Indiana-Illinois state line and on shipments between that territory on the one hand and Illinois and the eastern part of Western Classification Territory on the other.

Southern Classification applied in the territory east of the Mississippi and south of the Ohio and Potomac rivers.

Western Classification applied in the territory comprising the states west of the Mississippi River, also the Upper Peninsula of Michigan, the State of Wisconsin and the northwestern portion of Illinois.

Illinois Classification applied between points in Illinois.

While this was a step in the right direction, it did not completely solve the problem, for in addition to the different ratings that applied in each territory, there were different rules, packing requirements, commodity descriptions and carload minimum weights. There were very few interterritorial class rates, and most shipments moved on combinations of rates constructed over border points of the respective territories. Thus, a shipment from New York to New Orleans might carry a description that would assure it the correct rate for that part of the haul to the border of Southern Classification Territory, but that might not appear at all in the Southern Classification. Or a shipment from Atlanta to Dallas might be in packages that conformed fully to the requirements of the Southern Classification, but that would incur a penalty under the Western Classification.

The vast expansion of interterritorial commerce that took place in the early 1900's made this an unsatisfactory arrangement, working hardships on both carriers and shippers alike. During World War I, when the United States Railroad Administration controlled and operated the railroads, the Commission decided that if ratings could not be made entirely uniform, at least commodity descriptions, rules and regulations, packing requirements and carload minimum weights could be. An investigation was ordered, as a result of which the railroads were instructed to create committees representing Official, Southern and Western Classification territories for the purpose of planning the publication of a consolidated freight classification.

Out of these proceedings came the publication of Consolidated Freight Classification No. 1, which became effective December 20, 1919. Although it preserved the individual differences of rating peculiar to each territory, it was a significant advance toward uniformity. The Illinois Classification, which had continued as a separate publication, was incorporated into the Consolidated Freight Classification effective November 15, 1933.

Until the 1920's, the various class rate scales were graduated according to numbered or lettered classes, none of which bore a fixed relationship to the first class rate. Thus a third class rate might be anywhere from 60% to 80% of the first class rate for the same distance, while the fifth class rate might range from 32½% to 45% of the same first class rate. In a series of decisions which caused the revision by December 3, 1931 of all class rate scales east of the Rocky Mountains, the Commission prescribed fixed percentage relationships to the first class rate for each of the numbered or lettered classes. Thus, in Official Classification Territory, second class always was 85% of first, third class 70%, fourth class 50%, fifth class 35% and sixth class 27½%. In Western Classification Territory, second and third classes carried the same percentages as in Official, but fourth class was 55%, fifth class 37½% and there was no sixth class. The lettered classes started with A at 45% and scaled down to 17½% for class E. Southern Classification Territory classes were numbered from 1 to 12 and corresponded to Western as far as fourth class, fifth class being 45%, sixth class 40% and so down to twelfth class at 17½%.

FREIGHT CLASSIFICATION

The Commission still was not satisfied with the rail classification situation, for in addition to the wide differences in ratings in the various territories, a vast number of exceptions to these ratings had found their way into the railroads' tariffs to meet a variety of competitive situations. Although the Commission on several occasions had declared that the interests of the shipping public pointed up the desirability of uniform classification ratings and urged the railroads to remedy the situation, no voluntary action was forthcoming. Consequently, as it is permitted to do under section 13(2) (now para. 11701) of the Interstate Commerce Act, the Commission initiated an investigation on its own motion and placed it on the docket under No. 28310, Uniform Freight Classification. In a decision handed down May 15, 1945, the Official, Illinois, Southern and Western Classifications were found to be unjust and unreasonable for the future, in violation of section 1(4) and (6) (of the pre-1978 Codefied Act) and the railroads were ordered to prepare within a reasonable time a classification embodying the principle of basic uniformity of classification ratings for applications in all territories. The order directed them to weave into the new classification as many of the exception ratings as possible and to refrain from using the order as a means of increasing their revenues.

Complying with the order, the railroads appointed a committee, which held public hearings in major cities during 1948 and 1949 and considered in four separate dockets the reclassification of all articles then listed in the Consolidated Freight Classification. Finally, after seven years of preparation and study, Uniform Freight Classification No. 1 was published to become effective May 30, 1952. In it, the old system of numbered and lettered classes was dropped and all ratings were expressed as percentages. Thus Class 70 is 70% of Class 100, while Class 250 is 2½ times Class 100. It resulted in many increases and decreases in ratings and brought about several hundred requests for suspension by shippers who claimed to be affected adversely by the increases. However, so determined was the Commission to realize a goal of more than sixty years standing that it denied them all and suggested that adjustment of the items under attack be sought through the regular docket procedure of the Uniform Classification Committee.

INDUSTRIAL TRAFFIC MANAGEMENT

Originally, these uniform ratings applied only in connection with the uniform class rates which went into effect on the same date, as the result of the Commission's order in a companion case, Docket 28300. These rates applied in the territory east of the Rocky Mountains and with their establishment practically all of the LCL and many of the carload exception ratings in the same territory were cancelled. Then in 1956, the Commission extended the Docket 28300 principle to the entire country by its decision in Dockets 30416 and 30660, and the Uniform Classification governed all interstate class rates effective August 15, 1956. The Consolidated Classification is no longer in effect.

The Uniform Classification has as its current edition number 13. Numerous supplements have been issued and are to be consulted as a part of making sure any referrals are up-to-date. Examples from number 10 are part of this text, and are representative of the index and descriptions in any edition.

The Motor Carrier Classification

In 1935, when Part II of the Interstate Commerce Act brought motor carriers under regulation, the American Trucking Associations, as agent for most of the regulated motor carriers, published and filed with the Commission, National Motor Freight Classification No. 1. The ratings conformed closely to those published in the rail classification except for articles rated lower than 30% of first class and certain bulky articles on which the motor carriers elected to publish higher ratings. The "volume" minimum weights were generally the same as the rail carload minimum weights, but in almost all territories exceptions to these minimum weights were published separately because the size of the motor carrier equipment then in use and state laws as to maximum loading prevented shippers from loading that much. The rules of the National Motor Freight Classification differ from those of the rail classification and both will be summarized at the end of this chapter.

With the introduction of the rail Docket 28300 class rates and the Uniform Classification, the motor carriers brought out National Motor Freight Classification No. A-1, patterned along the same

FREIGHT CLASSIFICATION

lines as the rail Uniform Classification. At first, it had very limited application, but gradually motor carriers adopted class rate scales designed to be competitive with the rail Docket Nos. 28300, 30416 and 30660 scales and it now governs all motor carrier class rates. The National Motor Freight Classifications are published by the National Motor Freight Traffic Association, Inc., as agent for participating carriers.

The current publication of this Classification is NMFC 100F. Subsequent to the time of the last edition of this text, "Items," rather than "Rules," have come to be used in designating individual provisions. The changed terminology is reflected in the comparisons of the Classifications that are later tabulated in this chapter.

The motor carriers in New England have had their own distinctive classifications since the early days of motor carrier regulation. Originally two separate publications, issued by the Eastern Motor Rate Bureau and the New England Motor Rate Bureau, they were combined under the name Coordinated Motor Freight Classification when the two bureaus merged in 1954 as the New England Motor Rate Bureau. This classification regards density as its principal standard and confines itself to eight ratings: 2½, 2 and 1½ times first class; first, second, third, fourth and fifth classes. Illustrative of the importance and effect of density to this Classification is Rule 210 which states criteria that must be met to warrant "compressed" ratings. Under it, material weighing less than 5 pounds per cubic foot is rated Class 1; 5 to 10 pounds, Class 2; 10 to 15 pounds, Class 3; 15 to 20 pounds, Class 4; over 20 pounds, Class 5. Each article is assigned but one class rating, the class rate scales being constructed on a series of weight breaks that provide progressively lower rates in each class for each increase in minimum weight.

More than 200 motor carriers participate in the rail classification. Some of these are trucking subsidiaries of rail lines; others are independent carriers that join with the railroads to provide through services on rates published by the railroads. Conversely, a number of railroads participate in the National Motor Freight Classification for various purposes, including that of joining with motor carriers in through rates published by the motor carriers.

INDUSTRIAL TRAFFIC MANAGEMENT

UNIFORM FREIGHT CLASSIFICATION 10
INDEX TO ARTICLES

STCC No.	Article	Item	STCC No.	Article	Item
37 299 35	Nacelles or nacelle sections, aircraft	4705	32 959 59	Natural stone, ground or pulverized, noibn	47715
28 169 46	Nacreous pigment (pearl essence)	24745	19 411 50	Naval gunfire control systems, armored	73020
25 421 60	Nail bins, revolving, metal 44070, 44220, 44230		38 111 40	Navigating logs, vessel	11740
25 411 50	Nail bins, revolving, wooden 44070, 44220, 44230		38 111 35	Navigation sounding machines, vessel, ot radar	65930
31 999 60	Nail head protectors, leather	80085	20 861 20	*Near beer	56850
34 298 37	Nails and picture hangers combined †49771, 50210		20 143 30	Neatsfoot oil, noibn	72470
	Nails		29 919 15	Neatsfoot oil, petroleum	77240
33 992 10	Aluminum	5670	20 144 40	Neatsfoot stock	70095
33 992 15	Brass, bronze or copper, not plated †49771, 50220		38 421 10	Nebulizers	8095
	Cement-coated	21070	34 998 28	Neck rings, gas cylinder, iors 49670, 49781	
33 992 25	Copper-clad 49771, †49781, 50230		26 471 35	Neck strips paper	76405
33 152 40	Crate fastener, iors	54410	34 449 66	Necks, can, sheet steel 85710, 86010	
33 152 83	*Egg case	54430	20 119 10	*Necks, hog, fresh or chilled	67910
33 152 45	Horse shoe, iors	54420	34 298 49	Necktie racks	52710
33 992 50	Hungarian, noibn, brass, bronze or copper †49771, 50810		31 999 63	Neckyoke centers, leather or leather and metal combined	94180
33 152 25	Hungarian, noibn, steel, with ot steel or zinc heads	50820	37 994 11	Neckyokes, steel, or wooden in the white, ironed †4050, 32830	
33 152 30	Hungarian, noibn, steel, with steel heads †49781, 50830		37 994 12	Neckyokes, wood, in the white, not ironed †4050, 32830	
34 994 70	Hungarian, noibn, steel, with zinc heads	50840	37 994 10	Neckyokes, wooden, finished †4050, 32830	
33 152 50	Iors, noibn, plain, galvanized, japanned or tinned, or coated with brass, bronze, cadmium, cement or copper, or with lead-covered or lead-rimmed heads	54430	37 994 13	Neckyokes, wooden, in the rough	57810
			20 334 40	*Nectar, guava, unfermented, ot frozen	40210
			20 334 40	*Nectar, juice, unfermented, ot frozen	40210
			20 334 40	*Nectar, papaya, unfermented, ot frozen	40210
			20 371 30	*Nectarines, fresh, cold-pack (frozen)	40050
33 992 55	Nickel, nickel-copper, or nickel-iron-chromium alloy	70220	01 225 10	Nectarines, fresh, ot cold-pack	41830
33 992 25	Nickel-plated †49771, †49781, 50230		10 929 20	Needle antimony (antimony sulphide) concentrated ore, lump	27080
33 152 20	Picture †49771, 50210		10 929 25	Needle antimony (antimony sulphide) concentrated ore, pulverized or powdered	27090
33 992 50	Shoe, noibn, brass, bronze or copper †49771, 50810			Needles:	
33 152 25	Shoe, noibn, steel, with ot steel or zinc heads	50820	35 321 64	Miners', copper	91030
			35 321 64	Miners', iors	91040
33 152 30	Shoe, noibn, steel, with steel heads †49781, 50830		39 642 65	Noibn	70100
34 994 70	Shoe, noibn, steel, with zinc heads	50840	36 511 55	Phonograph	88530
33 152 10	String, boot or shoe, steel, coiled	13630	08 619 46	Pine (pine straw)	77480
			38 411 40	*Syringe, surgical	70100
33 992 60	Zinc	98400	36 511 55	Talking machine	88530
34 619 36	Name diagram plates, rwy, iors	82130	35 321 61	Well diggers', copper	91030
34 998 31	Name plates, metal †49771, †49781, 50350		35 321 64	Well diggers', iors	91040
29 119 82	Naphtha distillate	77240	41 114 74	Negatives, photograph, glass, old	46160
29 919 18	Naphtha solvent, coal tar 23415, 33800		28 134 75	*Neon gas	45580
29 919 31	Naphtha, coal tar, crude	72370	36 999 29	Neon sign leads or lead wires	34890
29 119 82	Naphtha, petroleum	77240	28 212 20	Neoprene rubber, crude	84270
28 141 49	Naphthalene, crude 2460, 33800		28 186 86	Neo-decanoic acid 2345, 33800	
28 151 39	Naphthalene, ot crude 24650, 33800		28 186 88	Neo-heptanoic acid 2345, 33800	
28 141 46	Naphthenic acid	77330	28 186 84	Neo-pentanoic acid 2345, 33800	
28 151 13	Naphthol, alpha 22110, 33800		28 186 87	Neo-tridecanoic acid 2345, 33800	
28 151 17	Naphthol, beta 22880, 33800		28 185 52	Neopentyl glycol	23980
28 799 80	Naphthyl methyl carbamate	6	35 228 23	Nest eggs, poultry equipment	36290
26 471 40	Napkin holders and paper napkins in combined packages	76635	35 228 45	Nest traps, hens', steel †35980, †36000, 36800	
26 218 20	Napkin paper	75700	35 228 28	Nests, hens' †35980, †36000, 36460	
26 471 35	Napkins, paper	76405	33 612 35	Net floats, aluminum	38783
26 471 40	Napkins, paper and napkin holders in combined packages	76635	24 941 62	" floats, cork	38784
			30 716 22	" floats, plastic, cellular or expanded	38785
26 472 10	Napkins, sanitary 74540, †74550		30 713 37	" floats, plastic, ot cellular or expanded	38786
23 924 10	*Napkins, table, cotton, or cotton and rayon or flax	28160	34 449 36	" floats, sheet steel	38787
32 411 15	Natural cement 21680, †77130		24 999 21	" floats, wooden	38788
13 211 10	" gasoline, suitable only for mixing, blending, or refining purposes	77290	41 111 17	" outfits, camouflage, military	73530
				Nets	
08 422 90	" gums, noibn 48040, 33800		39 496 58	Basketball	7690
08 423 25	" rubber, crude	84270	22 981 78	Camouflage, noibn	70110
14 219 90	" stone chips, noibn	47715	34 819 80	Camouflage, woven wire, interwoven cloth, glass fibre, or steel wool	70120
32 959 59	" stone dust, noibn	47715			
40 271 45	" stone waste, noibn	47715	22 981 81	Fish	70125
14 219 90	" stone, crushed, noibn	47715	31 999 33	Fly, horse	51250

For explanation of reference marks, see top of page 17; for abbreviations, see last page of this Classification.

Alphabetical Index to Articles, Uniform Freight Classification. Included are Standard Transportation Commodity Code numbers.

FREIGHT CLASSIFICATION

UNIFORM FREIGHT CLASSIFICATION 10   40150-40360

ITEM	ARTICLES	Less Carload Ratings	Carload Minimum (Pounds)	Carload Ratings
	FOODSTUFFS,BEVERAGES OR BEVERAGE PREPARATIONS,NOT NAMED IN OTHER MORE SPECIFIC GROUPS,NOT COLD-PACK NOR FROZEN,NOT FREEZE-DEHYDRATED NOR FREEZE-DRIED,UNLESS OTHERWISE SPECIFIED (see also Item 33800) (Subject to Item 39410)—Continued:			
40150	Jams,jellies or preserves,edible,noibn,in kits,pails,tubs or metal cans,in crates,in barrels,boxes,kits,pails or tubs,or in Package 1264	60	36,000	35
40160	Jelly,corn syrup,in kits,pails,tubs or metal cans in crates; or in barrels,boxes,kits,pails or tubs	60	36,000	35
40170	Juice,citrus fruit,frozen,in containers,in barrels or boxes,in bulk in barrels,in metal cans in crates or in Packages 575,1164 or 1415; also CL,in Package 1188	100	36,000	37½
40180	Juice,citrus fruit,other than frozen,noibn,see Note 36,Item 40182: In barrels,boxes or Packages 575 or 1415, also CL,in Packages 835 or 1164	60	36,000	35
	In insulated tank cars,Rule 35,except as to minimum weight	......	40,000	35
40182	NOTE 36.-Ratings also apply on fruit juice coloring,the gross weight of which does not exceed 5% of gross weight of shipment when shipped with fruit juices in the same outer container.			
40190	Juice,citrus fruit,with sugar,citric acid,vegetable gum and oils, certified food color and water,added,other than frozen,in glass in boxes,in metal cans in boxes or crates,in barrels or in Package 575	60	36,000	35
40200	Juice,fruit,artificial or natural,frozen,noibn,in barrels or boxes in pails or metal cans in crates,or in Package 575	100	36,000	37½
40210	Juice,fruit,artificial or natural,other than frozen,noibn,in barrels or boxes,in pails or metal cans in crates,or in Packages 575,592, 1264 or 1331	60	36,000	35
40220	Juice,grape,other than frozen,noibn,in barrels,boxes or Package 575; also CL,in glass in crates with solid tops,in Package 454,or in tank cars,Rule 35	60	36,000	35
40230	Juice,grape,with sugar or other sweetening and water added,other than frozen,in metal cans in boxes or crates,in bulk in barrels or in Package 575	60	36,000	35
40240	Juice,pineapple,other than frozen,in barrels or boxes; also CL,in Packages 838 or 873	60	36,000	35
40250	Juice,prune,other than frozen,in barrels,boxes or Package 1304	60	36,000	35
40260	Juice,vegetable,frozen,noibn,in barrels or boxes,or in pails or metal cans,in crates; also CL,in metal cans,loose	100	36,000	37½
40270	Macaroni,noodles,spaghetti or vermicelli,in barrels,boxes or double bags,or in Packages 58,852 or 1119	60	36,000	35
40280	Macaroni,noodles,spaghetti or vermicelli,prepared,with or without cheese,hominy,meat or vegetables,in containers in barrels or boxes or in metal cans in crates	60	36,000	35
40290	Macaroni,noodles,spaghetti or vermicelli,dry,not cooked,with cheese, meat,vegetables or other ingredients,in combined packages in boxes	60	36,000	35
40300	Malt and milk compound (not malted milk),in barrels,boxes,paper lined cloth bags,or in metal cans in crates,see Note 38,Item 40312	65	36,000	40
40310	Malt or milk and chocolate or cocoa compounds,beverage preparation, in double bags,barrels or boxes,or in metal cans in crates,or in Package 1100,see Note 38,Item 40312	65	36,000	40
40312	NOTE 38.-With LCL shipments there may be included paper cups and paper lids for such cups,wooden ice cream sticks and printed advertising matter,subject to packing requirements provided therefor,in an amount not exceeding 20% of the weight of the entire shipment,the entire mixed shipment to be subject to the Class 65 rating.			
40320	Malt extract,liquid:			
	In glass in barrels or boxes	70	30,000	40
	In metal cans in barrels or boxes or in bulk in barrels	65	36,000	35
40330	Malt extract,other than liquid malt powder (dehydrated malt syrup), or maltose (malt sugar):			
	In glass in barrels or boxes	70	30,000	40
	In bulk in barrels,boxes or 4-ply paper bags in containers other than glass or earthenware in barrels or boxes	65	36,000	35
40340	Meat substitutes,processed from vegetables,soybean products,peanuts or grain products,with or without seasoning,in glass or metal cans in barrels or boxes,or in metal cans in crates; also CL,in Package 1437	60	36,000	35
40350	Meats,cooked,cured or preserved with or without vegetable,milk,egg, fruit or cereal ingredients,noibn,in glass (Rule 5,Sec.2 (b) not applicable) or metal cans in barrels or boxes,in metal cans in crates,in Packages 223,1307 or 1400; also CL,in Package 1345	60	36,000	35
40360	Milk,cream,buttermilk,or dry milk solids,powdered or flaked,with or without vegetable fats not exceeding 1% by weight,in containers in crates or cloth bags,or in double bags,multiple-wall paper bags, barrels or boxes,or in Packages 3,197,864,1227 or 1307; also CL,in bulk in covered hopper cars,Rule 37,see Note 56,Item 41046,or in bulk in metal shipping containers	60	36,000	35

For explanation of abbreviations,numbers and reference marks,see last page of this Classification; for packages,see pages following rating section.

Bill of lading descriptions, ratings and carload minimum weights—Uniform Freight Classification.

45

Water Carrier Classifications

Class rates via domestic water carriers that are published jointly are subject to either the rail or motor classification according to the nature of the joint service. Local class rates, also, may be subject to either classification. For the most part, these have been subject to the rail classification in the past.

Freight Forwarder Classifications

The rail or the truck classification governs the class rates of domestic freight forwarders. Some forwarders are shown as participants in both publications.

General Comments

Each tariff that is governed by the ratings or rules of a classification will contain a definite statement to that effect and name the classification to which it is subject. By such a reference, it is possible to avoid the reproduction of voluminous rules in each of the many thousands of individual tariffs in effect, as well as assuring the uniformity of such rules in all tariffs. There are occasions when exceptions to classification ratings and rules are published and, if they do not appear in the tariff itself, reference to the tariff in which they are published must receive prominence equal to the reference to the governing classification.

In all the rail and motor classifications there are ratings that differ according to the released valuation of the goods shipped; others vary according to the actual valuation. The distinction between these two types of ratings is important and should be understood fully by shippers of articles so rated (see illustrations).

Ratings for released valuation limit the carrier's liability for loss or damage to the amount stated and, because of this change in the bill of lading contract, may be published only with the express permission of the Interstate Commerce Commission. Shippers using

FREIGHT CLASSIFICATION

released valuation rates should also be acquainted with the provisions of Section 2 of the bill of lading as it affects claims for loss or damage. The shipper is not obligated to declare the full value of the goods, but may elect to release the shipment to the valuation that will entitle him to the lowest rating. If the shipper does elect to move his shipments via released valuation, he has three options regarding the difference between the true value and the released valuation:

1. Self-insure
2. Buy commercial insurance
3. Do nothing, dependent on claim experience and/or the spread of difference between true value and released valuation.

Those ratings which call for the declaration of actual valuation are quite another matter, for with these the carriers' liability is not limited. The shipper must declare that the valuation of the goods falls within the stated ranges published in connection with the rating. Failure to do so will subject him to the penalties provided in the Interstate Commerce Act for false billing and make him liable to a fine of as much as $5,000.

The Interstate Commerce Act, Part I, section 1(4) (see section 10701 of the Codefied Act) and corresponding sections in Parts II, III and IV require of common carriers that they establish just and reasonable classification. The monumental task of performing this duty for the railroads is in the hands of the Uniform Classification Committee, while motor carrier classification is handled by the National Classification Board. Both groups are composed of men who have made a speciality of classification matters for many years and it is a tribute to their skill that relatively few formal complaints involve the results of their work.

While the recent increase in the number of carload and trailerload "Freight-All-Kinds" rates, the cost-related scales for shipments weighing less than a certain number of pounds and other developments are taken by some people to herald the breakdown of traditional classification principles, it is doubted that the classifications, as we now know them, will be discarded in the foreseeable

INDUSTRIAL TRAFFIC MANAGEMENT

Item	ARTICLES	CLASSES (Ratings) LTL	TL	Min. Wt. Factor (See Rule 115)
99400	**HIDES GROUP**, subject to item 99800: Hides, Pelts or Skins, dressed or tanned or not dressed nor tanned, NOI, dry, see Note, item 99402:			
Sub 1	Released to value not exceeding $1.50 per pound, in packages	100	85	12.2
Sub 2	Released to value exceeding $1.50 per pound but not exceeding $5.00 per pound, in packages	250	150	12.2
Sub 3	Released to value exceeding $5.00 per pound but not exceeding $7.50 per pound, in packages	300	200	12.2
99402	Note—The released value must be entered on the shipping order and bill of lading in the following form: The agreed or declared value of the property is hereby specifically stated by the shipper to be not exceeding _____ per pound. If the shipper fails or declines to execute the above statement or designates a value exceeding $7.50 per pound, shipment will not be accepted, but if shipment is inadvertantly accepted, charges initially will be assessed on the basis of the rating for the highest value provided. Upon proof of lower actual value, the freight charges will be adjusted to those that would apply if the shipment had been released to the amount of its actual value. Ratings herein based on released value have been authorized by the Interstate Commerce Commission in Released Rates Order No. MC-519 of October 12, 1962, as amended November 19, 1962, subject to complaint or suspension. (See page 2 for state authorities.)			
88150	**GLASSWARE GROUP**, subject to item 87500: Glassware, NOI, in barrels or boxes, see Notes, items 88152, 88154, 88156 and 88166, or in Packages 563, 945, 1004 or 1024:			
Sub 1	Actual value not exceeding 35¢ per lb.	70	37½	24.2
Sub 2	Actual value exceeding 35¢ per lb., but not exceeding $1.50 per lb.	85	55	24.2
Sub 3	Actual value exceeding $1.50 per lb., but not exceeding $2.00 per lb.	100	70	20.2
Sub 4	Actual value exceeding $2.00 per lb., but not exceeding $3.75 per lb.	150	85	20.2
Sub 5	Actual value exceeding $3.75 per lb., but not exceeding $5.00 per lb.	200	100	20.2
Sub 6	Except as otherwise provided, see Note, item 88172, if actual value exceeds $5.00 per lb., or if shipper declines to declare value	NOT TAKEN		
88152	Note—Shipper must certify on shipping order and bill of lading the actual value of the property as follows: "The actual value of the articles herein described as Glassware NOI is embrasive of any authorized accompanying articles in the same package, and is specifically stated by the shipper to be 'not in excess of 35¢ per lb.', or 'in excess of 35¢ per lb., but not in excess of $1.50 per lb.', or 'in excess of $1.50 per lb., but not in excess of $2.00 per lb.', or 'in excess of $2.00 per lb., but not in excess of $3.75 per lb.', or 'in excess of $2.75 per lb., but not in excess of $5.00 per lb.' as the case may be.			

Excerpts from National Motor Freight Classification illustrating manner of publishing *released value* ratings and *actual value* ratings. The figures in the extreme right-hand column are minimum weight factors which, when used in connection with the table provided in Item 997, produce the truckload minimum weight.

future. The ratings presently in use appear to offer the best means of arriving at price levels for the transportation of an important part of the country's commerce; they have been upheld on many occasions by the Interstate Commerce Commission; and they seem to continue to have the important endorsement of the transportation industry and shippers alike.

Using the Classifications

At best, classification is far from being an exact science; rather it is a matter of judgment born of long experience. Because of the

FREIGHT CLASSIFICATION

many factors which are considered in the development of a classification rating, no precise formula is available to provide the shipping public with a key to the basis for the Committees' judgment. Some things can be stated with a degree of certainty, however. For example, the use to which an article is put does not control its rating; instead, the Committee will be guided by its transportation characteristics and what it is represented to be. Thus a manufacturer who sells an article of hardware made of brass may not expect as low a rating as the same article made of steel, even though his article is sold in competition with the steel product. Nor may something represented as "Miracle Cleaning Compound" but which consists of nothing but trisodium phosphate take advantage of any lower ratings which may apply on trisodium phosphate.

In selecting a classification description for a given article, it is not necessary to secure a ruling from the classification committee unless it proves difficult to locate its description among the 10,000 or more listed in the Classification. These various articles usually are grouped with others of similar nature under generic headings, such as Automobile Parts or Accessories (See pages 529 through 534 of UFC 12). If the article shipped plainly belongs under one of these generic headings it is not listed by name, a "noibn" (Not Otherwise Indexed by Name) description may be used if provided. The Uniform Classification furnishes an aid in assigning such descriptions by listing certain articles in the index only, referenced by an asterisk (*) and the item number for the "noibn" description that it has determined as being suitable for the article. Thus, the entry

* Scoops, ice cream 91810

appears in the index and reference to Item 91810 shows that the general description to be used is "Tools, noibn."

In the absence of any definite indication as to what the description should be, the matter may be resolved by submitting to the Committee as much information about the article as possible, including trade name, use, material from which made, weight per cubic foot, value per pound and how it is packed for shipment. Upon consideration of these facts, the Committee may

(1) Select a description already in the Classification.

INDUSTRIAL TRAFFIC MANAGEMENT

(2) Rule that it is to be described by analogy to another article of comparable characteristics.

(3) Proceed as in (1) or (2) above as a temporary expedient and place the matter on their formal docket for the determination of description and ratings to be used.

Item numbers from the index (page 11 through 425 of UFC 12) are "matched" with those for bill of lading descriptions to determine such descriptions and other information. In using a bill of lading description, care should be exercised to avoid ignoring an obviously applicable description in favor of another one that is not as specific, merely because the latter carries a lower rating. To illustrate, hand blown Lamp Chimneys, although they are articles of Glassware that could be valued at less than 35 cents a pound, may not be described as Glassware to obtain the Class 70 LCL rating applicable thereon because they are listed specifically and take a rating of Class 125. Also, it is important to bear in mind that classification by analogy, as provided in Rule 17 of the Uniform Classification, is not the shipper's prerogative, but is reserved exclusively to the carriers and their classification committees.

Shippers desiring changes in descriptions, ratings, minimum weights or packing requirements may file an application with the appropriate classification committee, either by letter or by using the regular forms provided for the purpose. The application should include the following information:

1. Name and complete description of article, material from which made and use or function of article.
2. Photograph, line drawing or blueprint in triplicate.
3. Length, width and height of package, in inches.
4. Weight, in pounds, per package or piece, as ready for shipment.
5. Weight, in pounds, per cubic foot (length times width times height in inches, divided by 1728, and the result divided into the weight per package).
6. Value per pound, as ready for shipment.
7. Details, if packaged differently for carload than for less-than-carload shipment, or truckload than for less-than-truckload shipment, as the case may be.

FREIGHT CLASSIFICATION

8. Details, if article can be or is nested (see Rule 21, end of this chapter).
9. Details, if article is knocked down (see Rule 19, end of this chapter).
10. Present classification description, ratings, minimum weight and packing requirements.
11. Details of change desired.
12. Data as to weight shipped annually, broken down by origin point and destination territory, CL and LCL or TL and LTL, as the case may be.
13. Complete justification for change.

The application will appear on the next public docket of the classification committee. Hearings are conducted by the rail committee six times a year at Chicago and elsewhere when required. Motor carrier classification docket hearings are held five times a year at Washington, D.C., New York, Chicago and Atlanta and elsewhere upon occasion. It is not absolutely necessary, but it is desirable, that a representative of the shipper requesting the change make a personal appearance at the hearing to answer questions the committee may have in connection with the proposal. Rather searching questions are likely to be asked and the representative should be thoroughly familiar with the article and all aspects of the proposal.

Decisions on proposals are not made at the hearings, but only after the committee members have had an opportunity to study and discuss the matter among themselves. Proponents then are notified whether or not the application received a favorable disposition, or if a modification was recommended. If the disposition was unfavorable or the modification unacceptable, a new application with additional justification may be submitted for docketing. If the committee then reaffirms its previous action and the matter is thought to be of sufficient importance, it may be made the subject of a formal complaint before the Interstate Commerce Commission.

The dockets of the Uniform Classification Committee appear weekly in *The Traffic Bulletin;* the dockets of the National Classification Board are furnished by the Board to the public for a nominal charge and are published also as a special section of *The*

INDUSTRIAL TRAFFIC MANAGEMENT

Traffic Bulletin, while the monthly docket of the New England Motor Rate Bureau Classification Committee is distributed free of charge to all tariff subscribers. Each one should be reviewed, promptly upon receipt, for the detection of proposed changes that might affect a given firm's commodities. Those thought to be favorable can be supported, and those considered objectionable can be protested, either by letter or by personal appearance at the hearings. In the latter case, the Committee should be asked for an assignment of time at the most conveniently located hearing. Shipper proposals appearing on the rail classification dockets will be heard by the full committee only in Chicago, but arrangements may be made for a hearing by one committee member in New York or Atlanta when necessary.

The skill with which the classification job is done in any company can have a far-reaching effect on that company's sales and profits as well as its compliance with the law; therefore it should be entrusted only to trained personnel, with the best possible background, who are thoroughly familiar with the transportation characteristics of their companies' products, as well as those of their competitors. It will assist them in the performance of their duties if they have a knowledge of Classification Committee procedure and the history of previous Interstate Commerce decisions and Classification Committee rulings affecting their own and other kindred commodities. (See Chapter 14 for the Traffic Department's functions in this regard.)

With every company having a direct cost interest in the freight charges on its shipments, it pays to make certain that correct descriptions are used on the commodities it ships or receives and that packaging is such as to command the most favorable rating. For example, one shipper may describe his product as "Hardware, iron" rated Class 70, LTL, while his competitor may correctly use the description "Iron, Hangers, pipe" rated Class 50, LTL. (NMFC A-12 Items 95190 and 105500, respectively, are involved in this example). FOB origin prices being equal, the difference in freight charges soon would force the first shipper out of a given market or compel him to reduce his prices. On the other hand, the price of raw materials or supplies purchased FOB destination may include a

FREIGHT CLASSIFICATION

transportation factor which would be inflated if an incorrect description or improper package were used.

Personnel responsible for shipping a company's products, as well as those charged with auditing its freight bills, should be kept posted currently as to the correct descriptions to be used. Those companies having but a limited number of products may do this by preparing and using a bill of lading on which both trade names and correct bill of lading descriptions are printed in the "Description of Articles" column, somewhat as follows:

Boxes Muffo (Muffin Preparations with Canned Fruit)
Boxes Gellina (Dessert Preparations, N.O.I.B.N.)
Boxes Bonzo (Malt and Milk Compound)

If the list of articles is too long to make a printed listing on a bill of lading practicable, the articles may be similarly listed on a sheet or card which can be displayed prominently in the shipping room and elsewhere for easy reference.

An extension of this idea may be found in a company classification, which may be prepared along the general lines and arrangements of the rail and truck classifications. It should contain a section devoted to an explanation of its use, shipping rules pertinent to the company's needs and a digest of such rules of the carriers' classifications as may be applicable to the company's shipments. Another section should consist of an index arranged alphabetically by trade names or numerically by product, piece part or specification numbers. By means of a system of keys or codes, these individual entries in the index are linked to a third section containing bill of lading descriptions and, if desired, packaging requirements, ratings and carload or truckload minimum weights. A simple publication of this kind might be set up along the following lines:

Section 1—Rules

Explanation of use. To find the correct bill of lading description, locate the name or number of the product in the index in Section 2, note the key (or code) number shown in connection with it and use the description provided in Section 3 for this key (or code). Example: Armatures are shown in the

INDUSTRIAL TRAFFIC MANAGEMENT

index as taking key (or code) 151. Reference to No. 151 in Section 3 indicates that the description "Electrical Equipment, Armatures" should be used on bills of lading and that they carry a rating of 77½.

Mixed Packages. When different articles are shipped in the same container, the rate for the article taking the highest rating will apply to the total weight of the package. Do not pack a small weight of high rated material in the same container with a greater weight of low rated articles. When it is necessary to ship different articles in the same container, the description for the highest rated article should be shown on the bill of lading followed by "and other articles rated the same or lower."

Grouping Shipments. All freight shipped on one day to one consignee at the same address should move on one bill of lading to assure lowest freight charges.

Section 2—Index

Article	Key No.
Ammeters	13
Amplifiers	1
Anodes, Copper	215
Armatures	151

Section 3—Bill of Lading Descriptions and Ratings

Key	Bill of Lading Description	Rating.
1	Electrical Equipment, Sets, Amplifiers	125
13	Meters, Electric, noibn	100
151	Electrical Equipment, Armatures	77½
215	Anodes, copper (may be shipped loose if each weighs 25 lbs. or more)	55

In a simple publication such as this, only LTL ratings are shown. Since it would be designed primarily for use in shipping departments, any rules shown should be confined to those actually

FREIGHT CLASSIFICATION

affecting shipping functions and should be stated in terms that can be understood easily by shipping personnel.

Larger firms may find it desirable to expand on this format by adding packing requirements, ratings for both LTL, carload and those applicable via truck when different; also minimum weights. (It is not difficult to show truckload minimum weights even if they tend to vary from territory to territory and even from carrier to carrier.)

Publication of this sort must be under constant review for changes and additions, and for this reason it is best that they be issued in loose-leaf form. Each reissued page should bear a correction number or other identification by which it may be determined whether or not the publication is in current condition.

These three ways of assuring the use of correct commodity descriptions and obtaining proper freight charges require preprinted bills of lading, card lists, and company classifications. Combining and "personalizing" these aids to individual situations is also helpful. To illustrate: Partially preprinted bills of lading may be used in which the most frequently moved commodities are preprinted and blank lines left for the occasional movers. Also, partially preprinted bills of lading may be used along with card lists and company classifications where indicated.

The index of the Uniform Freight Classification carries the Standard Transportation Commodity Code (STTC) designation for all articles listed in the index—UFC 12. This code was originally developed to "input" transportation statistics to computers. Subsequently its uses have been broadened to include freight rates information and other purposes. A number of industries are using the Code in their classification work and in other ways. Its use is explored more thoroughly in Chapter 15.

Classification Rules

The carriers' classifications also set forth general rules governing the transportation of freight. This eliminates repeating those rules in each of the thousands of currently effective carrier tariffs.

INDUSTRIAL TRAFFIC MANAGEMENT

The rules of the various classifications govern class and commodity rates (for discussion of commodity rates, see Chapter 5) to the extent provided in the tariffs naming those rates. Thorough familiarity with these rules is an absolute necessity for all traffic professionals because of the prominent part they occupy in the transportation job. The following summary is not intended as a substitute for a careful reading of the rules themselves, but is designed to point out and explain the salient features of each. The listing shows the rules of the rail classification, with reference to the corresponding provisions, if any, in the truck classifications.

Rule 1. Gives shippers the privilege, rarely used, of tendering shipments with carriers' liability limited by common law instead of by the bill of lading contract, and provides an increase of 10% in freight charges for doing so.

Also provides for bill of lading forms to be used generally, except when shipments are tendered as above.

(Items 360 and 365 of National Motor Freight Classification do not carry the alternative privilege, but merely prescribe bill of lading forms to be used; Rule 20 of Coordinated Classification.)

Rule 2. States that bill of lading descriptions should conform to those in the classification and provides for inspection of shipments by carriers if they deem it necessary. See Chapter 7 for additional details.

(Item 360 of National Motor Freight Classification; Rule 20 of Coordinated Classification.)

Rule 3. Prohibits acceptance, either by themselves or as premiums accompanying other articles, of property of extraordinary value and other named articles.

(Item 780 of National Motor Freight Classification; Rule 150 of Coordinated Classification.)

Rule 4. Relieves carriers of obligation to accept freight liable to impregnate or otherwise damage equipment or other freight. See Chapter 7 for additional details.

(Item 780 of National Motor Freight Classification; Rule 150 of Coordinated Classification.)

Rule 5. Requires that articles be tendered for shipment in such condition, and so prepared, as to render transportation of them reasonably safe and practicable; and defines various packages and

FREIGHT CLASSIFICATION

provides basis for assessment of freight charges on articles found in transportation packed otherwise than required in the classification, but prohibits use of these provisions as a basis for quoting rates. See Chapter 7 for additional details.

(Items 680, 685, and 687 of National Motor Freight Classification contain provisions along generally similar lines as well as alternate forms of packing and packaging.)

Rule 6. Gives instructions for the marking or tagging of individual pieces of freight. See Chapter 7 for additional details.

(Item 580 of National Motor Freight Classification; Rule 40 of Coordinated Classification.)

Rule 7. Prescribes information to be shown on bills of lading and describes conditions under which delivery of shipments will be made under both Order and Straight Bills of Lading, as well as import shipments.

(Various items, e.g., 360 sec. 4, of National Motor Freight Classification and Rule 20 of Coordinated Classification are similar but make no reference to import shipments.)

Rule 8. Prohibits advancing charges to shippers, owners, consignees, their draymen or warehousemen unless specifically provided for in carriers' tariffs. This means that, for example, a warehouseman may not collect storage charges on a shipment from a carrier and ask the carrier to collect them, in turn, from the consignee.

(Item 300 of National Motor Freight Classification; Rule 130 of Coordinated Classification.)

Rule 9. Requires prepayment or guarantee of freight charges on shipments which, in the event of non-delivery, might not yield enough at forced sale to cover the amount of freight charges.

(Item 770 of National Motor Freight Classification; Rule 120 of Coordinated Classification.)

Rule 10. Prescribes the basis for freight charges on carloads consisting of a mixture of differently rated articles or articles carrying different carload minimum weights.

(Items 640 and 645 of National Motor Freight Classification and Rule 70 of Coordinated Classification prescribe bases for charges on LTL shipments as well as truckload.)

Rule 11. Provides that freight charges shall be computed on

gross weights or on estimated weights when specifically authorized.

(Item 995 of National Motor Freight Classification; Rule 60 of Coordinated Classification.)

Rule 12. States basis for assessment of freight charges on LCL shipments of one or more classes and for packages containing differently rated articles, less carload or carload.

(Items 640 and 645 of National Motor Freight Classification; Rules 70 and 80 of Coordinated Classification.)

Rule 13. Provides minimum charges per LCL shipment and per carload. This means that, if the weight multiplied by the rate results in lower charges, these minimum charges are applied.

(No corresponding provisions in National Motor Freight nor Coordinated Classifications. Carried in individual tariffs.)

Rule 14. Defines carload freight and states conditions under which carload rates will apply on shipments accorded services ordinarily not included in carload rates.

(Item 110 of National Motor Freight Classification and Rule 10 of Coordinated Classification define all shipments; otherwise no corresponding provisions.)

Rule 15. States conditions under which LCL shipments may be subject to carload charges and carload shipments may be subject to LCL charges. It specifically excludes from the application of carload rates shipments that have been accorded railroad pickup or delivery service, or an allowance in lieu thereof, unless carriers' tariffs provide otherwise.

(No corresponding provisions in National Motor Freight or Coordinated Classifications.)

Rule 16. Defines less carload freight.

(Item 110 of National Motor Freight Classification; Rule 10 of Coordinated Classification; LTL.)

Rule 17. Provides for classification by analogy when articles are not specifically provided for nor embraced by an "noibn" description.

Note: This may be done only by carriers or their agents, which also would include Classification Committees.

(Item 421 of National Motor Freight Classification; Rule 220 of Coordinated Classification.)

FREIGHT CLASSIFICATION

Rule 18. Provides that, when the combination is not specifically listed, articles which are combined or attached to one another will be charged for at the rating applicable to the highest rated article in the combination and, on carload shipments, at the highest carload minimum weight.

(Item 422 of National Motor Freight Classification; Rule 100 of Coordinated Classification.)

Rule 19. Defines "knocked down" articles by stating that KD ratings will apply only on articles taken apart in such manner as to reduce bulk by at least 33⅓ percent. See Chapter 7 for additional details.

(Item 110 of National Motor Freight Classification; Rule 10 or Coordinated Classification.)

Rule 20. States that parts or pieces constituting a complete article, when received on one bill of lading, will be charged for at the rating applicable to the complete article. Thus, for example, it is not permissible to ship a sidewalk sign consisting of a metal sign weighing seven pounds, a pipe upright weighing 12 pounds and a cast iron base weighing 30 pounds on the same bill of lading and describe them separately in order to obtain the lower ratings provided for the pipe and base.

(Item 424 of National Motor Freight Classification; Rule 110 of Coordinated Classification.)

Rule 21. Defines the term "nested." See Chapter 7 for additional details.

(Item 110 of National Motor Freight Classification; Rule 10 of Coordinated Classification.)

Rule 22. Defines "in the rough," "in the white" and "finished" as used in connection with classification descriptions of wooden articles. See Chapter 7 for additional details.

(Item 110 of National Motor Freight Classification; Rule 10 of Coordinated Classification.)

Rule 23. Prohibits loading of freight in the bunkers (ice compartments) of refrigerator cars.

(No corresponding rule in National Motor Freight nor Coordinated Classifications.)

Rule 24. Provides for manner of handling and basis for charges

INDUSTRIAL TRAFFIC MANAGEMENT

on quantities of freight in excess of that which can be loaded in or on a single car.

Note: This rule is recommended for careful study by shippers of carload freight because proper use of its provisions can operate to save considerably in transportation charges. Many exceptions to this rule appear in individual tariffs and in connection with individual rates.

(No corresponding rule in National Motor Freight or Coordinated Classification. Carried in individual rate tariffs.)

Rule 25. Provides for the interchangeability of the word "iron" with the word "steel" and states that where reference is made to the gauge of metal, it means U.S. Standard Gauge.

(Item 110 of National Motor Freight Classification; Rule 10 of Coordinated Classification.)

Rule 26. Does not appear in the Uniform Classification.

Rule 27. Requires shippers to load, and consignees to unload, carload freight and heavy or bulky LCL freight; also requires safe loading of cars and protection of detachable parts of articles shipped on open cars.

(Item 568 of National Motor Freight Classification, confined to heavy or bulky freight, which is defined. No similar provisions in Coordinated Classification.)

Rule 28. Provides that the word "rubber" shall indicate the natural or synthetic varieties; defines "synthetic plastics" and provides that references to aluminum, magnesium, zinc and metal, noibn, shall include alloys thereof.

(Item 110 of National Motor Freight Classification; Rule 10 of Coordinated Classification; rubber and plastics only.)

Rule 29. Provides basis for charges on carload and less carload shipments which, because of length or dimensions, require two or more cars or series of cars and specifies four cars as the maximum number in a series. Provision also is made for loadings which cannot be accomplished through a box car's side door measuring six feet wide by nine feet, four inches high.

(No comparable rule in National Motor Freight or Coordinated Classification.)

Rule 30. Defines dunnage as "temporary blocking, flooring or lining, standards, strips, stakes or similar bracing or supports not

FREIGHT CLASSIFICATION

constituting a part of the car" when required to protect and make carload freight secure for shipment. It specifically excludes from the term "dunnage" packing material such as excelsior, hay, sawdust, shavings, shredded paper, straw or packing cushions. Dunnage must be furnished and installed by the shipper at his expense. If the actual weight of the dunnage is specified on the bill of lading, carriers will allow up to 2,000 pounds of it to move in each carload free of transportation charges provided the carload minimum weight is met.

(Item 995 of National Motor Freight Classification and Rule 170 of Coordinated Classification. The latter is silent as to a provision for free transportation.)

Rule 31. Provides that ratings do not include the expense of refrigeration nor obligate carriers to maintain heat in cars, except to the extent that their tariffs provide otherwise.

(No corresponding rule in National Motor Freight Classification.)

Rule 32. Provides that transportation charges will not be assessed on the weight of ice in the bunkers of refrigerator cars unless removed and appropriated by the consignee, in which case it will be charged for at the carload rate applicable to the freight it accompanies.

(No corresponding provisions in National Motor Freight or Coordinated Classification.)

Rule 33. Provides for the alternative application of varying carload ratings and minimum weights.

(Item 595 of National Motor Freight Classification; see Rule 290 in Coordinated Classification as to Connecticut traffic.)

Rule 34. Deals with the matter of carload minimum weights that are made subject to the rule by specific reference or listed in the classification suffixed by the letter "R." In such cases, when a closed car longer than 40 feet 7 inches but not over 52 feet 8 inches is ordered for loading, the carload minimum weight is increased 40%, and if a car over 52 feet 8 inches is ordered, the carload minimum weight is increased 100%. Open cars are subject to a graduated scale whereby the carload minimum weight is applicable on cars up to 41 feet 6 inches and when longer cars are ordered the carload minimum weight is increased as per charts in Rule 34. The rule also makes provision for the furnishing of two cars when a single car of the length ordered is not available.

Note: This rule is of utmost importance to shippers of light and bulky freight and should be studied carefully for its effect on a given firm's shipping practices.

(No comparable rule in National Motor Freight nor Coordinated Classifications.)

Rule 35. Contains a number of basic provisions for the computation of carload minimum weights and freight charges on shipments in tank cars. It relieves the carriers of the obligation to furnish tank cars.

(No corresponding item in National Motor Freight Classification.)

Rule 36. Provides that, when applying a percentage factor to a given rate, fractions of less than ½ or .50 of a cent in the result will be dropped and those ½ or .50 of a cent or greater will increase the result to the next whole cent.

(Item 565 of National Motor Freight Classification; Rule 230 of Coordinated Classification.)

Rule 37. Specifies that publication of rates on shipments in covered hopper cars does not obligate carriers to furnish such cars. This rule is not applicable on Texas intrastate traffic.

(No comparable rule in National Motor Freight or Coordinated Classification.)

Rule 38. States that an exception rating must be used instead of the corresponding classification rating and that the establishment of a commodity rate on a given article supersedes both unless the tariff publishing and exception rating or the commodity rate specifically provides otherwise. It also permits alternation of class and commodity rates on Illinois intrastate traffic. Rule 38 does not apply on Florida intrastate traffic.

(Item 765 of National Motor Freight Classification, and Rule 180 of Coordinated Classification, except that they do not contain the provisions for Illinois and Florida intrastate traffic.)

Rule 39. Requires that explosives and other dangerous articles transportation comply with the rules and regulations prescribed by

FREIGHT CLASSIFICATION

the ICC and published in the tariff of the Bureau of Explosives or in the *Federal Register* or, if by water carrier, those prescribed by the Commandant, United States Coast Guard. It also requires that if the description in the Bureau of Explosives tariff differs from the classification, exception or commodity rate description, the former must appear first on the bill of lading with the latter following in parentheses. See Chapter 7 for additional details. (ICC means DOT here.)

(Item 540 of National Motor Freight Classification is comparable but refers to the Department of Transportation; Rule 40 of Coordinated Classification provides for marking and packing per the New England Motor Rate Bureau Explosives tariff.)

Rule 40. Carries detailed specifications for numerous forms of shipping containers made of materials other than fibreboard. See Chapter 7 for additional details.

(Items 200-297, inclusive, of National Motor Freight Classification; Rule 30 of Coordinated Classification.)

Rule 41. Carries detailed specifications for shipping containers made of solid or corrugated fibreboard including the form of certificate required to be imprinted thereon by the manufacturer of the container. See Chapter 7 for additional details.

(Items 200-297, inclusive, of National Motor Freight Classification; Rule 30 of Coordinated Classification.)

Rule 42. Provides that reshipping documents, invoices, assembly or operating instructions or x-ray photographs may be forwarded in packages with other articles at the rate or rating applicable on the articles they accompany.

(Item 428 of National Motor Freight Classification; Rule 200 of Coordinated Classification is along related lines but provides for forwarding without charge with truck loads.)

Rule 43. Prescribes the form of separate contract to be used when men accompany shipments of property, other than livestock, live wild animals or ostriches.

(No corresponding provisions in National Motor Freight nor Coordinated Classifications.)

Rule 44. Describes the method by which changes in individual items will be designated in supplements to the classification by use of lettered suffixes. Example: Item 1500-B cancels Item 1500-A of a prior supplement, which in turn had cancelled Item 1500.

(Item 381 of National Motor Freight Classification; Rule 50 of Coordinated Classification.)

Rule 45. Contains provisions for the acceptance of and assessment of charges on advertising matter, advertising signs, store display racks, stands and premiums when they accompany the articles they advertise or are given with. Ordinarily, the advertising materials may not exceed ten percent of the gross weight of those articles they accompany, and no more than one premium may be enclosed with each inner package. Freight charges then will be computed at the rate applicable to the article they accompany; any excess will be charged for at the rate applicable to the advertising matter or premiums.

(Item 310 of National Motor Freight Classification; Rule 90 of Coordinated Classification.)

Rule 46. Explains that the word "and" is used to couple the terms between which it appears and that "or" provides for alternation or use of either or both of the terms between which it appears, sometimes a very important distinction. It also states that a name appearing in parentheses is to be read as another description of the article preceding the parentheses. There is an explanation of the indentations used in the text of the classification to the effect that an entry so set away from the left margin must be read in relation to the heading to which it is subordinate. Example: Item 53630 reads "Angles, noibn" but because it is indented from the heading "IRON OR STEEL" in Item 53610, it must be read as "Iron or Steel Angles, noibn." This rule defines "rate" as the specific figure published in tariffs to be used in computing freight charges, and "rating" or "column" as signifying the numerals used in the classification or exceptions to identify the "rate" published in class or commodity column tariffs.

(Item 110 of National Motor Freight Classification; Rule 10 of Coordinated classification.)

Rule 47. Contains rules under which carriers will make collection of COD charges (not their freight charges) on LCL or any quantity shipments and provides a scale of charges based on the amount to be collected for performing this service.

FREIGHT CLASSIFICATION

(Item 430 of National Motor Freight Classification; Rule 190 of Coordinated Classification. These are not limited to LCL or AC shipments.)

Rule 48. Provides that where reference is made to tariffs, items, notes, rules, etc., such reference includes supplements to and reissues of those tariffs and reissues of those items, notes, rules, etc.

(Item 845 of National Motor Freight Classification; Rule 240 of Coordinated Classification.)

Rule 49. States conditions under which the Classification Committee will issue permits for the experimental use of otherwise unauthorized containers for the purpose of determining the merits of such containers. See Chapter 7 for additional details.

(Item 689 of National Motor Freight Classification, Rule 300 of Coordinated Classification.)

Rule 50. Provides that up to 25 pounds of empty shipping containers may be included in carloads for the purpose of reconditioning the car's contents at destination. See Chapter 7 for additional details.

(Item 426 of National Motor Freight Classification as to bags; no corresponding provisions in Coordinated Classification.)

Rule 51. Provides that when packing requirements call for wooden barrels, drums, pails or tubs, such containers may be constructed of fibreboard. Also provides penalties when specifications are not met. See Chapter 7 for additional details.

(Items 200-297, inclusive, of National Motor Freight Classification; Rule 30 of Coordinated Classification.)

Rules 52 and 53. Provide basis for computation of carload weights on shipments of Liquefied Petroleum Gas, Butadiene, and Isoprene.

(No comparable provisions in National Motor Freight nor Coordinated Classifications.)

Rule 54. Provides that, as to multi-level flat cars, rates and weights published in U.F.C. shall not be applicable. This provision is currently under suspension by the ICC.

(No comparable provisions in National Motor Freight nor Coordinated Classifications.)

Rule 55. Describes method of denoting matter brought forward from one supplement to another without change.

(No comparable provisions in National Motor Freight nor Coordinated Classifications.)

Rules 56-59, inclusive. Do not appear in Uniform Classification.

Rule 60. Provides minimum charge for the use of certain types of flat cars.

(No comparable provisions in National Motor Freight nor Coordinated Classifications.)

Rule 61. Specifies calculations to be made in determining weight per cubic foot.

(Item 110 Sec. 8 of National Motor Freight Classification; Rule 10 Sec. s of Coordinated Classification.)

National Motor Freight Classification Rules

The National Motor Freight Classification has several items not referred to above. They include:

Item 360. Certain sections establish charges for extra copies of shipping documents, other papers, and related matters. Exceptions include documents incident to bank and sight draft plans.

Item 420. Explanation of classes.

Item 535. Table of expiration dates in connection with certain temporary provisions.

Item 997. The purpose of this rule is similar to that of Rule 34 of the Uniform Classification in that it attempts to adapt truckload minimum weights to the size of equipment available. Articles in the classification are given individual "minimum weight factors" (see last column, illustration, page 48) instead of minimum weights stated in pounds. Application of the minimum weight factor to the proper table will indicate the truckload minimum weight in pounds.

4

SHIPPING DOCUMENTS

Of all the documents required in the transportation of goods, none is more important than the bill of lading, the origins of which extend far back into the annals of commerce when ships were the only carriers. The form and terms of the bill of lading have evolved over the years and varied according to the country in which used, but its basic purpose has remained unchanged. In the United States, standardization of its form and contract provisions took place on January 1, 1917, under the Federal Bills of Lading Act.

This act, in addition to standardizing, clothed the bill of lading with much of its high standing as a document by making provision for heavy penalties for altering, forging, counterfeiting or falsely making a bill of lading for criminal purposes. Convictions on such charges subjects the guilty parties to imprisonment for as long as five years, up to $5,000 in fines, or both, for each offense. The Interstate Commerce Act imposes upon common carriers the duty of issuing bills of lading and prescribes the general terms of the contract of carriage. The actual form of the bill of lading is set forth in the carriers' classifications.

The importance of the bill of lading lies in the several functions it serves in connection with the shipment it covers. Today, the bill is usually issued in quadruplicate although for many years it was issued in triplicate. The consignee may sign the forms for receipt of the goods described on the bill by using interleaved carbon paper. (Some forms are now printed on chemically treated paper so that a signature will appear on all copies without the use of carbon paper.) The forms consist of an original, shipping order (carrier's copy) and either two memorandum copies or an additional shipping order and a memorandum. The carrier must get the shipping order. Whether he gets a second copy is dependent on the payment system used by the shipper. Some shippers may give the carrier a second copy, which the carrier may have to attach to his freight bill, if the shipper uses a bank-freight-payment plan.

In any event the purposes and uses of the copies of the bill of lading are as follows:—

No. 1 Copy—Original. This copy, bearing signatures of shipper's and carrier's agents, serves as a description of the shipment and its contents and as a receipt by the carrier for the goods; thus it is acceptable as proof that shipment has been made and as evidence of beneficial ownership. At the same time, it is a contract and clearly fixes the carrier's liability (see Appendices 4 and 5). Under ordinary circumstances, it should be in the possession of the party having title to the goods while in transit, for it is required in support of loss and damage claims.

No. 2 Copy—Shipping Order. This copy is retained by the carrier and serves as forwarding instructions and as a base for billing.

No. 3 Copy—Memorandum. This copy is an acknowledgment that a bill of lading has been issued and, while it does not have the standing of the original as a receipt and contract, is valuable as a record, especially in cases where the original must be surrendered.

Although the Interstate Commerce Act requires carriers to issue bills of lading, they are almost always actually prepared by the shippers. Sometimes they are on blank forms supplied by the carriers, but more and more frequently they are on forms designed to suit an individual shipper's needs, carrying printed lists of his principal commodities and printed at the shipper's expense. The carrier signs the bill as a firm contract. Often, the three bill of lading copies are part of a manifold of six or more that also may include an order acknowledgment, warehouse release, shipping notice, invoice, ledger copy or other forms incident to the transaction. There are no restrictions as to what may be manifolded with the bill of lading as long as the bill of lading itself conforms to the illustrations published in the carriers' classification.

SHIPPING DOCUMENTS

What is a Bill of Lading?

When a shipper turns over his shipment to a common carrier he obtains a special type of receipt called the bill of lading. On its face, the bill of lading contains a description of the shipment, its value, routing information and the names and addresses of the consignor and consignee. This information is used by carriers in making up the waybill and other operating papers required in the routing and handling of the shipment as well as for invoicing.

On the back of the bill of lading are the printed terms of contract, setting forth in detail the responsibilities and liabilities of the carrier and of the owner of the shipment. The purpose here is to give reasonable protection to both shipper *and* carrier. The conditions on the back of the bill summarize the laws on which the bill of lading is based. The bill of lading is therefore at one-and-the-same-time a receipt, contract, basic operating paper and evidence of ownership of the shipment.

There are three types of bills of lading prescribed by the Interstate Commerce Commission for use by the railroads. The U.S. Government, the water carriers, and the air lines have their own special forms, as do certain specialized motor carriers and household goods carriers.

Although the ICC does not prescribe a bill of lading form for general commodity motor carriers, the National Motor Freight Classification publishes general rules and formats almost identical to those governing rail traffic. Currently there are several proposals pending that would eliminate "Order Notify Ladings" and establish a uniform format for motor carrier bills of lading. Attachment 2 shows copies of the currently applicable NMFC rules.

Railroad Bills of Lading:

Uniform Domestic and *Uniform Through Export* bills of lading are of two kinds:
 1. *Straight Bill of Lading*—this is not negotiable and con-

tains a statement that goods are consigned or destined to a specific consignee.

2. *Order (Notify) Bill of Lading*—this is a *negotiable* document that may be bought and sold. But there are restrictions on its use:
— It must be consigned to an "agency" station
— No stopping in-transit privileges are allowed with this type of bill
— No predelivery inspection of the shipment is allowed

Uniform Livestock Contract—Uniform order or straight bills of lading cannot be issued to cover the movement of livestock or live wild animals. These must be consigned on a straight livestock contract.

The Carrier's Duty to Issue Bills of Lading

The Interstate Commerce Act states that a common carrier must issue a receipt or bill of lading. The obligation lawfully rests upon the carrier's agent to direct the shipper's attention to any inconsistency or conflict in the provisions of the bill of lading. Although it is the carrier's responsibility to issue the bill of lading, the shipper may, and usually does, furnish his own forms so long as they conform to all essential particulars of the documents prescribed by the ICC. There are many reasons why a shipper would want to provide his own forms:

—Savings in clerical expense through the use of special machines or other devices, and pre-numbered forms.
—Less likelihood of error in describing the goods if the accepted classification description is already printed on the form.
—Necessity for additional copies.
—Uniformity of size.

The shipper's bill of lading is generally the "short" form. The difference between "long" and "short" forms is that the bill of lading liability terms are not printed on reverse of the "short" form; but there is a statement at the top of the short bill to the effect that the shipper acknowledges and accepts the usual bill of lading terms, nonetheless.

SHIPPING DOCUMENTS

STRAIGHT BILL OF LADING
ORIGINAL — NOT NEGOTIABLE

Name of Carrier: KEYNLEE MOTOR FREIGHT

Shipper No. _____
Carrier No. _____
Date: August 19

TO:
Consignee: Donald's Specialities
Street: 4286 Burt Road
Destination: Detroit, Michigan Zip Code: 48228

FROM:
Shipper: F. Mac Williams & Sons
Street: 1234 Ninth Ave.
Origin: Charleston, WV Zip Code: 25303

Route: Keynlee Motor Freight - Columbus, OH - Marshall Freight
Vehicle Number: ____

No. Shipping Units	Kind of Packaging, Description of Articles, Special Marks and Exceptions	Weight (Subject to Correction)	Rate	CHARGES (for Carrier use only)
14	Glassware NOI, In Boxes	532	85	

REMIT C.O.D. TO: ____
ADDRESS: ____

COD Amt: $ ____

C.O.D. FEE:
PREPAID ☒
COLLECT ☐ $ ____

Note — Where the rate is dependent on value, shippers are required to specifically state in writing the agreed or declared value of the property. The agreed or declared value of the property is hereby specifically stated by the shipper to be not exceeding

One Dollar   Pound
$_____ per _____

TOTAL CHARGES: $ ____

FREIGHT CHARGES:
FREIGHT PREPAID ☐ Check box if charges are to be collected.

RECEIVED, subject to the classifications and tariffs in effect on the date of the issue of this Bill of Lading, the property described above in apparent good order, except as noted (contents and conditions of packages unknown), marked, consigned, and destined as indicated above which said carrier (the word carrier being understood throughout this contract as meaning any person or corporation in possession of the property under the contract) agrees to carry to its usual place of delivery at said destination, if on its route, otherwise to deliver to another carrier on the route to said destination. It is mutually agreed as to each carrier of all or any of, said property over all or any portion of said route to destination and as to each party at any time interested in all or any said property, that every service to be performed hereunder shall be subject to all the bill of lading terms and conditions in the governing classification on the date or shipment.

Shipper hereby certifies that he is familiar with all the bill of lading terms and conditions in the governing classification and the said terms and conditions are hereby agreed to by the shipper and accepted for himself and his assigns.

SHIPPER	F. Mac Williams & Sons	CARRIER	Keynlee Motor Freight
PER		PER	
		DATE	

Form No. 3, The Traffic Service Corp, 815 Washington Building, Wash, D.C. 20005

Uniform Straight Bill of Lading.

Functions of the Bill of Lading: The bill of lading performs three distinct functions:—
—It is a receipt for the goods.
—It is a contract of carriage.
—It is a documentary evidence of title.

Terms of the Bill of Lading: These are summarized below under the section headings used for the terms printed on the back of the straight bill. (For complete text of terms see Appendix 1).

Section 1.—Common carriers are subject to three types of liability: (a) common law; (b) common carrier or bill of lading liability; and (c) warehouseman liability. This section of the bill of lading covers the carrier's liability and indicates that a common carrier shall be absolutely liable, without proof of negligence, for all loss, damage or injury to goods transported by it, unless the carrier can affirmatively show that the damage was occasioned by one of the exceptions contained in the terms of the bill of lading which include: 1. Acts of God; 2. acts of the public enemy; 3. acts of negligence of the shipper; 4. inherent nature of the goods; and 5. acts of public authority. The terms also prescribe a situation in which a carrier is liable only as a warehouseman and other situations in which the carrier is exempt from liability.

Section 2—"No carrier is bound to transport said property by any particular train or vessel, or in time for any particular market or otherwise than WITH REASONABLE DISPATCH."

Section 3—Except when such service is required as the result of the carrier's negligence, all property shall be subject to necessary cooperage and bailing at the owner's cost.

Section 4—Concerns application of storage and warehouse charges for freight not removed by the party entitled to receive it.

Section 5—No carrier will carry or be liable in any way for any documents or articles of extraordinary value not specifically rated in the classification or tariffs unless a special agreement is made. But if the goods are specially rated in the classification, they must be accepted regardless of value.

Section 6.—Every party shipping explosives or dangerous goods without previous full written disclosure to the carrier shall be liable for any loss or damage caused by such goods, whether it be to other goods or to the carrier's property. Such goods may be warehoused

SHIPPING DOCUMENTS

at the owner's risk and expense or destroyed without compensation.

Section 7—The first part of this section pertains to the "without recourse" clause (see below). The second part indicates the procedure an agent acting as consignee must follow if he desires to escape liability for lawful freight charges. The third part states that the carrier always has a right to require prepayment or a guarantee of the freight charges at time of shipment.

The "without recourse" clause refers to the fact that the bill of lading is the contract of carriage, and the shipper, being a party to this contract, is held liable for the freight charges even when the carrier delivers a "freight collect" shipment without first collecting the freight charges. The shipper may exempt himself from this liability by signing the "without recourse" clause found on the face of the uniform domestic bills of lading. If the shipper does not sign this clause in the bill of lading, the carrier will have recourse to the shipper for the freight charges even though it delivered the shipment and did not make any real effort to collect the charges from the consignee.

Section 8—Provides "If this bill of lading is issued on the order of the shipper, or his agent, in exchange or in substitution for another bill of lading, the shipper's signature to the prior bill of lading as to the statement of value or otherwise, or election of common law or bill of lading shall be considered a part of this bill of lading as fully as if the same were written or made in connection with this bill of lading."

Section 9.—This section stipulates the liability of carriers by water when they are parties to a through movement and route. The essence of water carrier liability is that it arises only out of the failure or negligence of the owner in loading cargo, in making the vessel seaworthy, or in outfitting and manning the vessel. The water carrier is not responsible for loss or damage resulting from fire, bursting of boilers, etc.

Section 10—"Any alteration, addition, or erasure in this bill of lading which shall be made without the special notation hereon of the agent of the carrier issuing this bill of lading, shall be without effect, and this bill of lading shall be enforceable according to its original tenor."

Preparing the Bill of Lading

The bill of lading, for clarity, should be typewritten and should as a minimum contain the following information: (See illustration on page 75.)

1. Name of shipper and mailing address (this may be printed on the form).
2. Point of origin.
3. Date of shipment.
4. Consignee.
5. Destination.
6. Delivery address—in detail, if necessary.
7. Car number or vehicle number (if applicable).
8. Seal numbers, if seals are used.
9. Complete route.
10. Designation of "prepaid" or "collect" for freight charges.
11. Description of shipment, and the number of, and kind of packages.

Description of articles—special marks and exception. The correct freight classification is essential in describing the article, or product shipped. When more than one product is shipped and more than one freight classification is applicable, there must be a separate entry on the bill of lading for each freight classification description, and if more than one type of package is involved, there must be a separate entry for each type of package (to permit the application of correct freight rates).

For "dangerous articles" the provisions of Section 173.427 of the

---

NOTE *Shipper's Load and Count Notation:* The Bill of Lading Act provides, "The carrier may also, by inserting in the bill of lading the words, "shipper's weight, load and count," or other words of like purport indicate that the goods were loaded by the shipper and the description of them by him, and if such statement be true, the carrier shall not be liable for damages caused by the improper loading or by the non-receipt or by the mis-description of the goods described in the bill of lading." Although this stipulation does not relieve the carrier from liability for loss or damage, it has the effect of placing the burden upon the shipper to prove that the amount specified was loaded and that a smaller amount was taken out of the car by the consignee; whereas in the case of a bill of lading not so qualified the burden is upon the carrier to prove that the amount specified in the bill of lading was either not in fact loaded, or was delivered.

SHIPPING DOCUMENTS

Dangerous Articles Tariff must be met. This provides "each shipper offering for transportation any dangerous articles subject to the regulations, shall describe that article on the shipping paper by the shipping name prescribed in Section 172.5 and by the classification prescribed in Section 172.4, and may add a further description not inconsistent therewith, abbreviations must not be used. The total

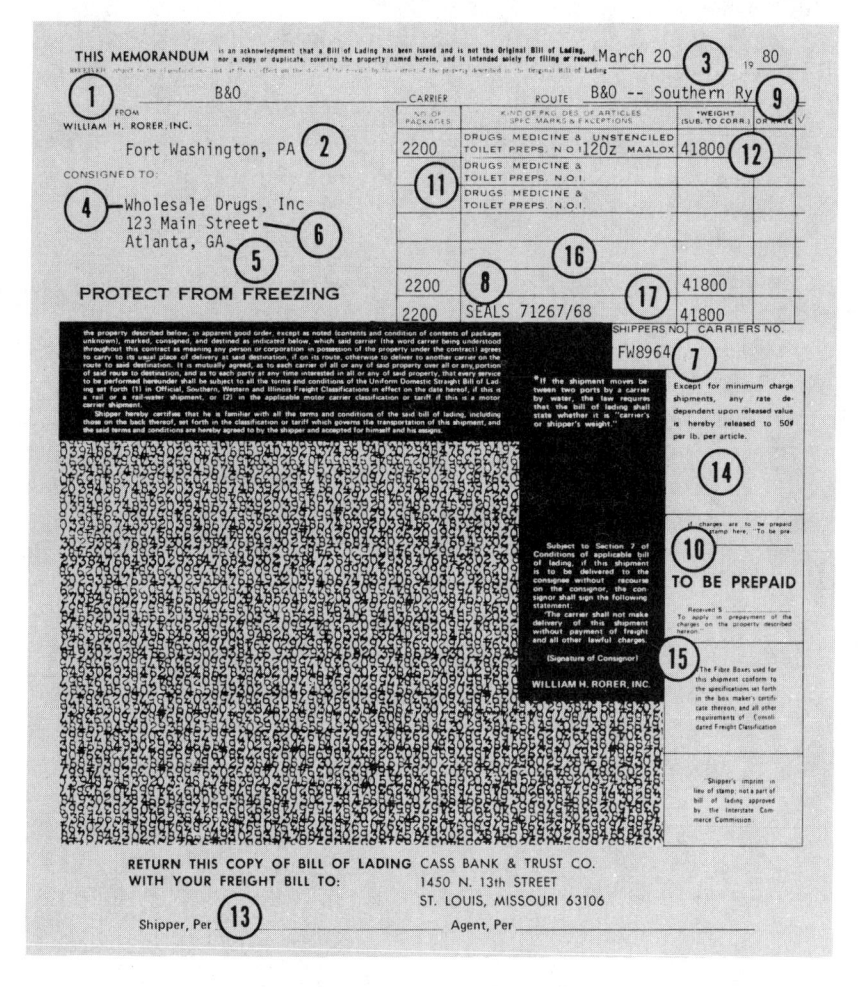

Straight Bill of Lading—Short Form.

quantity by volume weight or otherwise must be shown." The proper freight classification and dangerous articles designation should be determined from the specific instructions.

Stop Off Bill of Lading.

SHIPPING DOCUMENTS

12. Weights. For bulk shipments, the gross, tare and net weight should be shown, except when estimated or calculated weight is authorized. For package shipments, the gross weight (actual or agreed) for each individual entry on the bill of lading, plus the pallet weight, should be indicated, with a total of the product weights shown after the last item is listed, with pallet and dunnage weights (when used) shown separately. For agreed weights, the weight agreement stamp must be used on all copies of the bill of lading.

13. Signature. This is required on the Department of Transportation certificate for shipments of dangerous articles and acknowledges the conditions of the certification must be met; the signature is optional on other shipments, but is recommended.

14. The "released valuation" (at 50¢ per pound) provision, if applicable.

15. Other imprinted provisions, including the "fiber box conformation" and, in some instances, the designation of trip lease notation—"shell full" or "not shell full" designation—and other specific provisions required for individual locations.

16. Any additional information, detailed instructions for the carrier, or special provisions necessary for the proper handling of the shipment, should be included in the description section of the bill of lading, but information confusing to the carrier or confidential to your company should not appear on the carrier copy of the lading.

17. More than one car or truck may be covered by one bill of lading if they are part of the same shipment and there is a separate indication of seal numbers, weights and vehicle number.

Distribution of the Bill of Lading Copies:

Since practices vary widely, dependent on particular company and carrier requirements, there are no general rules as to the number and distribution of copies. Generally, however:—
- The carrier should always receive the # 2 copy.
- The original must be retained by shipper for a minimum of three years from date of shipment.

The Ocean Bill of Lading

The ocean bill of lading serves the same purposes as the domestic bill of lading, i.e., a contract of carriage, a receipt for the goods shipped, evidence of title, and so on; but it also has other purposes—for example, its use in customs entry. Ocean bills of lading are usually prepared by the shipper or his export agent at the departure port on blank forms supplied by the steamship company. The ocean bill of lading is issued both in straight and order form. The type of endorsement, whether "on-board" or "received for shipment" by the steamship company, is important for banking purposes. Most letters of credit call for payment against on-board bills, which means the goods are actually on-board the ship. If the ocean bill of lading is endorsed "received for shipment," it means that the carrier acknowledges the receipt of the goods at the pier for shipment on a certain vessel. For years there was much confusion and litigation over the responsibilities and liabilities and the rights and immunities of carriers under the ocean bill of lading contract. The leading nations of the world recognized the desirability of uniform standards in these matters, and by international convention the Hague rules were formulated. America's endorsement of these rulse is contained in the Carriage-of-Goods-by-Sea Act, which became effective July 16, 1936.

U.S. Government Bill of Lading

The United States Government, being the world's largest shipper, has found it necessary to print its own bill of lading to control the flow of goods and the payment of freight charges. The government bill of lading is, in a sense, a draft on the Treasury Department. When properly executed, it becomes an order on the Government for payment of charges legally due the carriers for their services.

The most important thing to remember about this bill is that government installations cannot receive "collect" shipments. Therefore, if freight charges are to be paid by the Government direct to commercial transportation companies, the shipment must move on a government bill of lading. Where a government bill of lading is not available, the carrier may accept a commercial bill of lading if a government office has so authorized, and all copies of the commercial bill must contain the notation "to be converted to a government bill of lading at destination."

Air Bill of Lading

Each airline has its own printed bills, but they are all substantially the same in form and all non-negotiable. There are no terms similar to those found on the back of the uniform domestic bill of lading.

Other Documents

Railroad Waybills: The waybill is strictly a carrier document and could properly be called the "historical record of shipment." It is prepared for each railroad shipment from the shipping order, which is the carrier's copy of the bill of lading. All the information contained on the shipping order that concerns the transportation of the shipment and the services to be performed by the carriers is transcribed to the waybill. It moves with the freight from origin to destination, regardless of the number of carriers over which the shipment is transported.

Freight Bills: For *railroads* the freight bill is the carrier's invoice. It is prepared at the point of origin on "prepaid" shipments and at destination from the waybill on "collect" shipments.

For *motor carriers*, the freight bill as a rule performs both the functions of an invoice and waybill. It is prepared at the point of origin on both prepaid and collect shipments.

To repeat, the bill of lading in its various forms is the single most important transportation document in use today. The efficient traffic manager should be aware of all terms and conditions of this document to ensure that it is prepared properly. The bill of lading starts the movement of goods, is the basis on which the carrier invoices for freight charges, and serves as industry's receipt for the goods as tendered. The bill of lading contract sets the liability of the carrier in case of claims. Once the movement is completed, the bill of lading becomes an important historical document.

Some carriers today reproduce bills of lading on photo-copying equipment, with an attached form of a freight bill. The final document then is a combination of a bill of lading and freight bill on one sheet. However, it is still advisable to retain the original bill of lading in a file to support any differences or a claim that might arise at a later date. Bills of lading are also now being produced by computer in coordination with invoices and packing lists.

Particular attention must be paid to section 10 of the bill of lading affecting any alterations, additions or erasures on the document. Any such changes must carry the signature of the carrier's agent to make the changes and the document valid. The bill of lading itself must carry the full signature of the carrier and its agent (or driver) as a receipt for the freight. If the carrier uses a local cartage company as its pick-up agent, then the shipper must be sure the signature received identifies the local cartage company as the agent of the line haul carrier. The following is one manner in which it can be done:—

>   CROSS COUNTRY CARRIERS
>   Local Delivery Co., Inc. Agent
>   per John Smith
>   September 25, 1980

Some carriers provide their drivers with rubber-stamp-ink-pad pocket size combinations which may take this form:—

```
        RECEIVED
      X-Y-Z TRUCKING CO.
   By ........................................
           Driver No. 24
   Date ..................... Pieces ....
```

SHIPPING DOCUMENTS

When freight is loaded by the shipper and tendered to the carrier with seals applied to the car or motor vehicle, the carrier's agent naturally cannot check the number, kind or condition of the packages inside. In such cases, the bill of lading is signed subject to a "Shippers Load and Count" notation, which means that the carrier will not entertain claims for shortage except under certain well-defined circumstances.

One bill of lading may be used to cover unlimited quantities of freight as long as the freight meets the defination of "shipment." Thus, several carloads or trailer loads may be listed on a single bill of lading even though they actually move away from the shipper's place of business at different times during the day. Preparing bills of lading in this manner is particularly important in order to take advantage of "follow-lot" or "overflow" provisions of carriers' tariffs. These rules usually provide that when a quantity of freight offered for shipment is greater than can be accommodated in a single car or vehicle, each car or vehicle used for the shipment must be loaded to full visible capacity or weight limit except that vehicle carrying the overflow. Each load will be charged for at the carload or truckload rate and the actual weight or carload or truckload minimum weight, whichever is greater, but the overflow will not be subject to a minimum weight. Obviously, these provisions can be met only with a single bill of lading for the entire shipment; the savings to be realized will be lost if a separate bill of lading is prepared for the overflow.

A well-prepared bill of lading is excellent insurance that a shipment will travel to its destination as desired by the shipper, that freight charges will be correct and billed to the proper party and that all concerned will have a clear understanding of the transportation phase of the transaction. These benefits are well worth the effort.

5

FREIGHT RATES

Freight rates are transportation prices—the prices paid not only for transporting the freight but also for any of the accessorial services needed. Basically, the price of transportation is a cost for a service no less so than the price we pay for repairs to an automobile or for a pair of shoes.

However, determining the transportation price (rate) requires skill and an understanding of transportation, an understanding that adjustments in rates are often to the advantage of both shipper and carrier, and an understanding that one can *negotiate* rates within the limits set by the Interstate Commerce Commission and the decisions of both the Commission and case law.

The transportation pricing structure of this country probably is the most comprehensive system of user costs ever devised. It has been developed to embrace every known article of commerce, grouped into about 30,000 categories, moving from any given point to any of upwards of 50,000 other points in the country. According to estimates, some 40 trillion prices are involved!

The price list of a manufacturer is not as complicated as the "price list" of a carrier. You can simply look up the cost of one or any number of items for sale on a price list; looking up a freight rate requires a number of steps. It involves classifying the item, calculating the volume to be shipped, and taking account of the packaging, the origin point and the destination, the rules of the classification and the tariff, as well as determining what carriers participate in the rate and the route.

In reality we are not dealing with price lists, but with tariffs

INDUSTRIAL TRAFFIC MANAGEMENT

which are the publications carrying the rates, charges, rules and requirements. These publications are on file with the Interstate Commerce Commission (or with a state regulatory agency where intrastate movements are involved). Once filed with the Interstate Commerce Commission, the rates are lawful and a carrier may not charge any rate other than that filed in the tariff. If the rate is challenged, then the ICC may suspend the rate until it has fully investigated the challenge and the rate, at which time the ICC may re-instate the rate or cancel it.

Kinds of Rates

Previous chapters noted the requirement that common carriers publish tariffs covering the services they are authorized to perform. The tariff must be published in the form specified by the ICC. There are four major categories of rates:
- Class rates
- Exception rates
- Commodity rates
- Accessorial charges

Class Rates have a close relationship to the classifications that govern them. As a result of the system of class rates established under ICC Dockets 28300, 30416 and 30660 and similar scales adopted by motor carriers, virtually every community in the nation enjoying common carrier transportation service is linked with every other such community by an underlying freight rate structure. Thus there exists a basis for the calculation of freight charges on just about anything that is likely to be shipped from one place to another.

A tabulation showing the history of the first class (Class 100) rail rate New York to Chicago is shown on page 85. Of particular interest are the increases and the times at which they have occurred during the period which represents about three-quarters of the century.

Class rates applicable via all-rail routes are published in most of the tariffs listed later in this chapter. Because it would greatly in-

FREIGHT RATES

HISTORY OF FIRST CLASS OR CLASS 100
ALL-RAIL RATE
NEW YORK—CHICAGO
1900—1971

Rate in Cents	Date Effective	Case
75	1900	
78.8	2-23-15	Five percent case, 32 ICC 325
90	6-27-17	Ex Parte 57
113	6-25-18	General Order No. 28
158	8-26-20	Ex Parte 74
142	7- 1-22	Reduced Rates, 1922
152	12- 3-31	Docket 15879
154	1- 4-32	Ex Parte 103
152	10- 1-33	Expiration of Ex Parte 103
163	4-18-35	Ex Parte 115
152	1- 1-37	Expiration of Ex Parte 115
167	3-28-38	Ex Parte 123
177	3-18-42	Ex Parte 148
167	5-15-43	Voluntary Suspension of Ex Parte 148
186	7- 1-46	Reinstatement of Ex Parte 148
209	1- 1-47	Ex Parte 162
225	8-22-47	Docket 28300 Interim Basis
247.5*	10-13-47	Ex Parte 166, Interim Decision
270*	1- 5-48	Ex Parte 166, 2nd Interim Decision
292.5*	5- 6-48	Ex Parte 166, 3rd Interim Decision
293	8-21-48	Ex Parte 166, Final Decision
311	1-11-49	Ex Parte 168, Interim Decision
322	9- 1-49	Ex Parte 168, Final Decision
334.9*	4- 4-51	Ex Parte 175, Interim Decision
351*	8-28-51	Ex Parte 175, 2nd Interim Decision
370.3*	5- 2-52	Ex Parte 175, Final Decision
373.8*	5-30-52	Docket 28300 Final Basis
374	12- 1-55	Ex Parte 175, Rate Conversion
396	3- 7-56	Ex Parte 196
424	12-28-56	Ex Parte 206, Interim Decision
444	8-26-57	Ex Parte 206, Final Decision
453	2-15-58	Ex Parte 212
454	10-24-60	Ex Parte 223
477	8-19-67	Ex Parte 256
501	11-28-68	Ex Parte 259B
531	11-18-69	Ex Parte 262
563	11-20-70	Ex Parte 265B
642	4-12-71	Ex Parte 267B

*Approximation—increases were granted as percentage surcharges on total freight charges and rates themselves were not converted.
Note: Interim adjustments from 1960 are not separately specified; this is to help in streamlining.

INDUSTRIAL TRAFFIC MANAGEMENT

crease the bulk of each tariff if it were to list every individual point between which it applied, the class rate tariffs are arranged to apply between groups of points. These groups, or rate bases, are roughly 40 miles in diameter and usually are named for either the most prominent or most central point. The class rate tariffs refer the user to the National Rate Basis Tariff issued by Tariff Publishing Agent Robert H. Lindsay in behalf of the agencies listed on page 93, to ascertain the names of the rate bases in which origin and destination are located. Certain of the trans-continental class rates in the Transcontinental Freight Bureau tariffs are published to apply between groups of much larger size and the National Rate Base Tariff is not applicable to those larger groups.

An excerpt of a page from the National Rate Base Tariff applicable to Illinois appears below. The first column specifies the railroad station; the second, the railroad and its station number; the third, the SPLC (Standard Point Location Code—more about this later in the discussions on computers); and the fourth, the city that controls as the rate base point.

Excerpt from page of National Rate Base Tariff.

FREIGHT RATES

The class rate tariffs spell out the applications of the rate base points through the use of tables having headline and side line cities. The numbers shown at the junction of each vertical and horizontal column are known as the rate basis number and are used to determine the scale of rates, applicable in either direction between those two rate bases.

These rate scales appear as tables of rates elsewhere in the tariffs. Examples of tariff sheets showing the application of rate bases are shown below. Examples of the tables of rate scales are shown on page 88. These are partial only and the tariffs themselves are much more extensive.

To illustrate the use of these on a representative move from Addieville, Illinois, to Chicago, the first reference is to the rate base tariff (third line) which shows Centralia as the rate basis city for Addieville. Thereafter, by matching Centralia and Chicago on the table for applying rate bases, we find the rate base number is 252. The next step is to relate the number to the appropriate rate scale on page 88. If we assume the material moving takes Class 100, the rate would be $2.36 per 100 pounds; Class 70 would be $1.65, and so on.

Freight Tariff No. I-1002

BETWEEN (See Item 100) AND (See Item 100)	Alexandria..Mo.	Aurora..Ill.	Barry..Ill.	Barstow..Ill.	Bloomington..Ill.	Brooklyn..Wis.	Burlington..Iowa	Cairo..Ill.	Carmi..Ill.	Carthage..Ill.	Centralia..Ill.	Champaign..Ill.	Chicago..Ill.	Clinton..Iowa	Danville..Ill.	Davenport..Iowa	Decatur..Ill.	DeKalb..Ill.	Dixon..Ill.	Dodgeville..Wis.
											RATE BASES APPLICABLE (For rates, see pages 16 and 17)									
Alexandria........Mo.	40																			
Aurora............Ill.	205	40																		
Barry.............Ill.	54	244	40																	
Barstow...........Ill.	117	127	163	40																
Bloomington.......Ill.	144	107	157	128	40															
Brooklyn..........Wis.	260	121	304	144	196	40														
Burlington........Iowa	42	169	91	79	130	221	40													
Cairo.............Ill.	315	356	264	396	249	445	333	40												
Carmi............Ill.	274	287	245	312	193	388	292	103	40											
Carthage..........Ill.	40	205	54	117	144	260	42	315	274	40										
Centralia.........Ill.	203	243	178	253	136	332	222	113	71	203	40									
Champaign........Ill.	192	130	186	175	48	244	177	239	155	192	126	40								
Chicago..........Ill.	243	40	282	161	127	130	206	365	279	243	252	127	40							
Clinton...........Iowa	149	104	195	40	131	110	116	380	321	149	267	179	138	40						
Danville..........Ill.	224	131	213	208	80	251	210	258	157	224	145	40	123	210	40					
Davenport........Iowa	117	127	163	40	128	144	79	366	312	117	253	175	161	40	208	40				
Decatur..........Ill.	165	151	135	168	41	210	170	205	119	165	93	47	167	174	74	168	40			
DeKalb...........Ill.	200	40	240	103	113	97	164	362	305	200	249	160	59	81	155	103	167	40		

Excerpt from typical class tariff showing application of rate bases.

87

INDUSTRIAL TRAFFIC MANAGEMENT

In Tariffs 1000 or 1011 and 1016, the rate basis number is the approximate short line mileage between the two groups. These were established some 25 years ago and there have been some abandonments and other changes since that time.

In Tariffs 1014 and 1015 it is the Class 100 rate that applied when the original tariffs became effective. The rate tables contain columns headed by the class rating percentages appearing in the classification and in the exceptions tariffs. Rates are expressed in cents per 100 pounds.

Motor carrier class rates are published in much the same manner as the rail class rates; but in some tariffs, group numbers are substituted for named rate bases. It is in the tables of rates that the motor carrier rates depart radically from the pattern set in the rail tariffs. In most, if not practically all cases, the rate scales are divided into varying weight breaks in addition to those specified for truck loads.

For example, the LTL rates will have separate scales for various weights, such as: Under 2,000 pounds, 2,000 to 5,999 pounds, and 6,000 pounds or over. Furthermore, the rates do not always reflect the actual percentage of Class 100 indicated by the rating, particularly the lower ratings for long hauls.

Excerpt from typical class tariff showing rate basis numbers and application of class rates to them.

FREIGHT RATES

The scales of the New England Motor Rate Bureau are entirely different from all others, not only because of the numbered classes and their lack of fixed percentage relationship one to another, but because each class is divided into weight breaks, with the truckload minimum weight theoretically representing the quantity that can be loaded conveniently in the equipment generally available. Shown below is a tariff excerpt illustrating these features. Class 5 has a 32,000 truckload minimum; Class 4 has 24,000, and so on. The less-truckload break patterns have but few common features.

The Interstate Commerce Commission has held on numerous occasions that class rates constitute the maximum reasonable basis for freight charges. Therefore, any other rates which exceed that basis may be attacked as unreasonable unless they have been established with clear justification. When the ICC Docket 28300 rates went into effect on May 30, 1952, many commodity rates and exceptions to the old class rates that were higher than the new class rates on the same articles remained in effect and Tariffs 1000 to 1011 each contained an item stating that the new class rates would not apply under such conditions. In the order that created the new class rates, the Commission recognized the fact that it would be

New England Motor Rate Bureau class rate tariff excerpt showing truck load minimums, less truck load weight breaks, and other data.

some time before tariffs could be cleared of the higher commodity and exception rates and gave the railroads five years in which to do so. Meanwhile, it denied all requests for reparation on shipments that moved prior to May 30, 1957. Since that date, the existence of a higher commodity or exception rate has been accepted as *prima facie* evidence of unreasonableness and usually produces an award of reparation. Tariffs 1014, 1015 and 1016, published as a result of the decision in ICC Dockets 30416 and 30660, alternate with exceptions and commodity rates and lower charges resulting from either apply.

Exception Rates. If the classifications were as accurate as they should be, there would be no need for exceptions. The fact is, however, that exceptions to the ratings and rules of classifications for many years have served a useful purpose in that they give carriers more flexibility in dealing with specific competitive situations. Thus, if motor carrier competition develops on a given commodity in a certain territory, the railroads may meet this competition with an exception rating without disturbing the level of rates on that commodity in the rest of the country. Or, if one commodity which, for classification purposes, is grouped with a number of others, such as in an "N.O.I.B.N." item, starts to move in a far larger volume than the other commodities in that group, there may be justification for reducing its rating by means of exceptions without making a similar reduction of the other commodities in the group.

One of the objectives of the Docket 28310 Uniform Classification investigation was to eliminate exceptions to the greatest extent possible. As a result, considerable headway has been made in clearing tariffs of those carload exceptions to the old Consolidated Classification that remained in effect. But the task has been complicated by the large number of commodities involved and the fact that over many years the resulting rate levels had become deeply embedded in the freight rate structure and marketing practices of the industries producing those commodities. In some cases it has been possible to convert them into commodity rate scales, while in others they have been cancelled, only to reappear as exceptions to the Uniform Classification.

Exceptions also are used to provide incentive ratings to encourage heavier loading of equipment. If, for example, the classification gives a stated commodity a Class 40 rating at a minimum weight of 30,000 pounds, the exceptions may publish a lower rating of Class 30 at a higher minimum weight of 50,000 pounds.

Exception ratings and minimum weights take precedence over and must be used instead of those appearing in the classifications. If it is intended to give the shipper the benefit of the lower charges accruing from the use of either, it is customary to bring the classification rating forward into the tariff, naming the exception rating together with an explanatory note providing for alternation. Since exceptions are used in connection with class rate scales, the rates are expressed in cents per 100 pounds.

Commodity Rates. Designed to meet specific conditions in a somewhat less general manner than exception ratings, commodity rates are published in a wide variety of forms.

There are general commodity rates which are published to apply on a single commodity or group of commodities from or to all points in a designated territory. Some of the older general commodity rate scales never have borne any relationship to the class rate scales and are set up between groups of origins and destinations to preserve traditional competitive relationships between manufacturing and marketing territories. Others at one time were based on the class rates then in effect, but subsequent changes in the class rate structure have obscured this relationship. Still others use the Docket 28300 groupings as their basis or are graduated according to mileage. Among these are commodity column rates, some of them applying on a single minimum carload weight, others on an incentive basis with different numbered columns for each of two or more graduated minimum carload weights; still others name one column of rates for a given minimum carload weight and another column of lower rates to apply on all weight above the minimum which is loaded in the same car. Because incentive rates are intended to encourage maximum utilization of railroad equipment, it usually is provided that Rule 24 of the classification will not apply to them.

By far the most varied of the commodity rates are the thousands of specific point-to-point rates that have been established to take care of individual movements. Justified by many different competitive situations, they are not likely to follow any set pattern, but the well informed industrial traffic manager will find it advisable to make a study of the history of those rates which affect his commodities.

For some years, carload and truckload rates have been published on freight-all-kinds, on which the same rate applies on any article in the shipment regardless of classification. These rates usually do not apply to bulk freight, certain light and bulky articles, live stock and a few other items. They may also be subject to provisions limiting any one article to from 25 percent to 60 percent of the total weight of the shipment or may apply only when a certain number of different articles are included in the shipment. Some of these rates carry provisions permitting the application of class, exception or specific commodity rates when those rates on a portion of the shipment are lower than the freight-all-kinds rate, but may require that the freight-all-kinds rate be used on a stated minimum portion of the shipment. Freight-all-kinds rates are of greatest benefit to freight forwarders and industrial shippers of large quantities of variety merchandise.

Commodity rates usually are expressed in cents per 100 pounds, but there are some commodities, such as brick and coal, with rates per net ton and an occasional rate per gross ton. Certain lumber rates are expressed in cents per 1000 board feet, pulpwood in cents per cord, and bulk liquids moving via motor carrier frequently are rated per gallon. Charges stated in dollars and cents per car formerly were confined to switching and other short-haul movements. The introduction of trailer-on-flat-car (TOFC) service changed that, however, for since 1958 literally thousands of per-car or -per-trailer rates have been published to apply on Plan I, II, II¼, II½, III, IV and V.

In recent years the publication of special commodity rates on multiple car shipments, "jumbo" cars, and so on where incentives are involved, has increased. Reduced multiple-car rates are conditioned upon the tender of two or more carloads at one time at one point by one shipper for movement over one route to one consignee at one destination. Variations on multiple car rates include unit train rates on which full trainloads of perhaps sixty cars or more may move, articulated car rates, and others. Quite often they apply to bulky commodities such as coal, grain, ore, and automobiles. "Jumbo" car rates have been made applicable to movements via large tank cars having capacities of roughly 18,000 to 40,000 gallons, large hopper cars, and so on.

Unless the tariff naming a commodity rate specifically provides alternation, the commodity rate on a given commodity from a given origin to a given destination must be used instead of the class or exception rate on that commodity from and to those points, even though the class or exception rate might produce lower charges.

The railroads of Canada have made use of a system of commodity "agreed rates" whereby shippers of a substantial quantity of a certain material who would agree that a major percentage of that tonnage would move by rail were accorded rates somewhat below those normally applicable. For a number of reasons this type of rate is not in general use in the United States, and only a few intrastate applications exist.

Today there are three major regional groups among railroads, the Eastern Railroads, the Southern Freight Association and the Western Traffic Association. Within each of these groups are several tariff publishing agencies as follows:

- Eastern Railroads
  Trunk Line—Central Territory Tariff Bureau

- Southern Freight Association
  Southern Freight Tariff Bureau
  Southern Ports Foreign Freight Committee

- Western Traffic Association
  Western Trunk Line Committee
  Southwestern Freight Bureau
  Texas-Louisiana Freight Bureau
  Transcontinental Freight Bureau
  Colorado-Utah-Wyoming Committee
  Pacific Southcoast Freight Bureau
  North Pacific Coast Freight Bureau

INDUSTRIAL TRAFFIC MANAGEMENT

In addition, Illinois Freight Association and a committee of east-west Virginia lines each publish tariffs separate from the three major groups.

Even though the publication of class rates by agencies had greatly reduced the number of class rate tariffs, there were more than 100 of them in effect prior to the publication of the ICC Docket 28300 rates. As a result of the manner in which the ICC Docket 28300, 30416 and 30660 scales were put into tariff form, the country's rail class rate structure was contained in 17 issues with two published by the Western Trunk Line Committee, two by the Illinois Freight Association, four by the Southwestern Freight Bureau, three by the Trunk Line Central Territory Tariff Bureau, one by the Southern Freight Tariff Bureau, one by the Trunk Line Territory Tariff Bureau, two by the Transcontinental Freight Bureau, one by the North Coast Freight Bureau and one by the Southwestern Freight Bureau.

Some agencies publish both interterritorial and intraterritorial commodity rates for their member lines, while others leave the intraterritorial and local rates for the individual railroads to publish. The trend, however, is for agencies to issue such rates, especially where duplicating rates are published by several railroads between the same points. Except for the class rates mentioned previously, there is no hard and fast rule that will enable a person to tell whether to look in an agency issue or an individual tariff for a given rate.

Motor carriers likewise have created tariff publishing agencies within their various regional associations. But whereas the railroad groups mentioned act in behalf of all railroads in the respective territories, the same cannot be said of the motor carrier groups. Consequently, there are many overlapping motor carrier bureaus and some carriers represented by no bureau at all. The principal motor carrier groups are:

- New England Motor Rate Bureau
- Middle Atlantic Conference
- Southern Motor Carriers Rate Conference
- Motor Carriers Tariff Association

- Motor Carriers Tariff Association
- Eastern Central Motor Carriers Association
- Central States Motor Freight Bureau
- Central and Southern Motor Freight Tariff Association
- Middlewest Motor Freight Bureau
- Southwestern Motor Freight Bureau
- Southwestern Motor Freight Bureau
- Rocky Mountain Motor Tariff Bureau
- Interstate Freight Carriers Conference
- Pacific Inland Tariff Bureau

Accessorial Charges. Grouped under this heading are the charges made by carriers for a wide variety of services and privileges made available in connection with the transportation of goods. They include charges for switching, loading, unloading, weighing, pickup, delivery and transit privileges such as stopoff for completion of loading or unloading, storage, inspection, grading, repackaging, milling and fabrication. Also included would be charges for demurrage and vehicle detention, as well as allowances paid to shippers and consignees for services performed by them, such as pickup and delivery and the furnishing of shipper owned or leased railroad cars or highway equipment.

There are several other types of rates that should be identified.

Freight Forwarder Rates, both class and commodity, were originally patterned after those published by the railroads and they were governed by the rail classification. Today, the prime regulated competitor of the forwarders is the motor common carrier. Consequently, forwarder tariffs are predominantly governed by motor classifications, and forwarders endeavor to maintain rate levels that do not exceed those of motor carriers.

The basic concept of freight forwarding is to assemble and consolidate "less-load" individual shipments into carload or truckload consignments, which are moved from forwarder origin terminals to forwarder destination terminals, where breakbulk and distribution are performed. In essence, separate and distinct parts are combined into a whole and then reduced again to separate and distinct parts. Since the forwarder pays his underlying rail or motor carrier the charges applicable to carload or truckload consignments on the terminal movement, he is able to publish rates of his own to the public on any volume of traffic less than those quantities. Forwarder class and commodity tariffs, as a consequence, name rates on both "less volume" and "volume" shipments. While the vast bulk of forwarder shipments weigh less than 1,000 pounds, a heavy concentration of forwarder tonnage and revenue is to be found in the higher weight ranges.

The forwarder through his use of negotiated piggy-back rates has promoted AK (all kinds) freight rates that allow shippers to mix certain commodities into volume shipments. The piggy-back concept of negotiated rates for the forwarder has also given him the opportunity for many volume shipments between major markets.

Forwarder class rates on "less volume" lots between terminal cities are generally competitive with motor carrier rates down to Class 50, but between other points they may be competitive at higher levels to compensate for the higher cost of assembly or distribution or both. Volume class rates usually are maintained to Class 35, regardless of location of origin and destination. Forwarder rates are not subject to the long- and short-haul provisions of Section 4 of Part 1 of the Interstate Commerce Act. This feature is often overlooked.

Assembly and Distribution Rates are published by many motor carriers for use in connection with shipments originating at or destined to points beyond the terminal areas of freight forwarders or freight consolidators. They appear as both class and commodity rates and usually are ten percent or more lower than the ordinary rates applicable between the same points, on the theory that multiple shipments delivered to or picked up at a consolidation point are less costly for the motor carrier to handle. The rules of "A&D tariffs" usually require that shipments be reforwarded within five working days via common carrier service and that records substantiating compliance with the rule be made available to the assembling or distributing carrier upon demand. These rates are of particular use to the industrial shipper when he maintains a regular consolidation operation either with his own facilities or through a professional consolidator.

MOTOR CARRIER RATES

Basically motor carriers are subjected to the same regulatory requirements of tariff publication as the railroads and are also bound by the principle of "same rates for the same haul." They also have their tariff publishing bureaus (the major ones were shown on page 94).

In addition there are many publishing agencies for intrastate operations, and for specialized carriers such as heavy haulers and perishables haulers. Motor carriers also publish "private" tariffs; these they publish in accordance with regulatory agency requirements. Private tariffs contain freight rates (generally commodity) on certain commodities and applicable only on the carrier's own lines.

Motor carrier rate adjustments can be processed by one or more of the carriers, or the shipper can file the proposed direct with the bureau. Most motor carrier bureaus provide standard forms for presenting proposals.

The data submitted in support of the proposal should be complete, factual and specific as to the rates sought. To ask for reduced rates without saying what they should be implies strongly that the proponent doesn't know how to judge the existing rates.

Some other rate forms found in both rail and motor operations are:

—*Arbitrary rates* are usually a fixed amount added to another rate. Generally they are used in areas where traffic conditions increase the cost of operations. Such rates do not find favor with the ICC.

—*Any quantity rates* are for articles that do not have a volume or load minimum weight in the classification. These rates may be applicable to items that rarely, if ever, move in volume lots.

—*Released value rates* are based on a value previously agreed upon by both shipper and carrier and declared in writing by the shipper on the bill of lading. Such rates reduce the liability of the carrier to a specified value for the article being shipped, for which the shipper then pays a lower freight rate.

—*Distribution rates.* Motor carriers consolidate a number of smaller shipments into a single volume shipment consigned to a specific terminal where it is broken down into small shipments for delivery. The carrier performs both the line haul transportation and the small shipment delivery. To the shipper the combined cost may well be less than the direct LTL cost.

HOW RATES ARE PUBLISHED

Even more wasteful, confusing and repetitious than each carrier publishing its own classification was the former practice of individual publication of each carrier's rate tariffs. After Interstate Commerce Commission regulations became effective, competing carriers adhered to the same scales of rates between given points. One of the main purposes of regulations was to end the discriminatory rate wars that had characterized the early days of the transportation industry.

Early in the twentieth century the railroads banded together into regional groups, which served as clearing houses for mutual problems, and agents were appointed within these groups to publish tariffs for their members. Thus, instead of Conrail publishing a local class rate tariff from New York to Chicago and a group of other competing railroads publishing joint tariffs with their various connecting railroads from and to the same points each became a party to an agency tariff which published a single scale of class rates that applied via all these routes. This resulted in a tremendous reduction in the number of tariffs and, while many tariffs still are issued by individual carriers, for the most part they cover local provisions or rates within a restricted territory. Lists of both railroad and motor carrier rate bureaus are on previous pages in this chapter.

There are other interstate groups (generally smaller) as well as those which publish intrastate rates in a number of states. Additionally, there are groups representing carriers offering specialized services, a list of which includes the following:
- Household Goods Carriers Bureau
- Western States Movers Conference

FREIGHT RATES

- Oil Field Haulers Association
- Eastern Tank Carriers Conference
- National Automobile Transporters Association
- Steel Carriers Tariff Association

In 1948, Congress passed, over the veto of President Harry Truman, the Reed-Bulwinkle Act which gave the carriers the right to act in joint approach on rate and allied matters and gave them immunity from anti-trust prosecution from such action. Today, this anti-trust immunity is again the subject of heated discussion and, accordingly, President Truman's veto message is appropriate reading. It is as follows:

PRESIDENT TRUMAN'S VETO OF THE REED-BULWINKLE BILL
To the Senate:

I return herewith, without my approval, S.110, a bill to amend the Interstate Commerce Act with respect to certain agreements between carriers, because it would permit an important segment of the economy to obtain immunity from the antitrust laws, and would do so without providing adequate safeguards to protect the public interest.

This bill would authorize exemption from the antitrust laws for any carrier acting in concert with one or more competing carriers in the establishment of rates and related matters, if the Interstate Commerce Commission approves the procedural agreements under which such action would be taken. The Commission would be required to approve these agreements if it should find that exemption from the antitrust laws would be in "furtherance of the national transportation policy."

Under this bill private carrier associations and rate bureaus, free from the restraints fo the antitrust laws, could exercise broad powers over most forms of domestic transportation, including railroads, trucks and busses, water carriers, pipe lines, and freight forwarders. Carriers could agree privately among themselves upon the rates to be filed with, or withheld from, the Interstate Commerce Commission. Acting through these bureaus, groups of carriers could exercise a powerful deterrent influence upon the filing by an individual carrier of proposed rates which might benefit the public.

The exercise by private groups of this substantial control over the transportation industry involves serious potential harm to the

---
*Source: *Congressional Record*, Vol. 94, Part 7, p. 8524, June 16, 1948

public. Transportation rates affect the cost of goods as they move through each phase of production—from raw materials, through finished products, to the consumer. Power to control transportation rates is power to influence the competitive success or failure of other businesses. Legislation furthering the exercise of this power by private groups would clearly be contrary to the public interest.

My disapproval of this bill does not signify opposition to carrier associations as such, or to all of their present functions. Many of their activities are useful and desirable. However, this legislation is not necessary for the continuation of such activities.

No legislation giving a major industry immunity from the antitrust laws should be enacted unless adequate alternative safeguards are provided for the public interest. This measure fails to provide such safeguards. Even the limited safeguards incorporated in the bill as originally passed by the Senate are omitted from the bill in its present form. It would require the Interstate Commerce Commission to approve any agreement which it finds to be in "furtherance of the national transportation policy." This is a vague and general standard and is manifestly neither adequate nor appropriate as a criterion for waiving the protection afforded the public by the antitrust laws.

Furthermore, the Commission would be placed in the position of applying this general criterion to the basic procedural agreements without being able to forsee fully the nature and effect of the joint actions which would be taken thereunder. Nevertheless, the exemption from the antitrust laws would extend to these subsequent actions without the necessity of further Commission approval. It would extend, moreover, even beyond the parties to the basic agreement to any "other persons" who participate in such actions.

Even though transportation rates are subject to regulation by the Interstate Commerce Commission, the public interest nevertheless demands that the general national policy of maintaining competition continue to be applied to this industry. Our present transportation policy contemplates a pattern of partial regulation, within the framework of which the pressures of competition will remain substantially effective. Regulation cannot entirely replace these competitive powers. It can guard against some of the potential abuses of monopoly power, but it cannot be an effective substitute for the affirmative stimulus toward improved service and lower rates which competition provides. By sanctioning rate control by groups of carriers, this legislation would represent a departure from the present transportation policy of regulated competition. This, I believe, would be a serious mistake, with far-reaching effects on our economy.

Antitrust cases are now before the courts challenging some of the very activities which would be covered by this bill. Pending judicial clarification of the issues raised in these proceedings, it would be inappropriate to provide the immunity proposed by this bill.

I have repeatedly urged upon the Congress the necessity for a vigorous anti-monopoly program. This bill would be inconsistent with such a program.

For these compelling reasons, I find it necessary to withhold my approval from the measures.

Harry S. Truman

The White House, June 10, 1948

Since the Interstate Commerce Commission is an arm of Congress and has been granted quasi-legislative and quasi-judicial powers, then the tariffs approved by the ICC, as long as they are in effect, have the effect of a Congressional statute. They are binding on both shipper and carrier, a principle established in AT & SF Ry. Co. v. Bouziden, 307 F. 2d 230 (1963), National Starch & Cheese Co. vs. McNamara Motor Ex. H9 N.E. 2d 766 (Ill 1963), I.H. Christopher Co. vs. Atlantic & D. Ry. Co. 289 ICC 716, 717 (1953) and Beaumont, S.L. & W. Ry. Co. vs. Magnolia Provision Co., 26F. 2d 72, 73 (1928).

HOW RATES ARE ESTABLISHED

One of the most challenging aspects of traffic work is keeping currently informed on the kaleidoscopic changes that occur in freight rates. Every week thousands of changes are proposed; thousands of these proposals are disposed of and thousands of tariff supplements are issued reflecting the decisions of dozens of rate committees throughout the country. The traffic manager who fails to keep posted is likely to be confronted with some unpleasant and difficult situations.

These constant changes in freight rates are brought about in a variety of different ways. In a great number of cases, they are the outcome of negotiations between shippers and carriers. In many others, they are proposed by the carriers themselves to meet the competition of other means of transportation or, as in the case of general rate increases, to improve their revenues. In still others,

INDUSTRIAL TRAFFIC MANAGEMENT

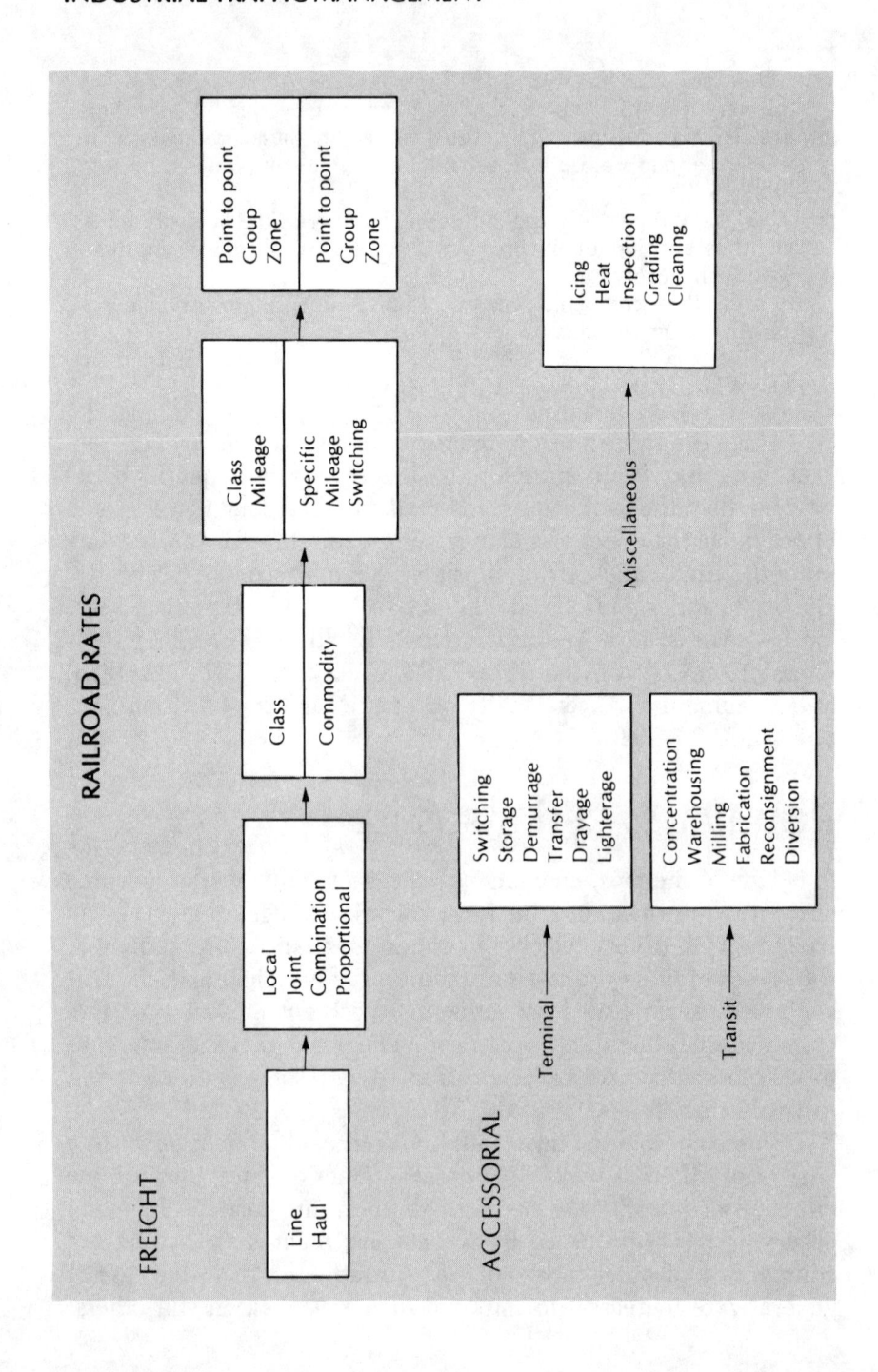

RAILROAD RATES

FREIGHT

Line Haul → Local / Joint / Combination / Proportional → Class / Commodity → Class Mileage / Specific Mileage Switching → Point to point / Group / Zone | Point to point / Group / Zone

ACCESSORIAL

Terminal → Switching, Storage, Demurrage, Transfer, Drayage, Lighterage

Transit → Concentration, Warehousing, Milling, Fabrication, Reconsignment, Diversion

Miscellaneous → Icing, Heat, Inspection, Grading, Cleaning

102

FREIGHT RATES

they result from orders of the Interstate Commerce Commission growing out of investigations such as Dockets 28300, 28310, 30416, 30660 and the various minimum rate orders affecting truck rates.

A shipper has every right to complain about—the proper traffic term is protest—any transportation rate or rule. But the burden of proof is on him. He must support his case with statistics on profitability, cost of operation, competition or any combination of these factors. He must establish the legality or illegality of a proposal through the previous decisions of the Commission and the courts. He need not sit back and blandly accept everything the carriers or their rate bureaus are publishing. The traffic manager may not win every "case" but the carrier will not win every "case" either.

The addition of section 5a to the Interstate Commerce Act in 1948, as a result of the Reed-Bulwinkle Bill, exempted common carriers from the operation of the anti-trust laws in connection with agreements developed between themselves and their various regional groups whereby procedures are established for the joint consideration, initiation and establishment of charges for the services they perform. Such agreements must outline all procedures to be followed and must receive the approval of the Interstate Commerce Commission. Basically, all procedures consist of consideration of tariff changes by a standing rate committee maintained as a part of the association's staff and, if necessary, reconsideration by an appeals committee composed of freight traffic officers of member carriers. All agreements of this nature must guarantee to member carriers the right of independent action.

The right of independent action by a carrier does not now allow the bureau to protest such action. Bureau members protests are limited to those individual carriers who are in direct competition with the rate resulting from the independent action of a carrier.

Generally speaking, proposals for tariff changes may be initiated by shippers, carriers or by the bureau itself on behalf of its members. The dockets giving a brief description of each proposal scheduled for consideration by the standing rate committee are published, usually weekly, and circulated among member carriers and such shippers as may request them. In addition, the rate committees of the railroads, motor and water carriers advertise their dockets in the weekly *Traffic Bulletin,* while a few motor carrier dockets are listed in *Transport Topics,* a weekly newspaper published by the American Trucking Associations. Rate bureaus will accept subscriptions to their dockets. The prices vary and interested persons should contact the bureaus for that information. The dockets will also carry information on the action taken on matters listed on the previous docket.

Thus the traffic manager may obtain advance information on proposed tariff changes. If he notes a proposal that will affect his company's operations, he may support or oppose it, as circumstances dictate, either by letter or by asking for public hearing on the matter. The importance of keeping posted in this manner cannot be overemphasized, for decisive action in advance of the actual publication of a tariff change often will forstall the difficulty of obtaining adjustments later.

The report of the standing rate committee will be final unless it is protested within a stated period. However, if it is protested, either by a member carrier or a shipper, it will be scheduled for hearing before the appeals committee, which is known by names such as "Freight Traffic Managers Committee," "General Rate Committee" or "Central Committee," and protestants may state their cases either by letter or in person. If the standing rate committee's report is upheld, this usually closes the matter, but if an individual carrier feels very strongly that a proposed provision should or should not be published, he may take independent action and either arrange for publication for his own account or withdraw from participation in the provision, as the case may be.

If the rate is to be published, ICC tariff rules require that the new rate or rule become effective on a 30-day notice to the Commission and the public. It is at this point that anyone who feels his interests are affected may file a protest and request a suspension with the Commission's Board of Suspension. There are time limitations on these and all other filings and traffic personnel responsible must be familiar with these limitations so that all filings are timely.

The appeal or protest must be supported by good evidence that

FREIGHT RATES

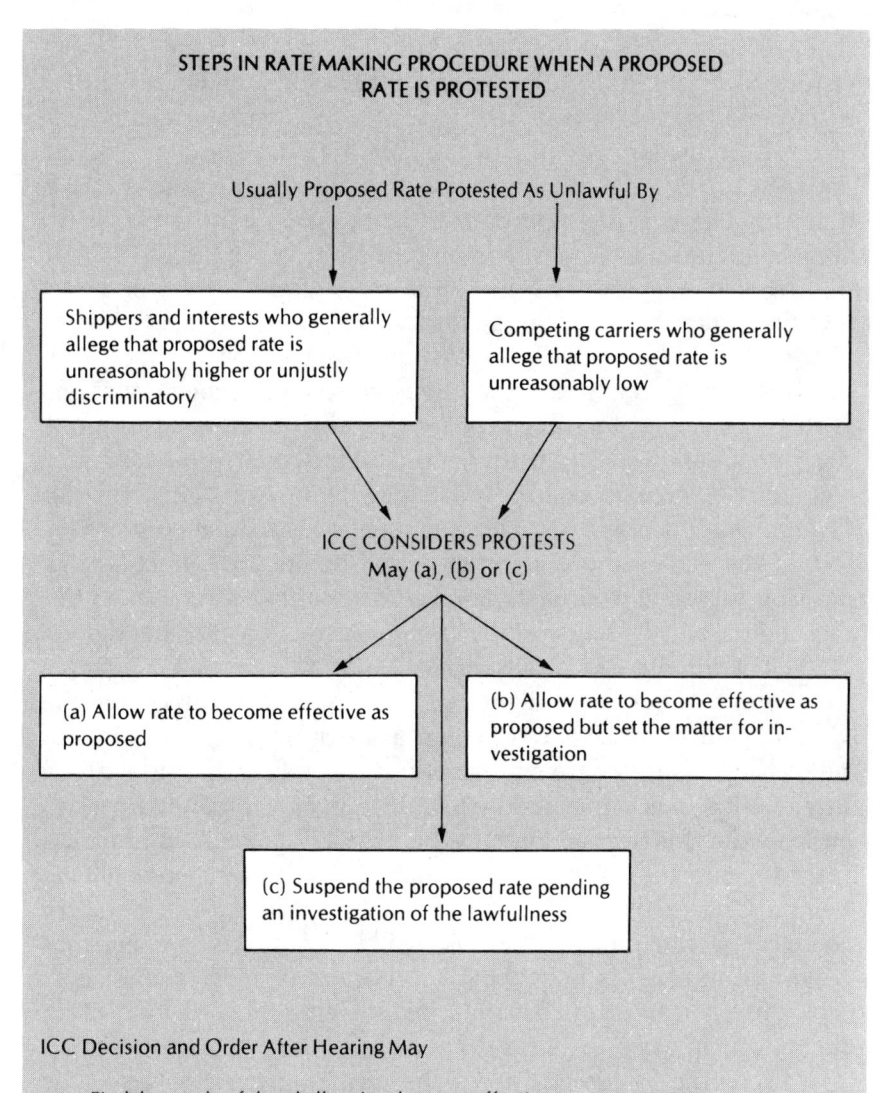

STEPS IN RATE MAKING PROCEDURE WHEN A PROPOSED RATE IS PROTESTED

Usually Proposed Rate Protested As Unlawful By

- Shippers and interests who generally allege that proposed rate is unreasonably higher or unjustly discriminatory
- Competing carriers who generally allege that proposed rate is unreasonably low

ICC CONSIDERS PROTESTS
May (a), (b) or (c)

(a) Allow rate to become effective as proposed

(b) Allow rate to become effective as proposed but set the matter for investigation

(c) Suspend the proposed rate pending an investigation of the lawfullness

ICC Decision and Order After Hearing May

   a. Find the rate lawful and allow it to become effective
   b. Find the rate will be unlawful for the future in which case the existing rate will remain in effect
   c. Find the rate will be unlawful for the future and (1), (2), (3) or (4):

1. Fix a maximum reasonable rate to be observed for the future
2. Fix a minimum reasonable rate to be observed for the future
3. Relate the rate to be observed for the future to other rates by a fixed differential
4. Fix a precise rate for the future

the provision will be unreasonable, prejudicial, preferential or otherwise in violation of the law. Whether or not the Board decides to suspend, its action may be appealed to the appropriate division of the Commission. Because of the short time that usually remains between the Board's announcement of its decision and the effective date, it is of utmost importance that the protestant act quickly. Some companies maintain representatives in Washington to help in this regard. Suspension, when granted, is for a period of seven months, unless the case can be disposed of sooner by the Commission or the publisher of the provision receives the Commission's special permission to withdraw and cancel the suspended provision. If the case cannot be disposed of within the period of suspension, the rate is allowed to go into effect unless both protestant and respondent agree to continue the period of suspension until the Commission's decision is announced. When a provision goes into effect at the end of the suspension period before the Commisssion's decision is ready, the investigation of the matter continues and the respondent may be required to cancel the provision at a later date.

Occasionally, the provisions of section 6 of the Act calling for thirty days' notice of a freight rate adjustment are waived and the Commission grants what is known as Sixth Section Permission to publish on a shorter period of notice. This can be as little as one day's notice, but it is generally for ten days or some similar period. Justification for such action must be strong; the proposed provision must be non-controversial, must be the result of an order of the Commission or the courts, or must be to correct a tariff error. In some cases, short-notice permission will be granted where there has been undue delay in publishing a duly recommended provision or when the person affected would be severely damaged by waiting the full thirty days.

From the foregoing, it will be seen that the machinery for publishing new or changed tariff provisions, though slow and apparently cumbersome is designed for the full protection of anyone whose interests are affected.

FREIGHT RATES

Rate Adjustments

At least one person in every traffic department should be assigned to observing the new rates and rules being proposed or being published that affect inbound and outbound goods. That responsibility can have a serious impact for good or bad on a company's profits. A carrier can hardly be expected to reduce its revenues without being asked, and even when it is asked, the request should be reasonable, logical and lawful. But when it is confronted with a carefully worked out proposal for a rate adjustment, any carrier that has a proper customer-relations policy will cooperate to the utmost in seeing that the proposal receives full and fair consideration.

How to Check Freight Rates

The tariff is published and several supplements to it have been published. The time has arrived to answer a request for a rate from the sales department so it can price articles and close sales. What are the steps you will take to find that rate?

1. Review the classification (Uniform for rail and/or National for motor) and find the proper nomenclature describing your product, its volume and less than volume class rating. Note this information.

2. Check in a tariff index to find the tariff (rail or motor) covering the commodity from the origin point to the destination. You are seeking a commodity tariff on your product.

3. Check the tariff carefully for the specific commodity and for a rate on that item from origin point to destination point. Note the rate and the weights affected.

4. Does that tariff carry an alternate provision which could reveal another rate in the same tariff or one in a class tariff? A class rate that might be lower? If so, search out the other provisions and read them carefully for there you will find a possible answer as to which rate you are allowed to use.

5. Suppose your destination (or origin) is not specifically nam-

INDUSTRIAL TRAFFIC MANAGEMENT

ed? You will then determine from references in the tariff where your communities are named and what larger points they are "attached" to for rate purposes. Is there a note in the commodity rate item (in the tariff) that brings your origin and destination under the larger umbrella? In this manner you can determine whether the rate shown to and from Chicago and Philadelphia might also be applicable to and from Fort Washington (a suburb of Philadelphia) and Hammond, Ind. If your answers are there, the problem has been an easy one.

6. If there are no specific commodity rates to be found covering your problem, you bring out your notes on what the classification told you. You must then refer to a Rate Basis Tariff to find the origin and destination points listed. You are now seeking named base points for your origin and destination, unless these are specifically named.

Once you have the base points from the Rate Basis Tariff and you go to the applicable tariff where you line up both points and find a rate basis number.

7. Using the class rate tariff and using a rating table in this tariff, you will line up the rating with the rate base number and where they meet, you will have a base rate for the classified commodity.

8. In any and every case you *must* check every supplement of every classification and tariff in use to determine whether any changes are in effect on your problem case. You must also be sure the tariffs are currently in use; that they have not been superceded by a re-issue of the tariff you are now using.

This is a very simplified example. There are steps that come into use as your experience grows. There are steps you will discard with experience and you will be quite safe. What can you do to avoid the tariff examination every time a question arises? If the origin and destination points come up with any regularity, then it pays to develop a card system (Rollodex or something similar) with the card containing the origin, destination, commodity, class rating, rates for each weight break, tariff reference and date of the entry on the card. Now you need only correct the card if a supplement or re-issue of the tariffs changes the important information.

You must exercise care in your rate search. A carrier is not responsible for any quotation that is wrong made by any form of communication. Therefore if you seek quotations from carriers, you must remember that the Commission and the courts have held that the shipper or receiver is responsible for determining the correct rate. Carelessness in searching for the correct rate can lose sales.

And in using freight rates you should remember that you cannot combine intrastate and interstate rates.

Freight forwarder rates present few of the problems of their rail and motor carrier counterparts. There are the class and commodity tariffs of the Freight Forwarders Tariff Bureau, to which most of the major forwarders are parties. Forwarders also publish their own tariffs, but there is little or no overlapping. Forwarder rates are not subject to long-and-short-haul nor aggregate of intermediates provisions, but there are differing patterns of rates as between individual forwarders. Whereas Forwarder "A" may be rail or truck competitive from Boston to Kansas City as low as Class 70, Forwarder "B" may be competitive only down to Class 85; yet their positions may be reversed from Boston to Minneapolis.

Auditing Freight Bills

One of the best means of training a traffic manager is by auditing freight bills, for from those bills he is in an excellent position to obtain a detailed picture of a company's raw materials and supplies and their sources; its products and markets; its routing policies and practices; and the transportation services available to and at its various locations. A good rate searcher can recoup his salary many times over, not only by the overcharges he detects, but by the economies he can effect by policing the entire shipping job.

Further, an alert auditor will observe opportunities to order and ship in more economical quantities; note suppliers' and shipping departments' deviations from proper bill of lading descriptions and prescribed routes; uncover possibilities for consolidations and stopoff arrangements which can reduce transportation charges; and report rates which appear to be out of line!

Carriers are required to collect their freight charges before relinquishing shipments at destination, but any reputable concern may be accorded credit arrangements upon request. Failure to pay bills within the credit period is an offense punishable by fine and can subject the carrier to prosecution for unjust discrimination if he neglects to remind his customers of their obligations and cut off the credit of delinquents. Shippers, also, have obligations in these connections and may be subjected to penalties where violations occur.

If freight bills can be audited and the remainder of their processing completed so that payment may be made within the credit period, considerable expense for filing of claims, processing of balance due bills, and adjustments of records and accounts may be avoided by making the corrections. In such cases, corrections are to be supported by statements of tariff authority or other reasons for the change and such information is sent to the carriers with remittances. If prepayment audits are not possible, the bills may be audited after payment and claims for overcharges thereafter presented to carriers.

Many shippers follow a practice of submitting freight bills to one or more of a large number of reliable freight audit bureaus for audit or reaudit. These firms may not have the specialized knowledge of a company's materials that its own auditors have, but their familiarity with tariffs plus their experience with other commodities often enables them to detect overcharges that otherwise might go unnoticed. These audit bureaus provide checks on the performance of a shipper's auditing procedures and have other functions. They also will perform the complete audit for concerns that do not do their own work. Their fees vary but may be up to 50 percent of the amount recovered.

Important to bear in mind in filing claims for overcharges is the statute of limitations. On interstate shipments, the statute is three years. This means that carriers may collect their charges and those who pay freight charges may file claims for overcharge within three years of the date shipment is delivered. Intrastate shipments are governed by the laws of the individual states, which provide limitations varying from two to ten years.

One of the most valuable services of the traffic department is the quotation of freight rates to other departments of the company. These quotations help the purchasing department in the selection of sources of supply and in making comparisons of laid down costs between FOB origin and FOB destination quotations. They can be used by the sales department in arranging distribution methods to the best advantage, in developing profitable marketing areas, in establishing prices to meet competition, and in determining amounts to be allowed on invoices covering goods sold on delivered prices. Manufacturing can use them too, in arranging for raw materials. Additionally, freight rates play an important role in the selection of new plant locations.

Because their accuracy may have great influence on the profit of a transaction, rate quotations should be in the hands of the most seasoned rate men, preferably those who have had considerable audit experience. An extensive tariff file is essential to a good quotation job.

Consolidations: Whether it be combining two LCL or LTL shipments into a stopoff carload or truckload or gathering a large number of individual small lots into scheduled pool loads, the success of the operation will hinge to a great degree on the accuracy of the rate information used. Not only must this information be readily available and up-to-date, but it must be under constant review by experienced rate searchers who can deal knowledgeably with the effect of every change in rates.

Sales Terms

There is a definite interrelation between freight rates and FOB terms, for the FOB terms indicate the purchaser's interest in transportation charges. Conversely, the transportation charges may have considerable influence upon the choice of FOB terms.

"FOB" is an abbreviation for "Free on Board" and implies loading on a conveyance. Several important questions may be answered by the manner in which the FOB terms are stated:
—Who pays the freight?
—Where is delivery to be made?

—Where and when is title and control of the merchandise to be transferred?
—Where is the point of shipment?
—How is the laid down cost to be determined?
—When is payment due?
—Who pays for packaging?
—Who is to select and make any necessary special arrangements with carrier?
—Who absorbs cost of loading into carrier's equipment?
—Who is to bear the risk of transportation?

As in any legally binding document, ambiguity in a contract of sale will be construed by the courts against the framer. But even if the matter never gets into the courts, improperly stated FOB terms will lead to misunderstandings and ill will between seller and purchaser and should be avoided.

Ordinarily, purchasing FOB destination would appeal to a buyer as relieving him of many bothersome details; but in an operation of any size this simply cannot be done any more than it is possible to buy everything FOB origin. Trade practices, the provisions of the Robinson-Patman Act, and the regulations of the Federal Trade Commission produce a wide variety of FOB terms, many of which cannot be altered at the desire of a single customer, no matter how important that customer may be. Some companies buy FOB origin whenever possible because they wish to control the transportation. This may be important to them. A complete examination of a delivered price as part of the FOB destination terms compared to the FOB origin plus transportation (and any accessorial cost, if applicable) is necessary for a decision.

In buying FOB origin, title and control of the goods pass to the buyer and invoice becomes payable when the carrier signs for the goods at that point. It follows that the buyer assumes risks of transportation, is entitled to route the shipment, and also must undertake the responsibility for getting the goods to destination and for filing claims for loss or damage regardless of who bears the freight charges. Unless qualified, FOB origin terms place the responsibility for freight charges on the buyer, but the terms may be worded in such a way as to provide for absorption of freight charges, or a portion of them, through allowance or prepayment.

FREIGHT RATES

"Freight allowed" and "Freight prepaid" have different meanings and should not be confused. "Freight allowed" means that shipments will move freight collect and, unless qualified, that an amount equivalent to the charges the buyer pays to the carrier will be deducted from the total cost of the goods when paying seller's invoice. By this device, seller may quote delivered prices without tying up capital funds in prepaid freight. "Freight prepaid" means that seller will pay transportation charges to the carrier and that buyer will remit the full amount of the invoice without deduction for freight. In either case, if seller is unwilling to absorb charges for premium transportation, the clause may be worded to provide this protection.

Origin FOB points should be designated by naming the city and the use of wording such as "FOB Shipping Point" or "FOB Factory" should be avoided. If the seller has more than one FOB point, the FOB clause should state whether the seller's or the buyer's option governs the selection of the shipping point. If it is the seller's option and material is not sold on a delivered price, provision should be made for the equalization of transportation charges with the seller's shipping point producing the lowest transportation charges to buyer's destination.

When goods are bought FOB destination, the seller retains title and control of them and the invoice covering them does not become payable until they are delivered and the contract of carriage has been completed. The seller selects the carrier and is responsible for the risks of transportation including the filing of claims for loss or damage. Unless the FOB clause states otherwise, he assumes transportation charges either by prepayment or allowance. The buyer may not assign, divert or reconsign the shipment. When the contract of sale contemplates delivery of the material in seller's trucks, or in the trucks of a carrier operating under contract to the seller, such arrangements should be FOB destination, even if a charge is made for the delivery service, for obviously the buyer cannot assume title or control of the goods nor the risks of transportation if the seller has not relinquished them.

The following list includes some of the basic clauses most frequently used in connection with price quotations and contracts of sale. A brief interpretation accompanies each.

INDUSTRIAL TRAFFIC MANAGEMENT

- *FOB (City and State at point of shipment).* This means that material will be placed on the cars or trucks of the carrier indicated in buyer's routing, at seller's expense, and that all transportation charges thereafter will be borne by buyer. Any local trucking charges at point of shipment will be absorbed by seller.
- *FOB (Seller's plant, factory, mill, warehouse or other designated facility, City and State at point of shipment).* This means that material will be placed on the cars or trucks of the carrier indicated in buyer's routing, at seller's expense, if it can be done at seller's facility indicated. If it is necessary to use local cartage to get the material to the carrier, or if the carrier indicated by buyer makes a charge for picking up at seller's facility, this expense, as well as subsequent transportation charges, will be borne by buyer.

These clauses may be modified to provide for delivered prices through allowance or prepayment of transportation charges by addition of one of the following auxiliary clauses:

- *Transportation charges to destination allowed.* (Or "prepaid," as the intent may be.)
- *On individual shipments of (weight, invoice value of pieces) transportation charges to destination allowed.* (Or "prepaid," as the intent may be.)

These mean that the seller will allow (or prepay and absorb) transportation charges via any route of movement and should be used only when this is mutually understood. If protection against charges for premium transportation is desired, one of the following may be used:

- *Lowest published common carrier (rail, truck, freight forwarder or water) transportation charges to destination allowed.*
- *Lowest published common carrier (rail, truck, freight forwarder or water) transportation charges to destination prepaid.* When buyer's routing results in higher charges, buyer will be invoiced, and pay, for the excess.

Both of these clauses may be further limited by prefacing them with *On individual shipments of (weight, invoice value or pieces).*

FREIGHT RATES

The danger in the above could be time-in-transit since the choice at the *lowest published common carrier, etc.* could put the freight into an operation that might defeat the need and value of the shipment.

To provide for equalization of transportation charges with seller's other shipping points or competitive shipping points, the following may be added to the basic FOB origin clauses:

- *. . . freight charges to be equalized with those applicable from seller's shipping point producing lowest transportation cost to buyer's destination.*
- *. . . freight charges to be equalized with (city and state).*

If more than one competitive point is involved, "whichever produces the lowest transportation cost to buyer's destination" should be added.

- *FOB Destination* or *FOB (City and State of destination).* This means that the seller will undertake to deliver the shipment to the city specified and will absorb the regular transportation charges incident thereto. In the absence of specific provision to the contrary, unloading at destination, as well as any local trucking charges in the city of destination, is to be arranged and paid for by buyer unless carrier's tariff provides that either or both of these services are included in the transportation charge.
- *FOB (Street number address, city and state)* or *FOB (buyer's plant, warehouse or similar designation, city and state).* This means that all transportation charges, including those for local delivery to designated address or location, will be absorbed by seller. Arrangements for and cost of unloading are for buyer's account unless carrier's tariff or FOB clause provides otherwise.

All of the above clauses may be modified to express the intent behind the transaction. The important thing to remember is that intent must be stated and stated clearly enough to remove any reasonable doubt. There is great need for standardization of FOB terms and their interpretations and it is hoped that the foregoing may be of some value as the author's contribution toward that goal.

6

FREIGHT ROUTING

Routing is one of the most important responsibilities of a traffic manager. No firm can compete successfully if it is obliged to tie up capital in abnormally large inventories to offset the uncertainties of poorly chosen transportation services. Careful control of the routing of all shipments is an absolute necessity if customer service standards are to be satisfied at the lowest possible cost. Every traffic department should have a basic routing policy, understood by suppliers, customers and carriers. Failure to route shipments makes it practically impossible to expedite them or to intercept them for diversion or reconsignment. The cost of delays, losses, damage and excess freight charges that results is far greater than the cost of providing effective routing control.

Tariffs carry routing "authorizations" showing what carriers are authorized to move shipments between any two or more points. In the previous chapter, we suggested that recording a selection of commonly used rates on a "Rolldex" or in a loose-leaf binder would save a lot of research time. Including the route or routes available at the rate shown between the points involved will make this simple reference tool even more useful. And as in checking rates, supplements to tariffs and re-issues of tariffs should be constantly checked for changes in routings.

To control the routing of a shipment, there are two things to look for in the tariff. The first is that the tariff may have an item stating that the rate quoted applies to all routes between the specified points of those carriers participating in that tariff. The second thing is that the tariff may include some very specific routing instructions affecting some carriers or may refer to a routing guide. The point here is that some carriers may not be parties to all the rates or may be party to a limited application of those rates.

Ownership of goods in transit is determined by the terms of sale. Shipments have time-and-place value in addition to their basic value. These are important considerations for the traffic manager in selecting the mode of transportation and in assessing a carrier's trustworthiness, its control of drivers and equipment, its schedule of operations and its insurance coverage. Shipments valued at millions of dollars are sometimes entrusted to carriers unknown to the owners of these shipments. Effective routing control can do much to reduce the risks of such "blind flying."

The Right to Route

The legal right of an owner of a shipment routed by rail to put in writing on a bill of lading the exact route of that shipment is set out in 49 USC 10763 (revised IC Act), originally Section 15(8) of the IC Act. This right exists only for shippers by rail. There is no comparable provision for any other mode of transportation in the IC Act.

While there is no provision giving a shipper the "right" to route when using a motor carrier, the Interstate Commerce Commission has stated that it would be unreasonable for a motor carrier to depart from the written routing directions of a shipper (The Murray Co. of Texas v. Morrow, 54 MCC 442).

Fundamentals of Routing

Railroads, motor carriers, airlines, water carriers and freight forwarders maintain sales networks, not only in cities along their routes, but frequently in markets where they hope to attract freight that may at some point move into the regions they serve. A smart carrier salesman often has an uncanny faculty for discovering potential business, and will, if given a little encouragement, work

FREIGHT ROUTING

out combinations of rates, routings and services that the shipper is unaware even exist. But because of the large number of carriers serving even the smaller industrial areas, it is manifestly impossible for even the largest firm to give business to all of them. Yet to keep his shipping options open, to have adequate service available in emergencies, and to ensure that he will at least get an interested hearing from his local carriers should he run into rate or service problems, the shipper usually will find it advisable to divide and distribute his business in accordance with a carefully worked-out formula—and to deal straightforwardly with those carriers whose services cannot be used, for the moment at least, to mutual advantage.

But the shipper need not divide his freight between the carriers absolutely evenly. Some shippers place carriers on "priority lists" (first choice, second, third) and rotate them monthly, quarterly, etc. Under this scheme, marketing areas are divided into specific carrier areas—for example, four carriers for the middle Atlantic-New England sales region, another four for the Southern region and so on. While a system such as this is primarily intended for motor carriers, there are comparable considerations for rail shippers.

Both customers and suppliers should be made aware of a company's routing policy so that they will not become confused as to "why they handle it that way." For example, if terminal facilities for handling either rail or truck shipments are limited, and the facts are made known to suppliers, they will be much more disposed to cooperate, even on shipments arriving FOB destination, than if standing orders carry flat instructions to use a particular mode of transportation. One large steel producer has incorporated in its sales agreements a statement to the effect that its facilities for shipping by truck are limited and that orders bearing motor carrier routing will be accepted subject to possible delay or change in route. This type of frankness will gain far more in good will than it will lose in sales and is likely to be accepted with greater tolerance than will a practice of disregarding customers' routings without explanation. Then too, if certain carriers are known to provide the most reliable service at point of origin, customers buying FOB that point will welcome the benefit of the shipper's experience and usually will arrange their routing accordingly.

A simplified version of the firm's traffic manual, given to each supplier and customer (and to the firm's own salesmen) is a good method of publicizing the routing policy. A fuller version of the traffic manual, with operating information, can be published at the same time and given to carriers.

As a general but by no means inflexible rule, it is good policy to route carload shipments so that the railroad on whose siding the shipper is located receives its longest possible haul. Exceptions may have to be made when speed is necessary and that particular railroad cannot move the shipment as fast as its competitors; or where delivery requirements of a consignee demand that the railroad serving it, or the team track on which it takes delivery, be accorded a line haul. Of course when shippers or consignees are located on more than one railroad or on belt railroads that connect with more than one of the railroads serving their respective cities, they have more rail routing options.

In selecting motor carrier routes, try to use those carriers maintaining terminals in or nearest the origin town. They are most easily reached by telephone for pickups and forwarding information and they usually can get their vehicles to the shipper's place of business more quickly and without having to travel empty for long distances which, in turn, can have an effect on the rates.

In distributing tonnage among carriers, don't distribute it so widely that the carriers favored do not receive enough to make the business worthwhile. Obviously it is *not* worthwhile if each receives only a few hundred pounds per pickup or delivery. Although common carriers theoretically are obligated by the terms of their certificates to handle all shipments within the scope of their authority and published rates, they can hardly be blamed for losing interest in an account they cannot serve profitably.

To be properly armed for his discussions with carriers, a traffic manager should have a monthly or quarterly report of the tonnage given each carrier. If data processing equipment is available, the report can be more comprehensive, but the essential information needed is:

FREIGHT ROUTING

- Tonnage per origin point per carrier
- Tonnage to specific destination areas (city or county or state) per carrier
- Average weight of shipment for the period of the report per carrier
- Total number of shipments per weight group (i.e., 150 shipments from 0-150 lbs., 300 shipments from 151-500 lbs., and so on) per carrier
- Total revenue per carrier from each shipping location.

This type of information is invaluable and is easy to get once a shipment is routed and the freight bill received. Make separate reports for each mode of transportation used. Beside their usefulness in negotiating rates and generally improving management control, such reports will show if one carrier is starting to get too much of a firm's business. Changes in routing (or the threat of changes), backed by these reports, will often bring rapid improvements in a carrier's service and claims record.

Routing by Rail

As we mentioned in "The Right to Route" section earlier in this chapter, the rail carrier is obligated to meet the shipper's routing specified on the bill of lading unless there are operational reasons beyond the railroad's control for cutting off that route, or unless there is an ICC order allowing carriers to change routings because of floods or other operational difficulties.

If a shipper names a route over which a combination of rates applies, it may turn out to be a costly choice. It is very important to understand the link between the rate and the route. If the shipper specifies a route over which the through rate does not apply, a combination of two or more rates may be assessed, and it would be a rare case indeed if this combination did not exceed the through rate. But even if a through rate applies, the shipper should carefully specify the route it covers. One tariff may contain both routes and rates. And the number of routes between several points can be astronomical. For example, the House Committee on Interstate and

Foreign Commerce, 84th Congress, determined that there were 4,717,664 routes between Dallas, Texas and Detroit, Michigan affecting class rates. These included using Memphis, Louisville, St. Louis and Chicago as junction points for connecting carriers. Specific routing is published in connection with most rates either in the rate tariff itself or in a routing tariff referred to by the rate tariff.

In checking out the routes available for a given rate, frequently it will be found that both direct and circuitous routes are named. This is where a set of railroad maps, such as those provided in the *Official Guide of the Railways,* or those available from Rand McNally, is a valuable adjunct to the routing job. Of course, circuitous routes often are useful, even if fast service is not obtainable, for sometimes they provide desirable stopoff privileges. Circuitous routes may also be helpful in distributing tonnage, solving inventory adjustment problems at origin or destination, equalizing mileage, and so on. But circuitous routes bear watching from both the service and the rate angles. Some carriers will "flag out" (not participate) in many rates involving circuitous routes because they are not profitable.

We have mentioned two sources for checking routes; the tariffs and the routing guides. Another useful routing aid is the rail "gateway" routing guide, usually published by an agent for the carriers within a specified territory and showing routes to and from a particular gateway. For the rail carriers, gateway routes are an assurance that the freight will move from one rate territory to another and is likely to be profitable to haul. Routing guides are also published by individual carriers to show the routes available exclusively on their lines.

Difficulties may arise in routing when the shipper shows a rate on the bill of lading but no route. The Commission has held that if the rate is a correct one, then a route has in fact been shown and the carrier must move the freight over the route covered by that rate. But if the shipper shows a route and no rate, the carrier must move the freight via that route *even if there is a less costly route in effect.*

However, if a shipper puts both the rate and route on the bill of lading and the rate is not applicable, the carrier may move the

FREIGHT ROUTING

freight over another route *even if the rate is higher.* The ICC has held that since the rate charged legally must be the lowest in effect, there is no damage to the shipper. In any event, the railroad agent is obligated to notify the shipper of any such differences and to ask for instructions.

The routes published in rate and routing tariffs specify the junction points between the railroads named in each route and sometimes these junction points differ for carload shipments. Junction points may be shown on the bill of lading, but care is required in their selection. Frequently, two or more junction points appear in routes involving the same two railroads and, if the shipper selects a little-used one, it may result in delay to his shipment.

Problems may arise, however, when junction points are *not* specified. A carrier generally takes the "long" haul where it can; that is, once it receives a car, it moves it the maximum distance it can on its own line, thus getting the largest share of the revenue on the car. But this long haul route may not be the best for the shipper and by inserting junctions he may choose faster and better routes. Specifying junctions can also keep cars out of congested gateways. The Peoria gateway, as an example, has proved very useful at times in keeping cars out of Chicago.

Where the originating and delivering carriers are different railroads, specifying the junctions can be used to assure each railroad of a long (and profitable) line haul. Some shippers feel that this is good business practice and that better rail service results. Of course, some companies may simply be able to by-pass junctions in bills of lading. But knowing how to select the "best" junctions and include them in the bills-of-lading routings is a valuable weapon in the traffic manager's perpetual battle for the best service at the least cost.

The routing of carload freight offers the traffic manager many opportunities to control his company's transportation costs, for carload shipping provides advantages not found in other forms of transportation. The smallest box car in common use has about 50 percent greater cubic capacity and from double to triple the weight-carrying ability of the average highway trailer. Recent years have seen the development of still greater capabilities in the

INDUSTRIAL TRAFFIC MANAGEMENT

form of box, hopper, gondola, tank and special purpose cars with capacities of 100 tons and more. Carloads by rail therefore are an ideal medium for moving bulk freight and large masses of goods and this superior carrying capacity frequently enables the shipper to benefit from incentive rates considerably below those offered by motor carriers. Many box cars carry special blocking, bulkheading and other damage prevention devices and some are equipped with extra-wide doors for ease in loading and unloading with mechanized equipment.

But rail short haul carload transportation is usually much slower than truck service and may be slower on the longer haul, too. Shippers must load and consignees unload rail cars while, in most cases, loading and unloading costs are included in motor car-

Auto rack cars, used to haul new automobiles from factory to dealer. Reinforced plastic side panels on these 61-foot, three-level cars serve to reduce damage to automobiles from flying stones, wind-carried grit and other abrasives.

FREIGHT ROUTING

rier rates. Even so, many truck rates are lower than rail rates between the same points for the same commodity. Most rail carload shipping is confined to companies with private rail sidings, but there are still many "team track" facilities for loading and unloading rail cars for those without private sidings.

Rail routings do offer the opportunities for heavier loading per vehicle than is possible with trucks. These increased weights per carload can reduce freight charges per shipment. Freight cars can be ordered in "jumbo" sizes with larger cube and weight carrying capacities than conventional cars. The "Big Boy" freight car shown here is an example of what is available from the railroads.

Among the other carload routing services offered by the railroads are:

—Stop-off privileges at several points to complete the loading of cars or to unload them. These privileges enable the shipper to move shipments under carload rates that would otherwise have to move at much higher LTL rates.

In addition to carload rates and privileges, certain railroads

A new "Big Boy" box car is shown with its little brother, standard size, at right. These cars are longer than a bowling alley and are designed to carry light, bulky loads. As illustrated, they can accommodate a tobacco loading of 124 hogsheads versus 36 for the normal or standard size car.

have attempted, with varying success, to establish "trainload" container rates and routes between East, Gulf and West Coast ports, under the "landbridge" concept. The idea here is to use the continental United States as a railroad landbridge between Western Europe and the Orient. Containerships are offloaded at U.S. ports, the containers are shipped by the trainload across country and reloaded on other containerships for onward movement east or west. This saves up to nine days on the all-water route through the Panama Canal and releases the high-cost container ships for more profitable shorter hauls. However, attempts to set up such services have run into considerable legal opposition from rival ports who maintain that they divert trade "naturally tributary" to particular ports. A limited application of the landbridge idea that has met with some success is the "Mini-bridge," under which trainloads of containers are shipped from East and Gulf Coast ports to Californian ports for onward movement to the Far East. A full-scale landbridge operation, with container trains shuttling between east and west coasts, has yet to be successfully established.

Stop-off rules vary with the territories or the railroads involved. The tariff should be examined closely for those rules and charges. Each stop may effect customer service and must be considered in an overall evaluation of costs and service.

—Transloading of one or more cars. This is generally offered only by Western railroads. The car or cars loaded with freight for two or more destinations are moved to the transloading point, where the railroad reloads the freight for each destination into a separate car and then moves the partially loaded cars to their destination under the same carload rates. Care must be exercised to see that the through rates from origin to each of the destinations apply via routes operating through the transloading point. Additionally there are other transloading rules to be found in the stop-off tariffs of the railroads.

It's important when routing by the carload to make certain that delivery can be made at stop-off points and final destination without additional switching charges. Usually, if the railroad on which the shipper's or consignee's siding is located does not receive the line haul on the car, it makes a separate charge for handling

FREIGHT ROUTING

between the siding and the point of connection with the line-haul railroad. In order to compete for the business, the line-haul railroad will frequently state in its switching tariffs that it will absorb the switching charges of connections and it then pays the charges of the switching road out of its line-haul revenue. But there is a growing tendency to break away from these traditional reciprocal arrangements, so switching tariffs of both origin and destination carriers should be given careful scrutiny before this type of routing is attempted. When loading or unloading takes place on a public team track, the railroad owning the team track always should receive the line haul, for most railroads do not publish switching charges between their team tracks and connecting railroads, and when they do, these charges are not absorbed by line-haul railroads.

Accurate routing directions on the bill of lading are quite important and have an effect on the service and the rate applied. Be sure the junction points shown are in fact served by the railroads selected and that the tariff shows those railroads to have interlining operations and rates.

Careful routing will cut the cost of moving perishables. The traffic manager must be conversant with the types of cars available and be sure that the refrigeration equipment is in working condition. Where many cars of perishables are loaded daily, some rail carriers have become competitive with motor carriers in hauling these commodities into major market areas. The ICC has allowed the railroads the same exemption from regulation as the motor carriers enjoy in hauling agricultural products. Rail carriers will now often negotiate charges "on the spot" and move large numbers of refrigerator trailers in special fast trains from agricultural centers to major market cities.

Piggyback

No discussion of rail routing would be complete without an explanation of the rail/road hybrid technically called "trailer-on-flatcar" (TOFC), but more familiarly known as piggyback. This con-

INDUSTRIAL TRAFFIC MANAGEMENT

sists of loading a highway trailer with goods and hauling it by over-the-road tractor to a specially constructed piggyback "yard" or "ramp." Here the trailer is loaded onto a railroad flat car, also built for the purpose, and the car forwarded in a train operating on a fast schedule to the piggyback ramp that is nearest the destination of the shipment. The trailer is then unloaded from the flat car, hooked up to a tractor, and completes its journey over the road.

The basic idea of piggyback is not new, having been used in one form or another for many years. In 1885, the Long Island Railroad placed into operation a so-called "pickaback" plan whereby farmers' wagons loaded with produce were transported, four to a flat car, from outlying points on Long Island to Brooklyn, whence they were ferried to Manhattan. In the late 1930's and early 1940's, a motor carrier operating between New York and Chicago used the Baltimore and Ohio R.R. to move his trailers between those two cities. But it was not until the late 1950's, however, that development of the piggyback idea by the railroads and accep-

Santa Fe's yards in Chicago use large cranes to load and unload piggyback trailers quickly.

FREIGHT ROUTING

tance by the shipping public combined to make news of the subject. Since that time there has been a steady increase each year in the points at which this service is available and in the number of trailers handled. Judging from the sales effort devoted to it, the railroads feel they have a service with which they can meet the flexibility of motor carrier operations and successfully compete for the business of smaller firms and those without railroad sidings. Shippers find that piggyback provides fast, dependable schedules at charges which are not so subject to the pressures of rising fuel and labor costs.

Piggyback service is offered under eight basic plans, each briefly described as follows:

Plan I—Offered to motor carriers only at flat charges agreed upon between railroad and truck line, this substitutes an over-the-rail haul for over-the-road haul, for all or part of an intercity movement. Shippers deal exclusively with the motor carrier at regular motor carrier rates. Shippers may prohibit the use of this substituted service, but may not order it used. However, they are not likely to know whether or not a given motor carrier used it.

Plan II—The railroads pick up freight at shippers' platforms and deliver it to consignees' receiving docks in railroad-owned or leased trailers. Rates for this service usually are the same as, or slightly lower than, motor carrier rates between the same points. Use of this type of service is elective with the shipper.

Plans II¼, II¾—These plans are similar to Plan II, but the railroad limits its over-the-road service to pickup or delivery, but not both. Under Plan II¼, the railroad picks up the trailer and moves it to destination ramp only; the consignee must provide the over-the-road tractor for delivery. Under Plan II¾, the shipper delivers the trailer to the railroads piggyback ramp using his own tractor.

Plan II½—Similar to Plan II, except that service is offered ramp-to-ramp only, shipper and consignee making their own arrangements for tractors to make pickup and delivery of loaded and empty trailers from and to railroad piggyback yard. Rates are substantially lower than those available via motor carrier.

Plan III—Trailers owned or leased by shippers are transported on

INDUSTRIAL TRAFFIC MANAGEMENT

flat cars furnished by the railroads and the railroads perform no terminal service except loading and unloading trailers onto and off flat cars. Rates usually are stated in dollars and cents per flat car up to a stated maximum basic weight of trailer contents, with any weight over that basic weight calculated at a rate per 100 pounds. The rates are set for two highway trailers or demountable trailer bodies per flat car. Rates are somewhat lower than Plan II½ charges. In some territories, Plans II½ and III rates are published to apply on single trailers at 60 percent of the two-trailer rates, one-half of the two-trailer maximum basic weight and the same rate per 100 pounds for any weight over the maximum.

As regards commodities, certain rates apply only to specifically-named commodities, while others are set on a freight-all-kinds basis. While the trend is away from restricting to 60 percent or some other figure the quantity of any one commodity which may be loaded, some tariffs continue to provide such restrictions. Where provided, the stated percentage may apply to the total weight or maximum basic weight, whichever is greater.

Plan IV—This differs from Plan III in that the flat car as well as the trailers must be owned or leased by the shipper and there are no single-trailer rates.

Plan V—Trailers move under joint railroad-truckline coordinated service and rates. A motor carrier solicits freight, loads either railroad-owned or its own trailers and delivers them to the railroad which piggybacks them to a ramp and hands them over to another motor carrier for delivery. Thus the railroads and the motor carriers together gain business they might never get separately. The service is available for both TL and LTL shipments.

The use of Plans II and V involves practically the same procedure as regular motor carrier shipments, although there may be some difference in the rules governing. Plan II class rates are subject to the rail classification with ratings that are competitive with motor carrier class rates. Plan V class rates may be subject to either the rail or motor carrier classification, depending upon whether the rates are published by the railroads or the motor carriers.

Plans II¼, II¾, II½, III and IV present additional problems in that they require the shipper to make arrangements for tractors to move the trailers to and from the piggyback yards at both ends of the rail haul. Usually there are rail truck subsidiaries or motor carriers who will provide this service, either on a single-trailer basis or, in the case of repetitive shipments, on contract. Arrangements must be made in advance, however, and it cannot be assumed that every consignee has such arrangements in force.

Plans III and IV also require that the trailers are owned or leased by the shipper. No difficulty is encountered if there is a two-way movement that will keep trailers under load in both directions, but since the return of empty trailers by piggyback usually costs the same as the loaded movement, any reduction in transportation costs realized on the outbound loaded movement probably will be more than offset by the expense of returning empty trailers. This problem is solved by trip-leasing trailers from any of a number of firms, which relieves the piggyback shipper of responsibility for the trailer once it is returned empty to the piggyback yard at destination.

Plan IV, with its requirement that the shipper also own or lease the flat car, likewise requires a two-way loaded movement if costs are not to outrun savings and thus does not lend itself to general use. It finds its greatest use among freight forwarders and shippers' associations. Territorially, some western carriers provide flat car lease charges for one-way moves and thus help in promoting a more general use of the plan.

The trend away from weight restrictions on single commodities of freight-all-kinds shipments under Plans II½ and III is making it less difficult for one-product manufacturers to obtain the low rates available through these plans. Many manufacturers in this category surmount weight-restriction hurdles through their shippers' associations. These groups consolidate members' shipments of different commodities to overcome the weight limitations. In most cases these shippers' associations were set up by, and are operated under, the guidance of experienced traffic professionals; and they can show sizeable reductions in their members' shipping costs, even where single-commodity restrictions no longer exist.

INDUSTRIAL TRAFFIC MANAGEMENT

Motor Common Carrier

There are in the continental United States over three million miles of federal, state, and county highways that can be used by motor trucks—trucks carrying freight to and from the railroads' piggyback yards, trucks picking up and delivering freight at the ocean and inland ports, trucks delivering and picking up freight at the airports, as well as the trucks of some 17,000 motor common carriers operating for their own account. The evolution of trucking has affected, and been affected by, the evolution of the other modes of transportation and today the truck is as much a servant as a competitor to its rivals.

The truck has also dramatically affected the demography of our society, although some will argue that it was the highway builder rather than the motor carrier who brought about the transformations we see today. Nonetheless, where the trucks can roll, the factories will follow. The speed and flexibility offered by highway transportation has revolutionized marketing and distribution practices. The share of intercity tonnage hauled by trucks increases year by year and no one appears ready to hazard a guess as to when this trend will level off.

The slow but sure increase in trailer sizes (now up to 45 feet)

A typical over-the-road long haul tractor trailer combination. Trailer is 45 feet long pulled by a cab-over-engine (COE) power unit.

FREIGHT ROUTING

brings the weight carrying ability of the tractor-trailer close to that of the smaller box cars. But the speed of the highway carrier and its ability to place a load of goods at almost any point accessible to a useable road overcomes differences in rates that might favor railroads. And lower truckload miminum weights often enable shippers of smaller quantities to take advantage of volume rates and to maintain lower inventories.

Motor carriers also offer a variety of equipment and vehicles "tailor-made" for particular jobs—"low bed" trailers for hauling large machines that cannot be disassembled, flat bed trailers for steel and large objects, refrigerator trailers for perishables (and, conversely, for protecting some cargo from freezing), household goods trailers (now often modified to protect computers and delicate electronic equipment), pole trailers for log carrying, special trailers for transporting passenger vehicles, trailers with light-weight demountable bodies that can be transferred as containers from trucks to the bellies of wide-body airplanes. The range is wide and offers the shipper routing options unknown not too many years ago.

For example, the tanks in tank trucks can be heated so that cer-

In certain states, motor carriers have found it possible to reduce costs by operating two trailers in tandem behind a single tractor.

INDUSTRIAL TRAFFIC MANAGEMENT

tain heavy oils will remain fluid, chocolate and sugar remain liquid. Milk can be transported in bulk in stainless steel tanks that meet health-department standards of cleanliness. The variations seem limited only by the transportation industry's frequent failure to see the potential of new technologies and products when they appear. But motor carriers offer a greater diversity of services than any other mode of transportation—and they offer those services door-to-door.

Part II of the IC Act, now incorporated into the Codified Revised Interstate Commerce Act of 1978 (P.L. 95-473), places the motor carrier under ICC jurisdiction. Basically, motor carriers offer two major services—LTL (less than truckload) and TL (truck load or volume). LTL is any shipment that weighs less than the prescribed weight for the truckload rate. TL is any shipment that is equal to or more than that prescribed minimum weight, or a shipment that fills the cube of a single vehicle.

Most motor carrier rates include loading and unloading by the driver from or to a point adjacent to the tailgate of the vehicle and

Containers being transferred from truck chassis to a container ship by land-based crane.

FREIGHT ROUTING

deliveries inside buildings may be arranged at reasonable additional charges published in carriers' tariffs. Because truck shipments are not likely to receive the impacts in transit to which rail cars are subjected, it is often possible to eliminate the elaborate blocking and bracing necessary in ordinary box cars and the resulting saving in labor and material can be considerable. Ordinarily, motor carrier shipments do not require the expediting so often necessary to insure prompt delivery of rail shipments.

Selecting a motor carrier route is quite different from selecting a rail route. Motor common carriers operate on certificates of public convenience and necessity granted by the Interstate Commerce Commission which may limit the commodities they handle, describe the area they may serve and, in the case of regular route carriers, define the highway routes they may use. Some carriers may serve all intermediate points along their routes, while others are limited to named intermediate points or none at all. There are certificates permitting service to named points not directly on the described routes, called off-route points, and some certificates allow transportation for compensation in one direction only. Every motor common carrier makes public its operating rights, usually in a directory issued by one or more of the rate conferences of which it is a member and referred to by the rate tariffs to which it is a party. In this manner, the public is charged with the knowledge of carriers' operating rights. Under Section 222(c) of the Act (11904(c) and 11906 of U.S. P.L. 95-473), shippers are made liable, equally with carriers, for any wilful violation of the law; hence it is important that the shipper inform himself, either by referring to a directory or by obtaining a written statement from the carrier, of the extent to which the carrier may legally serve him, and thus avoid any appearance of asking the carrier to depart from the terms of his certificate.

Wherever possible, it is advisable to choose a carrier that offers a single line service from origin to destination, particularly on shorter hauls. Needless interchange only serves to slow up a shipment and might increase freight charges, for there are some class rate scales that are higher for joint hauls than for single-line hauls between the same points. When it is absolutely necessary to route

INDUSTRIAL TRAFFIC MANAGEMENT

over two or more carriers, refer to the specific routing published with the rates, either in the same tariff or in routing tariffs referred to by the rate tariff. Unless these routes are used, combination rates over the point of interchange will apply or the initial carrier may find it impossible to turn the freight over to the second carrier specified. Trailer interchange between carriers is becoming more widespread, but is by no means universal; therefore, it will pay to ascertain whether the routing chosen for truckload shipments provides for through movement in the same trailer, thus saving time and minimizing the opportunities for loss and damage. While the motor carrier shipper does not have his right of routing spelled out in law, motor carriers will generally respect a shipper's routing as long as they have interline agreements with the continuing carrier named in that routing. However, by Commission ruling, a carrier must apply the lowest through rate or charges in the applicable tariff.

When selecting a motor carrier, it is important to ascertain the amount of cargo insurance it carries. The Interstate Commerce Act and many state acts provide that certificates may not be issued to a motor carrier or remain in force unless the carrier complies with

Typical over-the-road household moving van and sleeper cab tractor. This type of tractor is used for long hauls including transcontinental hauls of household goods and specialized equipment such as electronic data machines.

such rules as may be prescribed for insurance. The minimum cargo insurance has been set at extremely low figures. While the larger motor carriers have substantial coverage, some of the smaller lines cannot afford the premiums on large policies. A letter to the Insurance Section of the Interstate Commerce Commission requesting the name of the insurance company and the amount of coverage will bring an answer to any questions about the carrier's insurance coverage. If the value of the cargo offered exceeds the amount of insurance carried by the motor carrier, the carrier should be informed so that he may arrange for additional coverage, both for his protection and the shipper's. The amount of insurance carried in no way limits the carriers liability for loss or damage. for the carrier is 100 percent liable for the value of the goods, subject only to the exceptions set out in the bill of lading contract.

Household Goods Carriers

One type of motor carrier that deserves separate consideration is the household goods carrier who is basically a specialist in moving household furniture, although many of these carriers now offer ancilliary services such as transporting, packing and moving antiques and items of high value. The costs of moving services bought by the uninitiated can run extremely high so the traffic manager should keep firm controls. His expertise is also needed in scheduling the move so as to diminish the many inconveniences that company transfers can inflict on the employee and his family.

The average employee has no means of knowing or judging the reliability, the rates and the services of the carriers available. He is busy at his job and will often leave the moving arrangements to his wife, who is no more aware of the complexities and possible pitfalls than he is. He may suffer delays in pickup and delivery of his household treasures and will certainly experience some unpleasant surprises if he attempts to handle his own claims for loss and damage.

In the hands of the traffic department, household moves will

INDUSTRIAL TRAFFIC MANAGEMENT

be confined to routings via van lines of national reputation. Shipments are timed and coordinated to avoid such extras as warehouse storage or paying for expedited service on smaller lots. Shipments are released to specific cents per pound valuation, which produces the lowest rate and any excess is covered by the company's transportation insurance policy (or the household insurance of the individual if it is not a company move). The traffic department also assumes the responsibility for preparing and following up claims for loss or damage.

The traffic department should develop a standard "policy" covering company moves. As soon as the van carrier is selected, a letter should be sent to the carrier, with a carbon copy to the employee, with the following information:
- Name and address of the employee to be moved
- Address of destination
- Telephone number of employee at his office, present house, and destination house.
- Services to be performed
- Moving schedule
- Insurance coverage
- Any storage conditions agreed to
- Services to be performed at destination (for appliances, television, etc.)
- Any other information of value to the van line or the employee

The letter should conclude with a warning to the employee that items of extraordinary value such as gems, coins, precious metals, etc., should not be included in the move and that items of high value—paintings, sculptures, etc.—should be insured separately.

A separate letter to the employee should emphasize that someone from the household should be present when the carrier packs and unpacks the goods, loads and unloads the van, so as to note any damage or loss on the bill of lading. This will simplify any subsequent claims disputes.

Household goods carriers' rates run much higher than those of general commodity carriers, but since this type of service obviates expensive packaging and provides for inside pickup and delivery, their use often can be justified if the commodity to be shipped falls within the standard household goods description or the "special handling" category. A number of the larger household goods carriers have set up special handling divisions to transport computers, for example.

Freight Forwarders

Although freight forwarders are regulated under Part IV of Interstate Commerce Act, they are not carriers in the same sense as the other carriers discussed in this book. Freight forwarders consolidate smaller shipments into larger and therefore more economical shipments. They derive their profits from the spread between the rates charged the shipper and the total cost of their operations, which includes charges paid to the underlying carriers—railroads, motor carriers, shiplines, airlines, or combinations of different modes.

Their rates are not always competitive with those of other services, but they do offer definite advantages. One is speed; their rail shipments travel from origin to destination almost as fast as carloads do. Another is free pickup and delivery at the premises of shipper and consignee. Still another is assumption by the forwarder of full responsibility for the shipment from shipper's to consignee's places of business. A firm may welcome the opportunity to give all its shipments to a forwarder in one pickup each day instead of attempting to distribute them among several motor, rail, air or water carriers.

In cities where forwarders maintain terminals, they usually make pickups in their own trucks or in those of an affiliated company or a local carrier operating under contract. At more distant points, they designate motor common carriers as their agents and may have arrangements with several such carriers serving a single city. When shipments are tendered to motor carriers for delivery to forwarders, the forwarder should be shown on the bill of lading as the receiving carrier and the motor carrier should sign it as agent for the forwarder.

There are many freight forwarders operating in the United

States, some of them on a nationwide basis, others on a more limited scale, and still others specializing in the handling of a single commodity or group of commodities, such as paper. In order to provide a more frequent service at lower cost than each could furnish single, two or more forwarders will sometimes pool their available freight (joint load) for movement between terminals. Section 404(d) of the Act (Para. 10766 P.L. 95-473 49 USC) permits this and the practice is particularly advantageous at smaller terminal cities.

The introduction of Plans II½, III and IV piggyback service by the railroads was widely acclaimed by the freight forwarders as a boon to their operations and they use these services extensively. The resulting economy and flexibility has enabled them to publish more volume rates, including a large number of such rates with minimum weights comparable to truck rates.

Freight forwarders should not be confused with non-profit shippers' associations which, although they operate on freight forwarder principles, are not subject to regulation under the Interstate Commerce Act.

Parcel Post

Parcel post was originally designed primarily for small packages, but permissable sizes and weights have increased to a fair extent in recent years. Federal laws and regulations establish parcel post as fourth-class mailing matter. First-class is regular sealed mail and postal cards. Second-class is newspapers and periodicals. Third-class is unsealed envelopes containing announcements, advertising matter, books, catalogs and circulars, merchandise, seeds, plants, and so on, weighing less than 16 ounces. Complete information on parcel post regulations is outlined in the *United States Postal Manual.*

The weight and size of parcels that will be accepted for mailing depends upon the class of post office from which the parcel is mailed. Parcels mailed at a first-class office for delivery at another first-class post office are generally subject to size and weight limits

FREIGHT ROUTING

which should be checked with the local post office periodically (a subscription to the *Postal Manual* will save you the trouble; the *Manual* also describes the limitations for other than first-class post offices).

Post office service includes free delivery, but shippers must take packages to the post office. There are exceptions, dependent on the volume of business a firm can offer. The post office will pick-up daily, on schedule, if the volume is there. Additional services are available from the postal system, for specific fees. These services are subject to change but those described more fully in the *Manual* are:

- C.O.D. shipments
- Special handling (not special delivery but preferential handling and transportation)
- Special delivery—available at destination post office
- Insurance
- Guaranteed next-day delivery (there are limitations in that only certain post offices will accept such packages or mail).

Parcel post charges for each package and the charges of course vary according to the distance the package will travel from the origin post office.

Generally speaking, parcel post charges are set up on a zone basis, the zones being shown on charts and identified by the first three ZIP code digits. Zones established for postal rates and the approximate distances are as follows:

Local	
Zones 1 and 2	Up to 150 miles
Zone 3	Over 150 miles, not over 300
Zone 4	Over 300 miles, not over 600
Zone 5	Over 600 miles, not over 1,000
Zone 6	Over 1,000 miles, not over 1,400
Zone 7	Over 1,400 miles, not over 1,800
Zone 8	Over 1,800 miles

Checking parcel post rates basically means checking in the official zone chart for the office of mailing and the office of address. (The Official Parcel Post Key for the origin city was formerly used for this purpose.) Official zone charts may be obtained free by request to the postmaster at the office of mailing.

A loose-leaf *United States Postal Manual* is available that contains extensive information on parcel post and mail regulations. It goes into the subject from many different angles, some of them quite removed from average industrial traffic requirements. With mailing costs increasing, traffic people may be able to help their companies in the mailing of circulars, advertising, price information and other items as well as letters and parcel post. The *Manual* just mentioned, which describes second and third class privileges, and many exceptions, special provisions and services may point to some valuable savings.

Domestic Water Routing

At one time water routes were the backbone of intercity transportation in the United States and most of the centers of trade and industry were on navigable waters. With the coming of the railroads, the importance of water routes dwindled. But today industries that use vast quantities of raw materials are often prepared to trade-off the slowness of water transportation for the cheap bulk rates it offers, and specifically seek sites on waterways.

The domestic water transportation system includes:

—Inland water carriers, for the most part barge lines. Most of the tonnage moved by barge lines is bulk. Even along the Atlantic and Gulf coasts the Intracoastal Waterway system is basically a "canal" system that offers protection from open waters to the barges and tow vessels.

—Great Lakes carriers, again mostly bulk carriers of ores and coal.

—Intercoastal carriers covered by the Intercoastal Shipping Act of 1933, which permits only American-registered vessels to operate in this trade. Once important, intercoastal shipping in the U.S. is almost non-existent.

A lively segment of the domestic water trade consists in moving rail cars, trailers and containers on ships between various ports—a

FREIGHT ROUTING

sort of "fishyback" corresponding to the rail piggyback services. Rail cars are shipped between the Puget Sound area and Alaska as well as to and from points in British Columbia. Another type of service to and from Puerto Rico and East Coast ports uses roll-on/roll-off ships and multideck barges as well as conventional container vessels.

Like rail piggyback, roll-on/roll-off and container operations of this type provide an economical service between shippers' and consignees' places of business, with the freight remaining in the same vehicle without rehandling from origin to destination, and with a much smoother ride for the greater part of the journey. It thus minimizes packing expense, pilferage exposure, and loss and damage to the goods while in transit.

The barge lines that operate along the navigable rivers, canals and intracoastal waterways, together with the bulk carriers on the Great Lakes, account for the greatest share of domestic waterborne tonnage. A few barge lines have joint routes with rail and truck connections, but in the main their business consists of port-to-port operations. Modern barges are huge affairs, capable of loading some 1,500 tons of freight, and volume is a very important con-

5,000 H.P. M/V ASHLAND with 12 BARGE TOW
277,413 BLS.-41,000 TONS
Equiv. to 1,165-10,000 Gal. Tank Cars
Loaded Canebrake Terminal-Mississippi River
Up Bound Ohio River-May 27, 1967

Inland U.S. waterways barges pushed by a "tug" and used for bulk materials on the major rivers in the United States.

sideration in the use of this type of service. The most attractive barge rates generally have minimim weights of from 300 to 1,000 tons.

Domestic water carriers, with certain exemptions for bulk haulers and others, are regulated by the Interstate Commerce Commission as set forth in Part III of the Interstate Commerce Act (P.L. 95-473). Joint services of water carriers with rail and truck lines are subject to the conditions of the Uniform Straight Bill of Lading as prescribed by the rail and truck classifications. Bill of lading provisions limit carrier liability for the water leg of the move except when tariffs are applicable that provide liability, so insurance covering the water movement may be necessary as a safeguard.

Ocean-going carriers such as the intercoastal operators are subject to FMC jurisdiction. Thus port-to-port rates do not include marine insurance, for shipments are subject to the conditions of the

Three-deck roll-on/roll-off barge pulled by an ocean-going tug operating between Jacksonville, Florida and Puerto Rico.

FREIGHT ROUTING

ocean bill of lading. Separate insurance against the perils of the sea, etc., must be arranged and its cost taken into account in deciding on the use of these services.

Still other cost factors must be considered in connection with these port-to-port rates. Port tolls, wharfage and lighterage, as well as unloading and loading of cars and trucks and to be added (except in those comparatively rare cases where a vessel is able to tie up at the private dock or bulk loading terminal of shipper or consignee to receive or discharge its cargo). Rates for tollage, accessorial services, etc., are published in tariff form by the owners of dock facilities, such as port authorities, ship terminal companies and railroads. Only railroad charges are under the jurisdiction of the Interstate Commerce Commission.

Domestic Air Freight Routing

Air cargo has long outgrown the emergency-or-perishables-only category. Its growth has kept pace with advances in aircraft speed and size. Thus we now have freight versions of the Boeing 747, with cargo capacities of 275,000 pounds, which can accept 20-foot containers.

Air freight has opened new markets for products that were once limited to areas close to production points and the list of commodities that regularly move by air is expanding. Today, most industrial points can be reached by air or an air-truck combination. The air version of the twenty-foot container can now be made strong enough, within its tare weight limitations, to withstand the rigors of surface transportation, so that air-truck, air-rail and even air-sea intermodal transfers are now possible.

In 1980, the Interstate Commerce Commission declared that all motor carrier service that is incidental to Air Cargo Service is free of economic regulation. This is applicable to both an inbound movement and an outbound movement. The opportunity, then, to negotiate through rates from plant sites now exists and the traffic manager can look forward to the airlines becoming more interested in intermodalism for through routes and control of the freight available.

INDUSTRIAL TRAFFIC MANAGEMENT

The types of air freight routings available include trunk line and local-service carriers that are primarily passenger-oriented, but also offer various cargo services; "pure" air freight carriers; air taxi and commuter lines, some of which also confine their operations to cargo only; air freight forwarders and small-packages express services; and cargo-only charter operators.

The trunk line carriers operate over specific routes on specific schedules and generally fly passenger planes with room in the hold or belly for cargo. The local carriers or "feeder" lines also operate on schedule over fixed routes but within a confined short-haul area, generally feeding passengers and cargo into the bigger airports from the smaller communities.

Air-cargo-only carriers carry no passengers, but fly on schedules between major market areas, often using trucks to move cargo to and from major airports and nearby cities. And of course, some large trunk-line carriers have dedicated planes in their fleets to air cargo only and operate those on schedules.

Air-taxi operators fly smaller planes and occasionally use helicopters. If they are in the passenger commuter business, cargo is

Stretched DC-8 jetfreighters loading at Chicago.

FREIGHT ROUTING

a side line, but some all-cargo operators will fly as soon as the cargo is ready. Air-taxi lines usually fly shipments from the nearest airport of the consignor to the nearest airport of the consignee.

Flying Tigers' aluminum 8x8x10 containers, containing 631 cubic feet, were developed to fully utilize the 747 freighter's enormous cargo space. Weighing 850 pounds, the units are lockable and water-tight and can be easily equipped to accommodate garments on hangers.

Charter operators are normally only for the shipper with sufficient volume to fill an entire aircraft. Cargo charters are also available from the large trunk lines and the all-cargo carriers.

Air freight forwarders will combine a number of smaller shipments going to the same destination and offer this freight unitized or in containers to the trunk lines or cargo-only carriers. Many forwarder rates are the same as air freight rates, but forwarders will expedite and assume liability for shipments.

The main benefit of air freight is speed. Often the air shipper or his customers can carry small inventories, relying on speedy delivery of orders. Ideally, the whole process of filling an order, including payment for the supplier, is faster when using air freight, and some air carrier marketing departments will make quite extensive cash-flow analyses for a potential shipper to prove the point.

About 1975, many airlines established express package services, some with guaranteed delivery schedules. Each airline defines the dimensions and contents of the packages it will accept (perishables are usually excluded), and provides colorful wrappings or pouches. Some package services include pick-up and delivery, but most are airport-to-airport. Delivery of the small package to a passenger station at the airport up to 30 minutes prior to a flight time usually guarantees the package will go on that flight.

One large operator (Federal Express) in the field of fast air freight has set up a "hub" system with small, specially-modified cargo jets flying into the hub from all over the country. Here incoming packages are sorted for destination the same night so that they can be delivered the next day or no later than the second day.

United Parcel Service

Another small package carrier, and one that is quite competitive with the Postal System, is United Parcel Service (UPS). UPS is nationwide (including intrastate services) and has begun to expand into Canada and Europe. UPS will only handle packages with a maximum weight of 50 pounds per package and will not accept more than 100 pounds per day from one consignor to one consignee. Packages must not measure more than 108 inches in length and girth combined.

UPS also offers its "Blue Label" air delivery service which includes the use of containers in jet freighters. Blue Label service schedules second-day deliveries in major market areas, with fourth-day deliveries on long flights to outlying points.

UPS will serve any shipper upon payment of a small weekly service charge which entitles the shipper to an automatic daily call by a pickup driver, whether or not there are packages waiting for him. The shipper may also take his packages to UPS offices or call for pickup (which costs extra). Inside delivery will be made without additional charge and if delivery cannot be made on the first call, up to two additional attempts will be made. Undeliverable packages are returned to the shipper without charge. Rates are generally competitive with those of uninsured parcel post and include protection against loss and damage up to $100.00 per package, with additional coverage available at a reasonable charge. UPS billing and other documentation permit the prompt furnishing of information regarding the delivery of any shipment.

Pipe Line

Some pipe line companies are common carriers and some are privately owned. Their principal business is piping petroleum products, but they also handle other liquids and even some normally dry commodities, such as coal, in slurry form. Rates on petroleum products generally are published per barrel, subject to various substantial minimum tenders ranging from 5,000 to 25,000 barrels. Some pipe line companies maintain "tank farms" for storage at terminals for the accommodation of their shippers and receivers, and some are regulated under Part I of the I.C. Act.

Economic pipeline use generally requires having large volumes of material available for movement. The numbers of pipelines and the areas they serve are limited. Also, dealings in this field are technical, specialized, and heavily weighted toward engineering. Pipelines have been experimentally used to transport even packaged freight (in sealed capsules or "slugs" in oil pipelines, or in systems employing wheeled containers). But while such experiments are an interesting development in transportation technology, only in rare instances would they be practical.

Consolidation and Distribution in Routing

A freight consolidation program will often show a handsome profit even for the small shipper. Yet many traffic people take a dubious or suspicious attitude towards such programs—perhaps because they often involve holding shipments for several days to make up carloads or trailerloads, and that goes against the traffic manager's normal reflexes!

LTL shipments of course are the first group with consolidation potential. Check documents covering all shipments to be made within a 24- or 48- or 72-hour period going to a common destination. The total weight at the volume rate is the starting point of a consolidation calculation. Consolidating all shipments going to one destination for a given period means that only one master bill of lading and one manifest will cover all shipments in the consolidation. Result: A considerable reduction in paperwork. Now the question becomes which distribution tariff is most advantageous for the volume shipment—that of a warehouse in the destination city or that of the line haul motor carrier? Some simple addition could very well show the consolidation plus the distribution costs to be less than the total of all the LTL shipment costs.

Another consolidation possibility is a partial load from one plant with a stop-off for a load of the same commodity from another plant which makes up the car or trailer load. (This can be arranged even with a competitor shipping to the same general area for distribution.) The first shipper assumes the paperwork and pays the bills and the second pays the first his portion of the costs. The savings in consolidation are well worthwhile. During his firm's three-month intensive sales period, one shipper of consumer products, each year for 15 years, saved over $50,000 annually by consolidation. Consolidation of this type, or pool loading as it is sometimes known, works equally well for rail, motor or air.

Consolidation takes another form when shippers join in non-profit associations to pool their shipments from any one area to a number of different areas. At destination, arrangements are made with local common carriers to distribute the load, which is line-hauled by another common carrier (motor or piggyback) under a "freight-all-kinds" (FAK) rate found in many tariffs.

Shippers' associations are found all over the country, but they cannot solicit freight under provisions of the IC Act. Would-be members must seek them out, apply for membership and pay an entry fee—usually quite low in comparison to the freight savings and the annual return of a share in the earnings of the association.

Consolidation also works for inbound shipments. If raw materials can be accumulated and held at a common point, a shipper could consolidate at that point and bring them in at a volume FAK rate.

The distribution of these pool loads can also be arranged by working with a motor carrier who has, or will publish, a distribution tariff applicable at several of his terminals. His distribution rates will generally be lower than his local rates if he gets the line haul volume business. Another approach is to contact a public warehouse for its rates for breaking out a pool load and distributing it. Yet a third approach is to local carriers (or cartage haulers) who will receive the volume shipment and distribute it. Total cost is the key in judging these possibilities.

The shippers' association does all the arithmetic for a member, who need only compare what it offers to the normal common carrier costs.

Operators of private fleets can use destination distribution services to make better use of their over-the-road trucks. By arranging with a local cartage agent or a common motor carrier for distribution, the private trucker can deliver the volume shipment with the necessary freight papers and return home for his next load or report to a point for a return load to the plant. This will increase the utility of trucks and drivers.

INDUSTRIAL TRAFFIC MANAGEMENT

The American Institute for Shipping Associations, Inc. (1100 Seventeenth St., N.W., Washington, D.C. 20006) can supply a list of non-profit shipper associations in the United States. In 1980, there were about 130 of them doing over $100 million a year business and saving their members up to $20 million annually.

7

INDUSTRIAL PACKAGING & MATERIAL HANDLING

The traffic manager must understand packaging and materials handling. Both activities, properly practised, are essential to any program of damage-free transportation and also to reducing costs. Both are functions that, coupled with transportation, form the most important element of distribution.

Packaging in fact is one of the factors in the total cost of production in addition to having important effects on the cost of transportation. Hence the value of studying it for the traffic manager who will need to analyze the interaction between the two functions.

Materials handling in physical distribution is not an element in the cost of production, but a technique for moving goods speedily and safely in plants, docks, warehouses, etc. Here the traffic manager must be prepared to collaborate with the packaging and material handling engineers in the interest of maximizing company profits. An understanding, appreciation and respect for *their* problems is needed. What counts in physical distribution is not the lowest freight rate or the lowest packaging cost but the lowest overall cost to safely distribute the product to the ultimate consumers.

Packaging

It is estimated that 40 percent of all material shipped must be packaged—which makes the manufacturing of packaging materials a rather large industry but nonetheless competitive. Modern transportation technology and systems, the high cost of money and the importance of controlling inventory have all caused shipments to become smaller, and created the need for low-cost but efficient packaging.

The factors that go into the packaging "equation" are:
- Purchase cost
- Packing labor cost
- Sealing or closure cost
- Tare weight
- Cubic displacement
- Rates for chosen means of transportation
- Materials handling cost
- Warehousing cost
- Loss and damage experience

—and should be balanced to result in a package or container that provides:
- customer conveniences, such as reuse or ease of opening.
- Merchandising appeal—a miniature "billboard" or display, etc.
- Satisfaction of the customer prejudices that may be encountered in certain markets.
- Complete conformance to classification regulation of the carriers.

In packaging for transportation, there is a persistent tendency to spend too much for outer packaging. Overpacking, overprinting on the package and using large, heavily-loaded cartons (to reduce the total number of cartons in a shipment) are all resorts that send up packaging costs with no gain to customer or carrier.

One problem in packaging that has grown in importance recently is the question of what size of package is most useful at destination. Why package sixteen or eighteen items in a shipping carton when twelve-item cartons are most convenient in size and weight to the customer? The plant may have no difficulty in handling cases weighing 100 pounds but a retailer may not be able to handle anything over 40 pounds. Packaging for the customers' needs often reduces over-all warehouse or handling costs by eliminating broken lots at branch warehouses.

When considering a new package design, the traffic manager, together with the packaging engineer, should weigh the following factors:

- How will the carton, case, etc. be handled when it is packed (or repacked) at the manufacturing or customer's plant?
- Is it simple to close?
- Does it fit into surface or air transport containers?
- How will the inner package fit into the outer package?
- Is it adaptable to present packaging machinery or will new equipment be required?
- Is the test strength of the carton sufficient to meet the handling forces it is likely to encounter in transit?
- If manual handling is required, is the package shaped and sized properly?
- Can it be stacked both in transit and storage?
- Can the retail customer adapt it to "end-of-aisle" stacking displays?
- Is the shipping label large enough and in a position where it can be easily read?
- Are there overprinted legends and unnecessary matter on the case?
- Can the customer either dispose of the empty container easily, or re-use it or return it?
- Is it easy to identify the contents, sizes of the items and count etc., when the package is in inventory?

Packaging managers will take many of these factors into account but only with the traffic manager's "input" are all of them likely to be considered. A great deal of assistance in training personnel and in designing and using packages is available from container suppliers. Many "box" companies maintain staffs of expert designers broadly familiar with packaging problems. The larger companies usually provide laboratory test facilities and engineers for the evaluation and performance testing of various package designs and constructions. Many such laboratories are certified by the National Safe Transit Committee (NST) and can test and certify the adequacy of containers in conformance to the NST standards.

INDUSTRIAL TRAFFIC MANAGEMENT

Loss and Damage

Claims against carriers run into the millions of dollars each year. Frequently, the blame is laid to insufficient or inadequate packaging. There may be some justification for this but investigations by shippers and container manufacturers show there are other causes. Reducing loss and damage claims requires a claims prevention campaign by the traffic manager whose job it is to keep such drains on profits to a minimum.

Both the National Motor Freight Classification and the Uniform Freight Classification carry packaging requirements for many of the items they list. There are also rules in both publications that carry specific packaging requirements for corrugated fibreboard cartons as well as other outer shipping containers. The wise traffic manager will have his cartons tested annually to be sure they meet the bursting standards described in Item 222 of the NMFC and Rule 41 of the UFC.

The packing material used inside a shipping container should also be evaluated as part of the claims prevention program.

ITEM 222-2

SPECIFICATIONS FOR FIBREBOARD BOXES
Glassware, Articles in Glass or Earthenware or Fragile Articles
(Applicable only in connection with items 222 and 222-1)

Except as otherwise provided, glassware, articles in glass or earthenware, or fragile articles will be accepted in fibre boxes only under the following conditions:

(a) **Box Construction Requirements**: All other fibreboard boxes must comply with items 222 and 222-1 as provided in paragraph (d) or Note 1.

(b) **Weight Limit**: Except as provided in paragraph (d), glassware, articles in glass or earthenware or fragile articles must not exceed 65 pounds gross weight. Liquids in individual glass or earthenware containers exceeding one gallon capacity will not be accepted in fibre boxes.

(c) **Inner Packing Requirements**: Except as more specifically provided in paragraphs (d), (e) and (f), or Note 2, contents must be packed within container by or with liners, partitions, wrappers, excelsior, straw or other packing material which will afford adequate protection against breakage or damage and box must be completely filled.

Upgrading here could not only improve the safety of the package but might also reduce its weight. And other packaging factors must also be examined as part of an effective claims prevention program. These include the sealing process used, the materials handling equipment employed and the effect of automatic operations on the package. Is the package being "hit" with unnecessary force at some point in an automatic handling system? Are the lift truck operators competent enough to avoid hitting and dropping cases? Are packers, to save money, putting too much in a case, creating bulges that are easily burst or squeezed by other packages in a stack?

The traffic manager must also remember the differences in liability of a common carrier and the liability of a warehouseman in his handling of any claims.

The Shipping Package

A shipping package is basically an enclosure designed to protect its contents from loss or damage in handling, transportation or storage. It should therefore carry no more of a legend than is necessary by direct identification or by symbols, to show what the contents are, leaving enough space on the exterior for names and addresses of consignor and consignee.

The package usually consists of an outer container or shell and some inside material used to cushion or protect the contents. The inner packing may be shock-absorbing material, molded plastic forms, excelsior, torn newsprint or anything else that will serve to protect the packaged item.

The package or container itself serves many purposes. For example, it:
- Protects against rough handling and resulting shocks.
- Protects against dirt, dust, moisture and other contamination.
- Helps prevent pilferage or tampering.
- Helps safeguard hazardous materials.
- Simplifies and makes more efficient stacking, warehousing and inventory control when it is the standard shipping unit.
- Often defines a unit of sale.

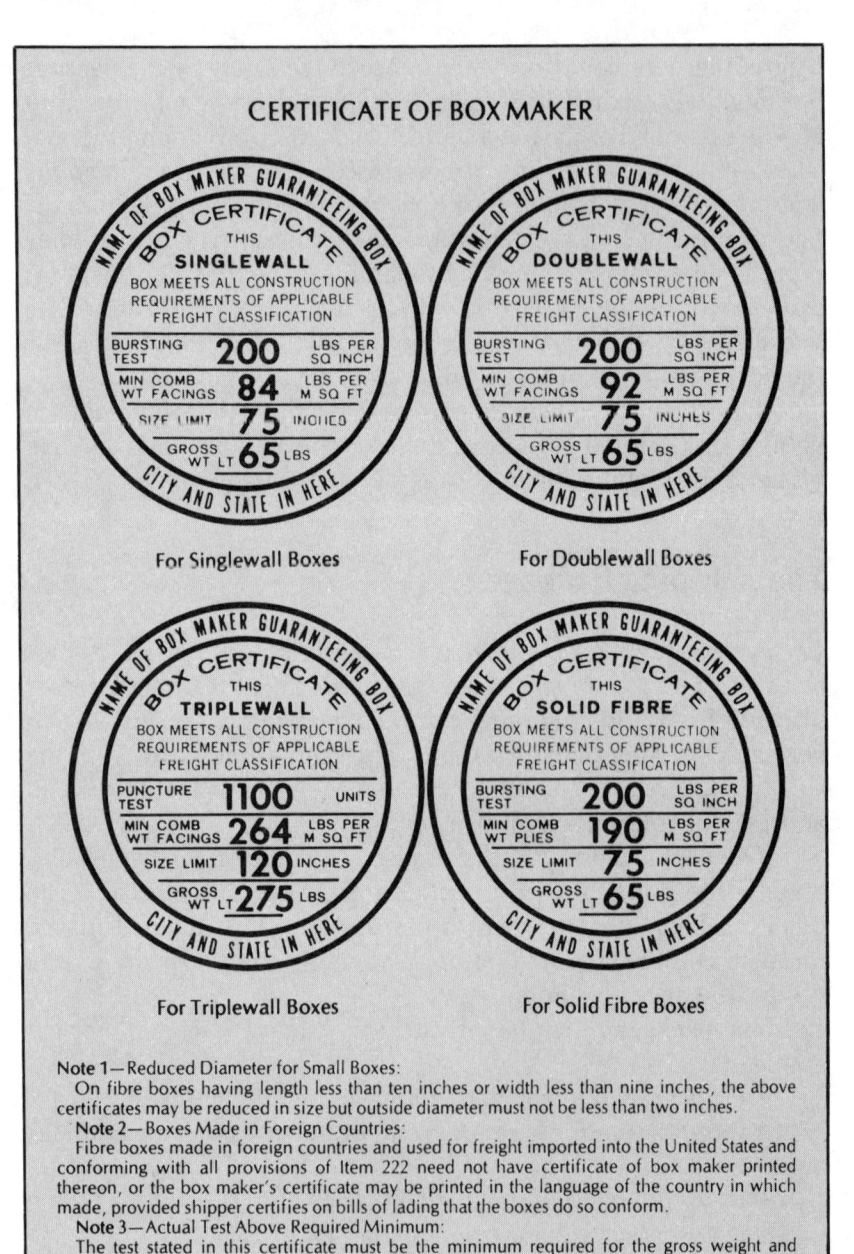

INDUSTRIAL TRAFFIC MANAGEMENT

CERTIFICATE OF BOX MAKER

For Singlewall Boxes

NAME OF BOX MAKER GUARANTEEING BOX
BOX CERTIFICATE
THIS SINGLEWALL
BOX MEETS ALL CONSTRUCTION REQUIREMENTS OF APPLICABLE FREIGHT CLASSIFICATION
BURSTING TEST **200** LBS PER SQ INCH
MIN COMB WT FACINGS **84** LBS PER M SQ FT
SIZE LIMIT **75** INCHES
GROSS WT LT **65** LBS
CITY AND STATE IN HERE

For Doublewall Boxes

NAME OF BOX MAKER GUARANTEEING BOX
BOX CERTIFICATE
THIS DOUBLEWALL
BOX MEETS ALL CONSTRUCTION REQUIREMENTS OF APPLICABLE FREIGHT CLASSIFICATION
BURSTING TEST **200** LBS PER SQ INCH
MIN COMB WT FACINGS **92** LBS PER M SQ FT
SIZE LIMIT **75** INCHES
GROSS WT LT **65** LBS
CITY AND STATE IN HERE

For Triplewall Boxes

NAME OF BOX MAKER GUARANTEEING BOX
BOX CERTIFICATE
THIS TRIPLEWALL
BOX MEETS ALL CONSTRUCTION REQUIREMENTS OF APPLICABLE FREIGHT CLASSIFICATION
PUNCTURE TEST **1100** UNITS
MIN COMB WT FACINGS **264** LBS PER M SQ FT
SIZE LIMIT **120** INCHES
GROSS WT LT **275** LBS
CITY AND STATE IN HERE

For Solid Fibre Boxes

NAME OF BOX MAKER GUARANTEEING BOX
BOX CERTIFICATE
THIS SOLID FIBRE
BOX MEETS ALL CONSTRUCTION REQUIREMENTS OF APPLICABLE FREIGHT CLASSIFICATION
BURSTING TEST **200** LBS PER SQ INCH
MIN COMB WT PLIES **190** LBS PER M SQ FT
SIZE LIMIT **75** INCHES
GROSS WT LT **65** LBS
CITY AND STATE IN HERE

Note 1—Reduced Diameter for Small Boxes:
On fibre boxes having length less than ten inches or width less than nine inches, the above certificates may be reduced in size but outside diameter must not be less than two inches.

Note 2—Boxes Made in Foreign Countries:
Fibre boxes made in foreign countries and used for freight imported into the United States and conforming with all provisions of Item 222 need not have certificate of box maker printed thereon, or the box maker's certificate may be printed in the language of the country in which made, provided shipper certifies on bills of lading that the boxes do so conform.

Note 3—Actual Test Above Required Minimum:
The test stated in this certificate must be the minimum required for the gross weight and dimension limit, except as provided in Note 4 of Item 222-1, and the combined weight of facings or plies must be the minimum prescribed by item 222, Sec. 3. When the actual test is in excess of

the minimum test required, the actual test may be stated below the certificate, but in such case all classes and rules in this classification as provided for a box having minimum test will apply.

Note 4—Notations for 'Size Extension' and 'Box within a Box':

On boxes made to comply with Item 222, Sec. 3, (Note 4), which permits increase in size when gross weight is reduced, 'size limit' and 'gross weight limit' may be omitted, but certificate must otherwise comply with requirements of Item 222-1 (a), and below certificate must be printed 'Conforms to Size Extension Formula.' On boxes made to comply with Item 222, Sec. 3, (Note 6), inner and outer box must bear certificate complying with Item 222-1 (a), and below the certificate on outer box must be printed Box Within a Box.'

Note 5—Special Box for Glass—100 Pounds—100 inches:

On boxes made to comply with item 222-2 (d), certificate must show the actual 'size limit' and 'gross weight limit' authorized by Paragraph (d) and the certificate must otherwise comply with requirements of this section, and below certificate must be printed, 'Special Box for Glass.'

(b) **For Double Thickness Scoreline Boxes:** Double thickness scoreline boxes which conform to requirements of Item 222-5 (k), must bear a certificate in the following form, size, type and wording:

```
┌─────────────────────────────────────────┐
│           BOX CERTIFICATE               │
│  THIS DOUBLE THICKNESS SCORE LINE BOX   │
│  MEETS ALL OF THE CONSTRUCTION          │
│  REQUIREMENTS OF APPLICABLE FREIGHT     │
│  CLASSIFICATION                         │
│  BURSTING TEST        GROSS WT LT       │
│                                         │
│  _____           _____          │
│  LBS PER SQ IN           LBS            │
│  (Name of Box Maker Guaranteeing Box.)  │
│       (City and State in here.)         │
└─────────────────────────────────────────┘
```

(c) **For Special Numbered Packages:** Fibreboard boxes, including those referred to as 'containers,' not authorized by this item but which are authorized by item 680, Sec. 1 (c) for some particular article must bear the appropriate certificate in the following form, size, type and wording:

```
┌─────────────────────────────────────────┐
│          PACKAGE CERTIFICATE            │
│  THIS BOX MEETS ALL CONSTRUCTION        │
│  REQUIREMENTS OF APPLICABLE FREIGHT     │
│  CLASSIFICATION                         │
│  FOR PACKAGE NO.  BURSTING TEST LBS. PER SQ. IN. │
│       000                 000           │
│                                         │
│     (Name of Box Maker Guaranteeing Box)│
│                                         │
│        (City and State in here.)        │
└─────────────────────────────────────────┘
```

For triple wall boxes, and double wall box specifications which refer to puncture test units, substitute the words 'Puncture Test Units' for 'Bursting Test Lbs. Per Sq. In.' in the above certificate.

Where numbered packages authorized different tests of fibreboard for bodies and caps, tests of the body only need be shown.

On boxes made to comply with Item 222-2, Note 1, certificate must be a rectangular form shown above, enlarged to include certification for packages 1304 and 1305. Such certificate must show bursting test, size limit and weight limits.

(d) **Special Identification for Unmarked LTL Shipments:** when used for LTL shipments, boxes which bear certificate prescribed in Paragraph (a), (b) or (c) which boxes do not show descriptions of contents must be marked with an identifying symbol or number and identifying symbol or number must be shown on shipping order and bill of lading.

Note 6—Shipments in fibreboard boxes manufactured on or before October 1, 1977 and bearing the form of certification required by Section 4 of Item 222 as of January 1, 1976 will be accepted for transportation if tendered to carriers parties hereto before March 31, 1979.

- Provides product identity and definition of quality, quantity, size, etc.
- Can be an aid to merchandising in various ways, particularly by means of the "miniature billboards" that attractively printed shipping containers provide.

Regulations

To handle his job, the packaging manager must of course be thoroughly familiar with all rules and regulations affecting the design and specification details of shipping containers and methods of closure and reinforcement approved. Since the traffic manager is also vitally concerned with all rules and regulations affecting the shipment of merchandise by various carriers, and is expected to be an expert in the interpretation of these rules, this is an area of strong mutual interest and one providing an opportunity for very effective collaboration.

Good sources of general information on specific types of materials and containers are the publications put out by packaging industry groups or companies. An excellent example is the *Fibre Box Handbook*, compiled by the Fibre Box Association, 224 South Michigan Avenue, Chicago, Illinois 60604, and usually available from the members of the Association. This handbook offers a useful short course in the language of the container manufacturer. It defines terms and gives clear illustrations of cartons in various stages of construction. The Association also publishes a manual, *Certification and Other Markings for Fibreboard Boxes*, which covers rail, motor, air, express, parcel post, and federal shipments.

Estimates have been made that over 95 percent of packaged freight moves in corrugated fibreboard containers. We will therefore use the corrugated container and rules governing its use as examples in the discussions that follow.

It's a common assumption that the provisions included in carrier regulations constitute an all-inclusive and safe guide to judging the adequacy of a given shipping container. This is not true. The provisions of Rule 41 of the Uniform Freight Classification (UFC) and Item 222 of the National Motor Freight Classification (NMFC) are the bare minimum—an attempt to write a rule that takes into account such packaging variables as shipping characteristics of the products to be packed and the hazards they may be exposed to in transit.

It is extremely important that the packaging manager, and others in the organization who may influence his judgment, recognize that in applying a minimum rule of this character, adjustments must always be given consideration. For example, it may become necessary to increase container protection for easily-damaged contents or to overcome unusually severe hazards. This may have to be done either by making the box itself stronger or by using special inner packing; or it may become necessary for the ruling agency to make exceptions, permitting the shipper whose merchandise requires little or no protection to use a lower-strength container.

It is incorrect to conclude: "Rule 41 says I can ship 65 pounds gross weight in a box whose dimensions add up to 75 inches, made of board testing not less than 200 pounds per square inch bursting test; therefore, such a pack is suitable for my product, regardless of what it is!"

Such fallacious thinking is common and management pressure will be encountered to implement its conclusions. While usually not taken to the extremes quoted above, it results in a continuing push to reduce safety margins between actual needs and the limits of the rule.

Figure 1 of Item 222-1 of the NMFC shows the certification required of the box manufacturer. Rule 41 of the UFC is almost identical to it.

Classification Rules

The chart on the next page shows the specific provisions affecting various types of packages and shipping containers in the NMFC and the UFC:

Remember to examine all supplements to the Classifications

INDUSTRIAL TRAFFIC MANAGEMENT

and note any changes or additions that will affect your packaging system. Remember also that you can effect those changes yourself by applying to the Classification Committee. The Committee will set a six-month test period to determine if a new package should be included in the list of those permitted. An applicant must first fill out the National Classification Board's "Test Shipment Permit" and then make periodic reports to the board giving date of shipment, carrier, pro number, number of cases, total weight, consignee's location, article shipped and the condition of carton at destination.

Airlines have had their own packaging requirements which are less stringent than the surface carriers.

Packaging Specifications

Following carriers' packaging rules will not guarantee that you will have an adequate container for the commodity packed. The

Container	Rule UFC	Item NMFC
Bags, multi-wall and single-wall	40 - Sec. 10	200
Low density fibre container	40 - Sec. 7¼	201
High density fibre container	and Sec. 7¾	
Barrels	40 - Sec. 4	210
Boxes general	40-Sec. 1	226
Fibreboard boxes	41	222 thru 222-6
Buckets & pails	40-Sec. 8	225
Rubber	40-Sec. 7½	230
Carboys	40-Sec. 9240	
Crates	40-Sec. 2	45
Drums	40-Sec. 7	255
Polyethelene drums	40-Sec. 5½	256-257
Molded pails	40-Sec. 11	258
Metal barrels & drums	40-Sec. 5 & 6	260

Provisions affecting packages and shipping containers.

packaging must take into account many factors and, by applying certain basic principles of protection, design and analysis of the economic feasibility of alternative approaches, arrive at a practical answer.

Taking Rule 41 at face value, it might appear that the only characteristics of the product the packaging engineer must consider are weight and size. Studying the matter further, one will find that the classification contains certain special provisions for a few specific commodities because of special characteristics they possess. However, nowhere in the classification will one find a list of any of the additional characteristics that must be considered, or what to do about them. Here is such a list of commodity characteristics; but no claim is made for its being all-inclusive.

Selected Characteristics Influencing Container Specifications

Weight	Urgency of need (intrinsic value)
Size	Physical form—liquid, granular,
Proportion	solid, etc.
Design	Susceptibility to contamination
Shape	Raw Material or finished product
Strength	Set up or knocked down
Fragility	Appearance and other merchandising
Value in dollars	features
Finish	Manufacturing setup
Trade customs	Storage and distribution setup
Storage life	Unusual hazards, explosives, poisons,
Sales practices	etc.
Reuse features	Types of transportation to be used

The foregoing factors are concerned with the merchandise itself and the conditions of practices surrounding its handling. While these dictate how a product should be packed, it is specific "outside" hazards that usually produce loss or damage and set the standard for the degree of protection a package must provide.

INDUSTRIAL TRAFFIC MANAGEMENT

Handling and in-transit hazards can be broadly classified into seven basic types.

> **Basic Types of Handling Hazards**
> 1. Piercing blows or pressures.
> 2. Crushing blows or pressure from heavy loads.
> 3. Impacts creating shock.
> 4. Impacts creating rupture or breakage.
> 5. Contamination.
> 6. Marking or imprinting through abrasion or pressure.
> 7. Theft or tampering.

A single blow or impact may, and usually does, induce several types of "side effects" that could result in multiple damages. For example, a container dropped several feet and landing on a corner may ultimately suffer damages described in hazards 2, 3, 4, 5, 6 and 7.

The knowlegeable traffic manager will keep records to find out which of the seven hazards are causing the damages *his* shipments suffer. The weaknesses revealed in the package can then be corrected. The results will also indicate whether the protection built into the package is of the right kind, in the right amount, and in the right place.

Packaging for transportation can be summarized as the application of 15 fundamental principles of protection:

1. Enclosure—The contents should be enclosed by suitable materials and the complete package should be structurally designed.

2. Compatibility—All packaging materials should be compatible with the contents of the container.

3. Retention—The container should be strong enough in *normal* use not to lose its contents or expose them to contamination.

4. Restraint—The container should be so designed and built as to eliminate or control all undesired movement of its contents.

INDUSTRIAL PACKAGING & MATERIAL HANDLING

5. Separation—Adequate dividers or other devices should prevent parts or surfaces of the packaged items hitting against one another.

6. Cushioning—Adequate resiliency or cushioning should be built into the package to prevent damaging shocks from reaching the contents.

7. Clearance—There should be sufficient clearance between the contents and the container walls to reduce damaging contact from any cause.

8. Support—The container and its interior packing should provide the necessary support to resist all the stresses and loads of stacking.

9. Position—The container should be designed so that the weakest parts of contents receive the greatest protection.

10. Non-Abrasion—The container should be designed to prevent rubbing and marking of labels, codes, etc., during handling.

11. Distribution—The weight-bearing surfaces of the container should be of sufficient size to spread the load imposed by the weight of its contents.

12. Suspension—Contents should be "hung" inside the container, if possible, in such a way as to prevent abrasion, shocks, etc.

13. Inspection—The container should have inspection or ventilation openings when needed.

14. Closure—The container should be adequately closed, sealed and reinforced.

15. Instructions—The container should carry adequate identification, instructions and information regarding contents and their proper storage and handling.

On page 167, an attempt has been made to tabulate all of this into what might be called a packing "equation." It is not a mathematical formula for in packaging, there can be many answers to a given problem. While Rule 41 has been used in this example, any other packaging rule or specification could be inserted in its place with equal validity.

The great majority of containers used by industry fall into the #3 class specification, i.e., they exceed minimum requirements. But it is the job of a trained specialist, the packaging engineer, to determine what specifications most effectively satisfy all parts of the formula. The traffic manager can often be of great assistance to him in deciding on the right package for the job.

The traffic manager who uses both rail and motor carriers must check the packaging specifications in both classifications to determine whether they are identical for his products. If they are not, he must carefully select specifications that will allow him to use a common package and packaging system for both modes. In addition to the rules of the UFC and the items of the NMFC, he must also check the commodity listing to note the packaging required for a particular commodity or the variables allowed. But relying only on the rules in the classifications and on commodity descriptions is not always sufficient; packaging requirements may also be included in the tariffs.

The novice can in fact be misled in reading classification rules. It often helps to re-write the rule in question by removing all words between commas. This will clarify the essentials. Then the qualifications, lists, exceptions etc. between commas should be written out below the basic rule. It will then become clear what additions to the basic rule must be taken into account.

At the same time a review of the punitive provisions in the rules is important. These tell you what penalties you face or extra charges you must pay if your packaging does not meet the minimum requirements of the classification. The UFC, in rules 2 through 40, spells out carrier rights, defines terms and specifies markings; the NMFC lists its almost-identical rules under one item (110), broken into a number of sections.

The traffic manager can obtain a lot of useful information from the various manufacturing associations of wooden boxes, corrugated boxes or steel containers. However, shipments to or for a governmental agency must meet the packaging specifications that are generally part of the bid request. Commercial packaging practices are rarely applicable to government freight. Government packing specifications should be discussed with a container manufacturer who will have the expert knowledge to meet them.

Some carriers also have their own container requirements—and this can be a controversial subject. United Parcel Service (UPS) has its own specifications for example, even though it

INDUSTRIAL PACKAGING & MATERIAL HANDLING

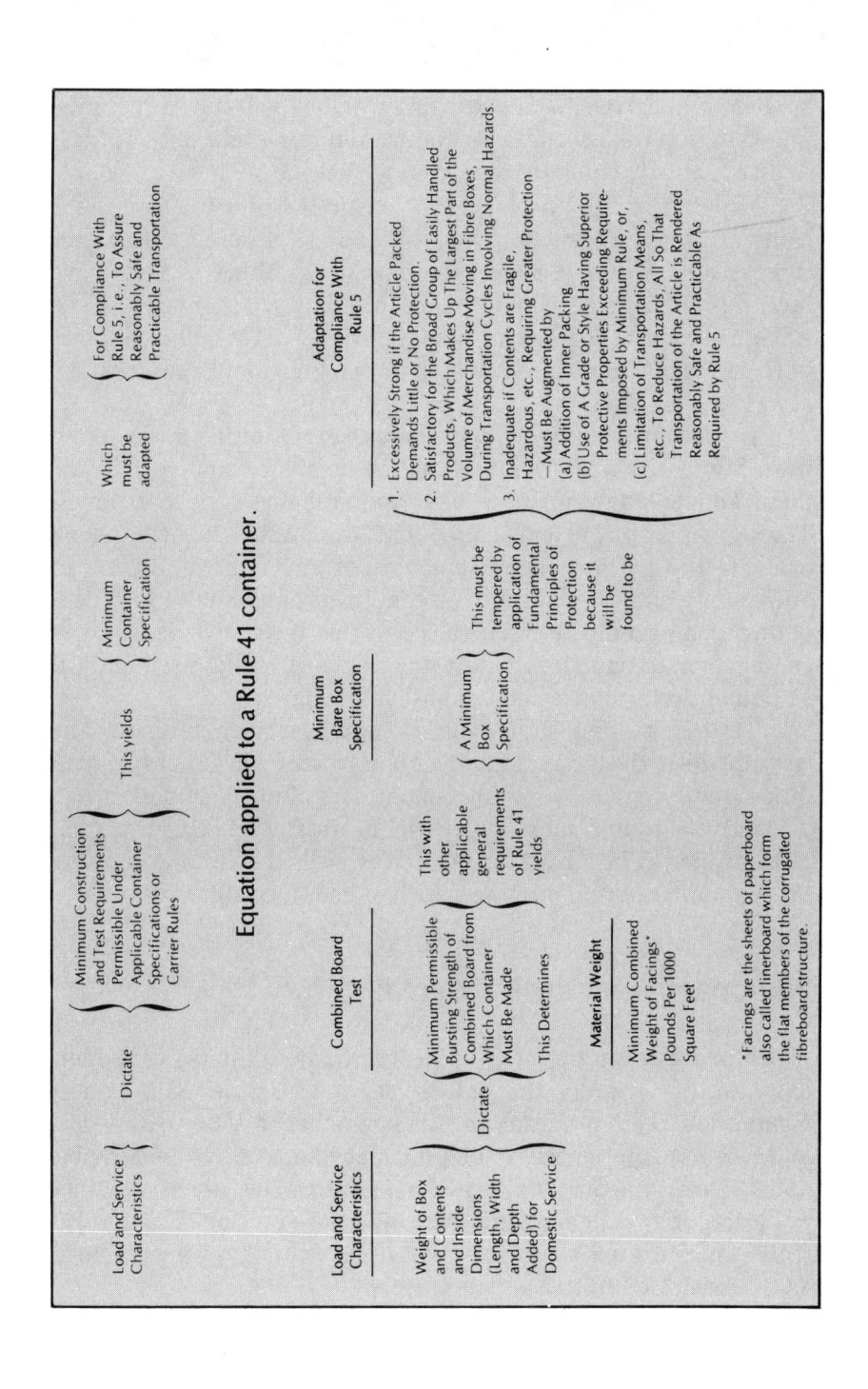

Equation applied to a Rule 41 container.

Load and Service Characteristics Dictate {Minimum Construction and Test Requirements Permissible Under Applicable Container Specifications or Carrier Rules}

This yields {Minimum Container Specification}

Which must be adapted {For Compliance With Rule 5, i.e., To Assure Reasonably Safe and Practicable Transportation}

Load and Service Characteristics

Combined Board Test

Weight of Box and Contents and Inside Dimensions (Length, Width and Depth Added) for Domestic Service — Dictate {Minimum Permissible Bursting Strength of Combined Board from Which Container Must Be Made} — This Determines

Material Weight

{Minimum Combined Weight of Facings* Pounds Per 1000 Square Feet}

This with other applicable general requirements of Rule 41 yields

Minimum Bare Box Specification

{A Minimum Box Specification}

This must be tempered by application of Fundamental Principles of Protection because it will be found to be

Adaptation for Compliance With Rule 5

1. Excessively Strong if the Article Packed Demands Little or No Protection.
2. Satisfactory for the Broad Group of Easily Handled Products, Which Makes Up The Largest Part of the Volume of Merchandise Moving in Fibre Boxes, During Transportation Cycles Involving Normal Hazards.
3. Inadequate if Contents are Fragile, Hazardous, etc., Requiring Greater Protection
   — Must Be Augmented by
   (a) Addition of Inner Packing
   (b) Use of A Grade or Style Having Superior Protective Properties Exceeding Requirements Imposed by Minimum Rule, or,
   (c) Limitation of Transportation Means, etc., To Reduce Hazards, All So That Transportation of the Article is Rendered Reasonably Safe and Practicable As Required by Rule 5

*Facings are the sheets of paperboard also called linerboard which form the flat members of the corrugated fibreboard structure.

is a common carrier. UPS packaging specifications have been questioned by several associations of manufacturers. Until this dispute is resolved, always compare the UPS requirements with the NMFC.

Expert testing of a new package is well worth the cost and is easily arranged. Over the years, a broad list of sound testing procedures and equipment has been developed. Many of these have been thoroughly studied and standardized by such agencies as The American Society for Testing and Materials (ASTM), Philadelphia, Pa., and The Technical Association of the Pulp and Paper Industry (TAPPI), New York, N.Y.

Laboratories equipped with the necessary testing devices, and staffed with trained engineers and technicians competent to use them and to analyze the results, are available in several areas. Through appropriate testing programs, it is possible to measure and evaluate the performance characteristics and relative abilities of both empty and packed shipping containers of all kinds. Such data permit comparison between (a) container types and designs, (b) container materials, (c) modifications of the product itself and (d) container performance and established standards.

Testing shipping containers is a highly technical business. The tests are deceptively simple, but certain factors, such as proper conditioning to standard moisture content, are of major importance in securing accurate results. Superficial or unskilled testing can yield gross errors, as in any other field. It is well, therefore, to be sure that container testing programs are handled by qualified people.

National Safe Transit Program

Much work in the pre-shipment testing of shipping containers has been done under the program of the National Safe Transit Committee. This program had its beginning in the excessive loss and damage experienced in shipping many porcelain enameled products. Determined to do something constructive about it rather than accept threatened rate increases, the Porcelain Enamel Institute inaugurated a program that led eventually to the National Safe Transit Committee and its program.

After a study of container laboratory testing equipment and techniques in use at that time, the committee adopted three test procedures reflecting the principle hazards encountered in shipment: (1) vibration tests, (2) drop tests, and (3) impact tests. The results of these test procedures were correlated with the actual hazards encountered in shipment, and standards were established to guide the performance of certification tests by the laboratories using them. The tests themselves are today conducted in accordance with standard test procedures established by the American Society for Testing and Materials and the Technical Association of the Pulp and Paper Industry.

After verifying the capability of the technical staff and the availability of authorized equipment, the National Safe Transit Committee (NSTC) awards certificates to qualified laboratories. These certificates permit them to perform tests and to certify the performance of various types of containers packed with specific items. This program has been successful, as evidenced by the fact there are some 120 certified laboratories and an even larger number of shippers cooperating in the program.

The appliance manufacturers also have widely adopted the program and NSTC procedures have been accepted and used on almost every type of commodity. Additionally, the program has been given approval by the various transportation agencies. Passing the NSTC requirements for certification is a major step in assuring shippers that their packaging will keep in-transit damage to a minimum.

The Committee describes the philosophy underlying its program as follows:

> If you will test your packaged products by these test procedures, and identify them by Safe Transit labels, experience has shown that your damage loss and your packaging cost will be acceptable minimums. It is up to each shipper to decide whether or not he will use these test procedures. The program is entirely voluntary and implies no connection with tariffs, freight rates, claim procedures, or any other existing transit regulations.

It should be noted that the procedures are used to determine the ability of the packaged product to withstand the dynamic load conditions encountered in transit and in handling. They do not

INDUSTRIAL TRAFFIC MANAGEMENT

evaluate the packaged products' protection from other types of damage such as corrosion, moisture, static compression loading, piercing, contaminating odors, failures to observe proper safety or security precautions, etc. The packaging engineer should determine whether his packaged product requires special protection and whether additional tests are required to assure a good overall performance against such hazards. He should also determine whether it meets carrier classification requirements in all other respects in addition to subjecting it to the standard NSTC test program.

Export Packaging

Specifications for export packaging are not set in the same way as those for domestic packaging. Insurance specifications are the main source of packaging requirements for export shipments, combined with the steamship lines' right to refuse freight that, in their opinion, is not well enough packaged to meet the many handlings given shipments destined overseas. Damage from salt water and interior moisture or "sweating" are particular concerns of export shippers, many of whom line their containers with waterproof paper, polyethylene, etc. for protection. Each shipper must develop an export package that meets the specifications and will deliver the freight to foreign customers in good condition.

Commodities shipped domestically in fibreboard containers may require further protection for export shipment. If careful handling and good service is available and well-designed heavy-duty containers are being used, they may perform satisfactorily on short trips to favorable ports. If poor or unknown conditions are to be anticipated, take precautions such as adding steel strapping, "overpacking" with domestic fibreboard, using "V" board weatherproof fibre boxes or nailed wood or wirebound boxes strapped, or packing directly in special export containers, etc.

Special grades of fibreboard containers were developed during World War II for shipping supplies overseas to the Armed Services. Known as "V" boxes, these containers are of board made especially to withstand outdoor storage and exposure. Lighter grades of the same types are identified as "W" boxes.

INDUSTRIAL PACKAGING & MATERIAL HANDLING

Tied in closely with packaging and loading in damage prevention programs is the use of the impact recorder, a small mechanism that shows the force and direction of each impact in transit and the time the impact occurred. Impact recorders are designed today for use in carrier vehicles, such as freight cars or trailers, and inside the packages themselves. The use of impact recorders inside packages is helpful in finding at which points the packages receive the most strenuous handling. Thus, impact recorders placed inside containers as merchandise comes off the assembly lines will show the force of impacts during warehouse handling, at both origin and destination, and in transit. Some recorders are designed to show if the position of the container is changed, which is a useful check on "This Side Up" freight. Impact recorders in carrier vehicles will highlight high-speed "humping," impacts, etc. en route. The use of an impact recorder is indicated if, upon delivery, damage is consistently detected and there is a need to prove that the packaging and loading are satisfactory but the handling en route is not. There are many different types of impact recorders on the market, designed for various uses.

Other damage-prevention measures include using cushion underframe box and flat cars and loading devices such as movable bulk heads, air bags, and so on. Many cars are equipped with both dunnage and cushion underframes to protect freight against impact. The cushion underframe cars use combinations of hydraulic pistons and fixed and movable friction plates to retard or "soak up" impacts. Movable bulk heads and air bags help in tightening loads so that shifts in lading do not occur and materials arrive at destination in sound condition. This equipment is widely available and should be used whenever possible.

A recent arrival in the packaging field is the shrink-pack, in which a number of packages are covered with a plastic film and given a brief flash of heat. The film shrinks tightly around the packages and a compact shipping unit is the result. This works with pallet loads of cases as well as smaller units. Shrink wrapping also reduces damage and pilferage.

Another recent development is the use of urethane or similar plastic foam blown into a case. The foam swells into an enveloping "cushion" around the item, affording excellent protection. This

INDUSTRIAL TRAFFIC MANAGEMENT

packaging innovation is particularly effective for fragile and highly valuable items.

Materials Handling for Distribution

There is much more to materials handling for distribution than pallets and forklifts—which are but two items in a vast arsenal of labor-saving devices. There is no "standard system" that a traffic manager can look for and apply to his problems. Each warehouse and each shipping dock must be looked at as an individual problem requiring an individual solution, even though the equipment employed is more or less standard.

The employment of materials handling equipment should reduce handling costs by eliminating as much manual handling as possible and by speeding the flow of packages from the end of the production line to loading dock or warehouse.

The efficiency of a materials handling system can be roughly measured by determining what the equipment handles (in weight

Using a forklift truck as a tractor for towing.

or count units) against time, resulting in some measure of "throughput" or man-hour evaluations. This only has a significance when compared to something, such as manual labor or to previously-used equipment.

Modern materials handling technology comprehends a vast range of equipment from the two-wheel hand truck—which still has great utility—to giant container fork lift trucks and computer-controlled stacker cranes and automated warehouses.

The first step in improving a handling system is to make a careful survey and cost analysis of the present system. Per unit costs may then be compared, old way vs. new way. It should follow that the unit cost will go down if the quantity handled increases. It has been said that one of the most important prerequisites for the successful materials handling engineer is to have and use a lot of common sense. He must be an efficiency expert, able to study an operation and figure out ways and means to do it better and cheaper. Someone once put it this way: "doing the mostest with the leastest. . . ."

First of all, data must be accumulated on each separate move

Forklift truck with push-pull attachment.

INDUSTRIAL TRAFFIC MANAGEMENT

(between successive points of manufacture and storage) as follows:
  (a) Description of material including unit weight, dimensions and over-all volume.
  (b) Details of movement: origin, destination, distance, clearances.
  (c) Number of employees utilized and time for each employee.

Care must be exercised that any plan developed to save costs does not impair necessary service to factory production and customers and does not lower safety standards.

Selection of the proper materials handling equipment (conveyors, overhead traveling cranes, industrial trucks either hand or power) must not be done hastily. Each type of equipment must be analyzed as to its advantages, limitations, cost and general adaptability. For example, automatic product-lift grab trucks are used to an ever-increasing extent where it is desirable to eliminate the handling and rehandling of pallets and pallet loads. Sometimes it is desirable to start out with a pilot operation on a trial and error

Forklift trucks with high telescoping masts.

INDUSTRIAL PACKAGING & MATERIAL HANDLING

basis. Frequently the design of a vendor's package must be changed so that incoming materials are received in proper unit packages suitable for power handling. Always remember that one big objective of materials handling is developing unit loads so as to use maximum cubic space; the use of storage to the roof if at all possible can mean good efficiencies. Another goal is loading and unloading shipments to and from carriers' vehicles and equipment mechanically. Constant alertness is needed to assure prompt replacement of obsolete equipment.

Loading

Loading is the science of placing and bracing freight in railcars, trucks, ships, or planes to get it to destination safely while complying with the carriers' and federal regulations. Poor loading and improper bracing of the load are major causes of damage.

When a shipment does not fully occupy the floor area of a box car, it must be blocked and braced with adequate dunnage material properly applied. If this is done, and that fact established beyond any reasonable doubt, it can be said that any damage in transit must have resulted from exceptionally rough handling by the carrier. Since the most damaging shocks are lengthwise in a car, the packages or pieces making up the shipment should be placed in the car in a stable arrangement using, when necessary, bulkheads, gates or steel bands nailed to the sides and floor of the car. The shipper using railroad equipment should examine the rules of the UFC for any special allowances on dunnage used to block and brace cargo in box cars.

The Association of American Railroads publishes a series of pamphlets containing rules concerning the loading of specific kinds of freight in or on railcars. Its Washington office issues a list of all AAR publications. Costs are small and every traffic department should have copies of the pamphlets that cover its types of loading operations. Some of these rules are mandatory.

The AAR rules governing open-top cars are mandatory. Loadings in this category include steel products, pipe, road and farm machinery, lumber, Department of Defense material, and so on. It is imperative that this kind of freight be handled and routed so that it does not leave the cars when they are in transit.

Rules involving closed-top cars are not mandatory except where interchanges between carriers may be involved. The possibility of the load "leaving" this type of equipment while in transit is not as great. Many types of freight fall into this category and approximately 40 pamphlets describing the loading rules that apply are available from the AAR. The pamphlets cover furniture, roofing felt, small cylindrical steel containers and so on. Closed-top car rules which are mandatory are in General Rules Circular 42D. Pamphlets are published by the Freight Loading and Container Section of the Freight Claim Division, 59 East Van Buren Street, Chicago, Ill. 60605.

The traffic manager should develop loading plans for his particular commodities and for the types of containers available. These plans and the proper use of material handling equipment, will reduce loading time and leave him available for other duties. But first the traffic manager must make sure his loading plans have been tried and proven.

Freight cars should be sealed after loading, with seal numbers noted on all records, including the bill of lading. Trucks should also be sealed but this should be done with the driver present. Truck shipments moving under a "shipper's load and count" tariff provision should be sealed. In all cases the consignee should check that the seal numbers at delivery are those recorded by the consignor. Any variations should be brought to the attention of the carrier before the cargo is touched.

One of the primary considerations in good loading is to plan each load in advance to be sure that there is at least sufficient weight in the car to meet the carload minimum weight required to apply a carload rate. If the shipper fails to do this, he will be shipping short weight loads, and he will be paying for "air." This is costly. For example, if an article is subject to an LTL rate of $1.00 per cwt. and a volume rate of 50¢ per cwt., minimum weight 30,000 pounds, the shipper should see to it that his shipment has 30,000 pounds in it. If it has 25,000 pounds he will pay for it "as 30,000 pounds," which means he is paying for 5,000 pounds of air as follows:

INDUSTRIAL PACKAGING & MATERIAL HANDLING

25,000 lbs.	@ LTL of $1.00 cwt.	$250.00
25,000 lbs. "as 30,000#"	@ CL of 50¢ cwt.	$150.00
30,000 lbs.	@ CL of 50¢ cwt.	$150.00

Of course, 25,000 pounds for $150.00 is 40 percent less than it cost LTL, but the cost per 100 pounds figures out to be 60¢, or 20 percent more than it would be if another 5,000 pounds were shipped.

Unless the tariff states that the shipper will load the vehicle, the carrier's driver must load his vehicle from the packages which the shipper stacks at the tailgate of the truck. Shippers are not obligated to aid the carriers' drivers unless the cargo is heavy or bulky and cannot be handled by one man under the best of conditions.

Air freight is never handled by the shipper at the plane. Unitizing and containerization are the usual practice in handling air freight. Unitizing air shipments often consists of simply stacking packages on a pallet covered with corrugated board or with film and strapped and covering the load with a net or plastic film.

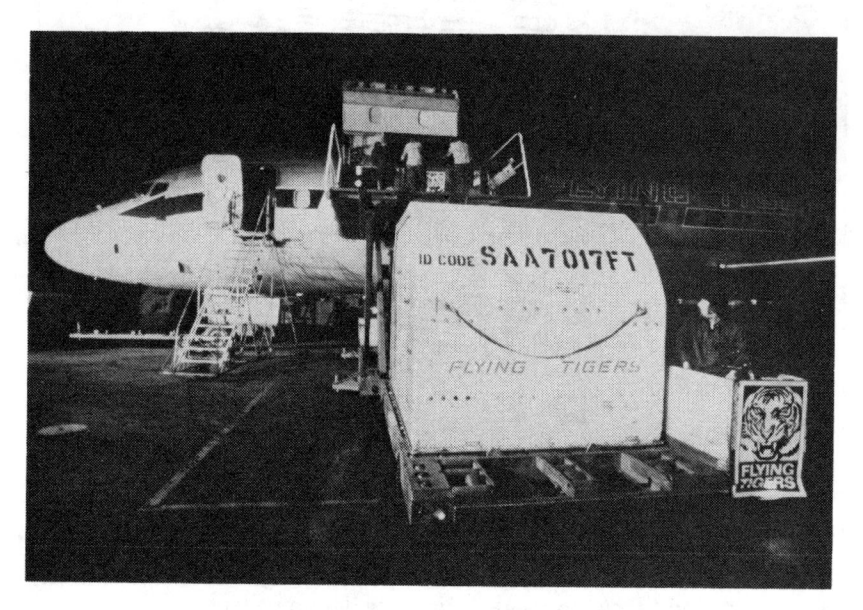

Airline container to be loaded on a DC-8 all-cargo plane.

INDUSTRIAL TRAFFIC MANAGEMENT

Courtesy National Wooden Pallet and Container Association.

The four basic pallet patterns are: Block, Brick, Row and Pinwheel. There are also a number of variations of each of these patterns some of which leave voids in the center of loads. Block patterns are used for containers or items of equal length and width including cylindrical containers. Block patterns are slightly unstable by nature thus requiring bonding if considerable movement is involved. Brick, Row, and Pinwheel patterns are employed for containers of unequal length or width. All three are customarily arranged in interlocking design which provides a stable load by placing adjoining layers at 90° to each other.

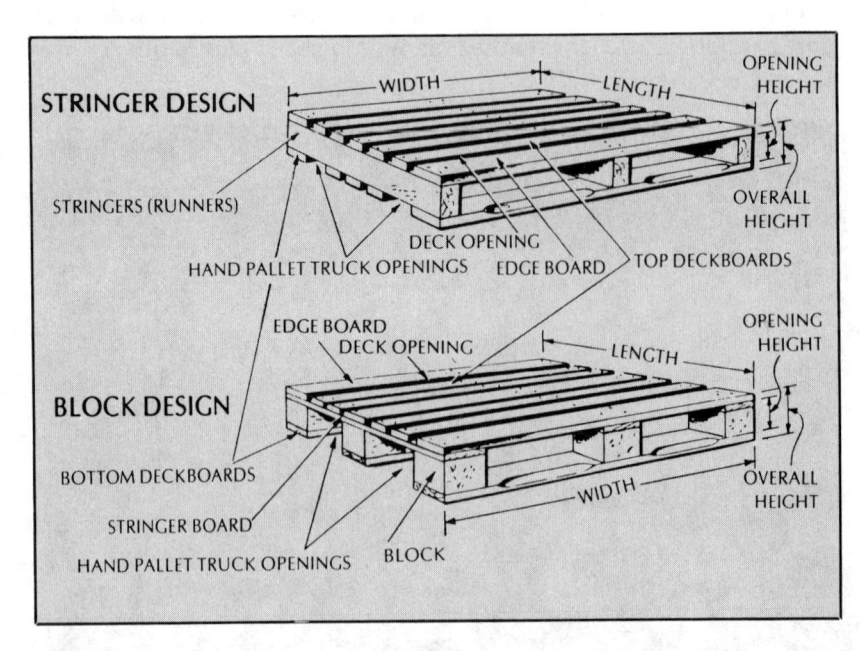

Principal parts and commonly used construction features of the *stringer* "two-way" and *block* "four-way" entrance design pallets.

If pallets are used in rail or truck shipments, tariffs should be examined for any weight allowances or free return of pallets. This can be a factor in deciding whether to buy reusable or disposable pallets. Disposables dispense with the accounting procedures and other problems of pallet "pools," such as receiving damaged pallets in return for sound ones, etc. An important advantage of all forms of palletization is the speed with which vehicles can be loaded and unloaded when freight is palletized. This should also influence rate negotiation since it improves the utilization of equipment and drivers.

Containerization

Containers have been used in transportation in various forms from time immemorial. Amphora, hogsheads, tuns, barrels, crates are all containers. But it is only in recent times that container systems, using large metal van containers and "dedicated" handling equipment, have made their appearance.

Van containers are today moving in the air, on railroads, on ocean-going vessels and barges and on trucks. When used for surface transportation, they are generally standardized in twenty foot and forty foot lengths. Containerization reduces handling time and costs when freight must be interchanged and the carriers' equipment is not interlined. It also reduces pilferage and in-transit damage.

Intermodal exchange of containers between railroads and trucks (trailer-on-flatcar-piggyback-and container-on-flatcar), between ships and trucks, railroads and ships (the "mini-bridge") and even between air and trucks have become part of the common carriers' service capabilities.

Containerization saves time for the carrier and thus reduces his cost. This is—or should be—reflected in the rates paid by the shipper. The benefits of containerization for both shipper and carrier can be summarized as follows:
- Reduces packaging cost.
- Reduces labor cost because mechanization means less handling.

INDUSTRIAL TRAFFIC MANAGEMENT

• Reduces pilferage—a factor which is of increasing importance in both domestic and foreign trade.
• Reduces damages which means fewer claims and lower insurance costs.
• Improves carriers' equipment utilization; trailers on flat cars are in pay-load use by railroads five times more than are box cars.
• Increases efficiency which in many cases will reduce the overall cost.

There are also reasons why containerization has had its difficulties:

• Original investment is large because of the high cost of new equipment and facilities.
• Maintenance is expensive.
• Lack of standardization, including clamps, lifting and other devices, has been a problem.
• Lack of coordination between the transportation companies involved sometimes prevents setting up satisfactory interchange arrangements.
• Unions have often objected to proposed container services because they see the large cuts in labor that they bring as a threat to jobs.

Container crane loading a "pure" container ship.

INDUSTRIAL PACKAGING & MATERIAL HANDLING

Containers loaded on roller decks in an all-cargo plane.

8

CLAIMS

For a traffic manager to establish an expertise in freight claims, he must first understand those bill of lading terms that set out the carrier's liability. This liability is based on the common law that we inherited from the English when they came to settle in the New World.

Common law is the accumulation of hundreds of years of precedents and court decisions rather than law resulting from the legislative process, in which a particular measure is debated and passed. Much of the common law affecting carrier liability derives from an appeals court which, in England in the 1700's, reviewed several issues before the court and laid down a number of rules that established the varying standards of care that must be exercised under different types of liability.

It established those degrees of care necessary for:

... the bailee (warehouseman) who keeps goods for the use of the bailor (owner of the goods), to whom he is liable only for gross neglect.

... the common carrier for hire who is chargeable with the highest degree of care; "the carrier is bound to answer for the goods at all events," and "the law charges this person to carry goods against all events but acts of God and enemies of the King...."

The general principles laid down by the court are the basis of our law today, with some changes and an addition or two limiting the carrier's liability so that he is not responsible for loss or damage resulting from:

183

- Acts of God
- Acts of the public enemy
- Acts or default of the shipper
- Acts of the public authority
- "The inherent nature" or vice of the goods

Bill of Lading Terms

The terms of the bill of lading, while no longer printed on the back of the document, are to be found in the Uniform Freight Classification and the National Motor Freight Classification. The following are those parts of the bill of lading contract affecting liability and claims:

Section 1—This section of the bill of lading covers a carrier's liability and says that a common carrier shall be absolutely liable without proof of negligence for all loss, damage or injury to goods transported by him, unless the carrier can affirmatively show that the damage was occasioned by one of the exceptions contained in the terms of the bill of lading which include: Acts of God; acts of the public enemy; acts of negligence on the part of the shipper; inherent nature of the goods; and acts of the public authority. The terms also describe a situation in which a carrier is liable only as a warehouseman and other situations in which he is exempt from liability.

Section 3—provides that except when such service is required as the result of the carrier's negligence, all property shall be subject to necessary cooperage and bailing at the owner's cost.

Section 5—provides that no carrier will carry or be liable in any way for any documents or articles of extraordinary value not specifically rated in the classification or tariffs unless a special agreement is made. But if the goods are specially rated in the classification they must be accepted regardless of value.

Section 6—provides that every party shipping explosives or dangerous goods without previous full written disclosure to the carrier shall be liable for any loss or damage caused by such goods, whether it be to other goods or to the carrier's property. Such goods may be warehoused at the owner's risk and expense or destroyed without compensation.

Section 8—provides that "If the bill of lading is issued on the order of the shipper, or his agent, in exchange or in substitution for another bill of lading, the shipper's signature to the prior bill of lading as to the statement of value or otherwise, or election of common law or bill of lading liability, in or in connection with such prior bill of lading shall be considered a part of this bill of lading as fully as if the same were written or made in connection with this bill of lading."

Section 9—This section stipulates the liability of carriers by water when they are parties to a through movement and route. The basis of water carrier liability is that it arises only out of the failure or negligence of the owner in loading cargo, in making the vessel seaworthy, or in outfitting and manning the vessel. The water carrier is not responsible for loss or damage resulting from fire, bursting of boilers, etc.

One section of the bill of lading that should be noted by those who ship under a released valuation rate (see description in this chapter) is section 2(a). That rule allows the return of freight charges for that portion of a shipment that is in claim if the goods were rated at a released value rate.

Terms of Shipment

Who takes what action in filing freight claims depends on the shipping terms under which the goods were moving. If the shipment is a freight prepaid movement, the consignor must act if any loss or damage occurs. In other words, the shipper assumes all obligations for filing the claim and collecting it. The consignee's responsibilities lie in supplying supporting evidence and documents on the claim.

On the other hand, if the shipment is a freight collect movement, the consignee files the claim and the consignor's obligation is to support the consignee with any evidence or documents in the shipper's possession.

In either case, if originals of documents are retained, photostats may be used if they are certified as true copies of the original and the certification is signed by the supporting consignor/consignee.

Carrier vs. Warehouseman's Liability

All carriers are allowed a "free-time" period within which to deliver freight. This free time is specified in the tariffs and if it expires without the carrier being able to deliver the freight, having made a proper attempt to do so, his liability changes from that of a carrier to that of a warehouseman. Generally speaking, the liability of a warehouseman is much more limited than that of a carrier. After the liability change takes place, a claimant must prove the carrier (now liable as a warehouseman) was negligent if the freight is lost or damaged. The carrier, under these circumstances, need only show he exercised "ordinary care," whereas a common carrier's full liability in most circumstances is absolute.

The transfer of liability, in the case of railroad carriers, takes place after the car has been actually placed or a tender of delivery made to those who are to be the "beneficial receivers" of the freight. A motor carrier's liability changes when the carrier is unable to deliver at the point designated, or when the consignee refuses the freight, or when a delay is created by a consignee.

Released Valuation Liability

Under certain circumstances, a carrier may limit his liability. This may *only* be done with the written approval of the Interstate Commerce Commission in a Released Rates Order. There are a small number of such orders and rates.

The net effect of this move is to reduce freight rates for shippers in exchange for reducing the liability of the carrier. An example of this is the limit of $1.35 per pound liability for loss or damage to pharmaceuticals. Thus if there is a claim, the claim can be no greater than $1.35 per pound for the portion of the shipment under claim, plus the freight charges for that portion.

The bill of lading must carry a notation advising the carrier to apply the released valuation rate. If that notation is not on the bill of lading, the full carrier liability rate, which is the higher rate, will be applied.

Claims

While the Interstate Commerce Commission has promulgated rules for claims procedures, the Commission has no authority to adjudicate claims that cannot be settled by the parties to the claim. The only recourse open to a claimant is the arbitration procedures of the Transportation Arbitration Board (described later in this chapter) or the courts of jurisdiction.

Freight claims generally divide into two categories, those that involve overcharges on freight bills and those that involve loss and damage to the shipment, either apparent on delivery or concealed.

Most claims are preventable by both shippers and carriers. Shippers can help prevent claims by proper packing—packing that is better than the minimum requirements of the classifications—by legible marking of consignee's name and address, by preparing bills of lading fully and clearly and identifying all instructions or information needed by the carrier to make proper delivery. The carrier can help prevent claims by fully counting the load and by questioning the shipper on any matter that is not clear either in the marking or on the bill of lading.

Overcharge Claims

These result from overpaying freight bills for any one or a combination of the following reasons:

—Errors by the carrier in calculating charges or in copying information from the bill of lading.

—Weights are not recorded properly.

INDUSTRIAL TRAFFIC MANAGEMENT

—Improper bill of lading descriptions corrected by the carrier.
—Differences in tariff interpretation.
—Duplicate payment of freight bills.
—Applying wrong rates.

An overcharge claim should be filed on a standard form for presentation of overcharge claims (shown on page 189). All parts of it must be completed, including the tariff authority used to arrive at the rate. The claim should be filed by the party paying the freight charges since that person has been "injured" by the carrier. In many cases, it is an auditing firm that files the claim, on behalf of the consignor or consignee.

In movements involving more than a single carrier, the claim should be filed against the carrier that presented the overcharge freight bill. The claim should be supported not only by identifying the error made and the specific tariff, supplement and item number, but also attaching the bill of lading and freight bill involved. (If your system requires that you keep the original documents in your files, photostats can be used if the claimant will certify that these are "true copies of the original" and sign that statement.) The claimant should also include an indemnity agreement attached to the claim (form shown on page 190). Any other documents applicable to the claim should also be attached, such as weight certificates or invoices.

It is common practice to seek interest on an overpayment, from the date of payment until the date of reimbursement at the rate of interest in effect at that time. While a statutory basis for this does not exist, the Interstate Commerce Commission has held that this is not rebating. Payments within thirty days are not subject to interest.

All claims should be filed promptly, as soon as the error is discovered. The law allows for the filing of an overcharge claim within three years from delivery of the shipment. Shippers must observe state laws on intra-state shipments which may vary from state to state and from the federal law. An action at law—taking the carrier to court—must begin within three years from date of delivery or six months from the carrier's declination of the claim, which ever is later.

CLAIMS

Standard Form for Presentation of Overcharge Claims

Approved by the Interstate Commerce Commission; Freight Claim Division, American Railway Association; National Industrial Traffic League, and the National Association of Railway Commissioners.

(Claimant's Number) c/o 317

Mr. R.E. Miller, Auditor 177 X St., Dayton, Ohio
(Name of person to whom claim is presented) (Address of claimant)

Cross Country Carriers April 7, 1972
(Name of carrier) (Date) (Carrier's Number)

PO Box 91, Centerville, Ill.
(Address)

This claim for $ 38.44 is made against the carrier named above by Ohio Department Store
(Amount of claim) (Name of claimant)
for Overcharge in connection with the following described shipments:

Description of shipment Glassware

Name and address of consignor (shipper) Eastern Supply Company, Philadelphia, Pa.

Shipped from Philadelphia, Pa. , To Dayton, Ohio
(City, town or station) (City, town or station)

Final Destination Dayton, Ohio , Routed via Cross Country Carriers
(City, town or station)

Bill of Lading issued by Cross Country Carriers Co.; Date of Bill of Lading March 4, 1972

Paid Freight Bill (Pro) Number 38677 ; Original Car Number and Initial

Name and address of consignee (Whom shipped to) Ohio Department Store, 177 X St., Dayton, Ohio

If shipment reconsigned en route, state particulars:

Nature of Overcharge Classification and rate
(Weight, rate or classification, etc.)

DETAILED STATEMENT OF CLAIM.

NOTE.—If claim covers more than one item taking different rates and classification, attach separate statement showing how overcharge is determined and insert totals in space below.

	NO. OF PKGS.	ARTICLES	WEIGHT	RATE	CHARGES	AMOUNT OF OVERCHARGE
Charges Paid:	25	Boxes Glassware	825	8.33	$68.72	
		Total	825		$68.72	
Should have been:	25	Boxes Glassware, NOI, actual value exceeding 35¢ per lb., but not exceeding $1.50 per lb.	825	3.67	$30.28	
		Total	825		$30.28	

Authority for rate or classification claimed NMFC A-7, Item 88150, Sub. 2; ECMCA Tariff 31-D
(Give, so far as practicable, tariff reference [I.C.C. number, effective date and page or item].)

IN ADDITION TO THE INFORMATION GIVEN ABOVE, THE FOLLOWING DOCUMENTS ARE SUBMITTED IN SUPPORT OF THIS CLAIM.*

(X) 1. Original paid freight ("expense") bill.
(X) 2. Original invoice, or certified copy, when claim is based on weight or valuation, or when shipment has been improperly described.
( ) 3. Original Bill of Lading, if not previously surrendered to carrier, when shipment was prepaid, or when claim is based on misrouting or valuation.
( ) 4. Weight certificate or certified statement when claim is based on weight.
  5. Other particulars obtainable in proof of Overcharge claimed: †

REMARKS: Invoice calls for 100 dozen Glass Candlesticks @ $4.55, total $455.00, or approximately 55¢ per pound.

The foregoing statement of facts is hereby certified to as correct:

Ohio Department Store A P Williams
(Signature of claimant)
Traffic Manager

† Claimant should assign to each claim a number, inserting same in the space provided at the upper right hand corner of this form. Reference should be made thereto in all correspondence pertaining to this claim.
* Claimant will please place check (x) before such of the documents mentioned as have been attached, and explain under "Remarks" the absence of any of the documents called for in connection with this claim. When for any reason it is impossible for claimant to produce original bill of lading if required, or paid freight bill, claimant should indemnify carrier or carriers against duplicate claim supported by original documents.
† Claims for overcharge on shipments of lumber should also be supported by a statement of the number of feet, dimensions, kind of lumber and length of time on sticks before being shipped.
Claims based on rates quoted in letters from traffic officials should be supported by the original or copies of such letters.

Form No. 1 The Traffic Service Corp., Washington, D.C.

Standard form for presentation of overcharge claims, illustrating manner of preparation.

INDUSTRIAL TRAFFIC MANAGEMENT

Indemnity Agreement

Loss or Damage Claims

A claim for loss or damage carries a requirement that the claimant take steps to minimize the loss. The carrier is responsible to make the claimant "whole," as though no loss or damage had occurred to the shipment. The owner of the freight does not have to prove negligence on the part of the carrier. However, a claimant is required to bring the dollar value of his claim down to the lowest possible figure, either by repair or replacement.

If damaged material cannot be used and must be returned to origin, or sent to some other location for repair or salvage, the cost of the return transportation should be compared to the amount that can reasonably be expected to be recovered. If damaged freight is returned, it should be routed by the same carriers that transported the original shipment. If the cost of transporting it back to origin, or to some other location, exceeds the total value of the material, the carrier should be so advised and should be requested to dispose of it, with written confirmation. If the damaged material is worthless, the carrier should be requested to dispose of it, with written confirmation as to how this was done. In this connection, a motor carrier freight claim rule provides that:

> "Freight damaged in transit may be returned without charge to initial or intermediate points for repairs or credit, provided return movement is made by the same route as when forwarded, reference to original billing to be shown on return billing."

Some shippers of damaged "name brand" materials prefer that these not be sold as salvage and special handling may be needed. Damaged items that may affect the health of the public must usually be given special handling. Shipments of drugs and medicines, for example, would be subject to the rules of the Food and Drug Administration. Thus, if labels are spoiled because one or two bottles in a case have been broken, it is possible that the entire case may have to be destroyed.

In determining the extent or measure of damages, the Cummins Amendment to the Interstate Commerce Act makes the carrier liable for "the full actual loss." The measure of this is the value at time of loss. If the shipment is going straight into inventory as happens when an intracompany shipment moves from plant to a distribution center, the value of the damaged goods is the cost of production plus administrative costs at the point of destination. This, again, comes back to the requirement that the carrier make the owner of the goods "whole"—that is in the position of one who has received an undamaged or full-count delivery.

If the shipment is subject to a released value rate, then the measure of damages is no greater than the value declared (in writing on the bill of lading), plus the freight charges for the lost or damaged portion (section 2a of the bill of lading), if the freight charges have been paid; if freight charges have not been paid, then they need not be paid for that portion of the shipment under claim.

If the articles can be repaired or reconditioned the carrier is also responsible for all costs, plus all incidental expenses involved while the damaged cargo is being repaired. This might include the loss of the use of equipment during repair time. In all cases, invoice prices of the item in question will be the governing factor in setting the costs of repair or replacement.

All claims for loss or damage against common carriers must be filed within nine months from date of delivery or within nine months after reasonable time has been allowed for delivery. If the carrier does not pay or agree to settle, this opens the door for a suit by the claimant against the carrier. The suit must be started within two years after the carrier has declined the claim. The alert traffic manager must watch his calendar carefully so that this statute of limitations will not be invoked and the courts refuse to hear his claims case.

A claim can be filed against the originating or delivering carrier; the owner of the goods does not have to prove that the loss or damage occurred at a particular point or time. It has been held by the courts (Keystone Motor Freight Lines v. Brannon-Signalgo Cigar Co. 115 F.2d 736) that a cartage carrier performing only pick-up or delivery service is *not* performing a service comparable to a switching railroad and therefore can be considered a carrier against whom a shipper can file a claim. (A switching railroad is looked upon as an agent for the line haul railroad and a shipper can only file a claim against that railroad if he has unquestionable proof that the loss or damage took place on that line—Louisiana So. Ry. Co. v. Anderson, Clayton & Co., 191 F 2d 784.)

Preparing Claims

The cost of preparing a claim is itself a factor in deciding whether to file it. Some companies establish minimum values for the filing of claims; below a certain figure, they do not file. These limits should be reviewed annually to determine whether the values should be lowered or raised. Some companies allow all overcharges to be filed and rely on later auditing of freight bills to catch the overcharges. However, most companies maintain what they consider acceptable limits on loss and damage. In any event, a record should be kept to evaluate the carrier's service levels.

If a company has its freight bills audited by one of the specialized freight bill auditing firms, then the claims, regardless of dollar value, will be filed by the audit firm for the claimant. The usual arrangement with auditing firms is that they retain 40 to 50 percent of the overcharges they recover from the carrier, which is the cost to the claimant for the service.

It is illegal to deduct money for overcharges or loss and damage claims from freight charges on shipments subject to ICC jurisdiction and most state laws also forbid this practice. The Supreme Court of the United States has rendered several decisions on this point. It has held that carriers may not accept service, advertising, property, or releases of claims against them in payment of transportation. They are required to collect the established rates and charges on each and every shipment, from all alike, in cash or its equivalent, regardless of any litigation, claim, debt or claims on any other shipment or shipments. (Possible exceptions exist to this ruling, but only in a few states and in special situations where litigation on the subject has not been clarified.)

One "claim" that may occur is that of an undercharge. The carrier, in these cases, has often made an error in the rate or arithmetic. The carrier then submits a "due bill" for the difference showing his authority for the correct rate. The shipper must pay the difference if the "due bill" is correct even if he is faced with difficulties in subsequently invoicing the buyer of his products.

A claim should be filed promptly since any delay in filing may prolong the period of investigation by the carrier. It can be filed on the standard form shown on page 194. This format is not mandatory, but the form used should contain all the essential information. If, as in the case of overcharge claims, duplicate documents are submitted, an indemnity agreement must be included. The for-

INDUSTRIAL TRAFFIC MANAGEMENT

Standard Form for Presentation of Loss and Damage Claims

Approved by the Interstate Commerce Commission; Freight Claim Division, American Railway Association; National Industrial Traffic League, and the National Association of Railway Commissioners.

Mr. J. J. Jones, Claims Agent
(Name of person to whom claim is presented)

Smithville, New Jersey
(Address of claimant)

L-259
(Claimant's Number)

Cross Country Carriers
(Name of carrier)

July 20, 1972
(Date)

PO Box 91, Centerville, Ill.
(Address)

(Carrier's Number)

This claim for $ 29.07 (Amount of claim) is made against the carrier named above by Smith Packing Company (Name of claimant)

for Loss and Damage (Loss or damage) in connection with the following described shipments:

Description of shipment 100 Boxes Tomato Catsup in Glass Bottles

Name and address of consignor (shipper) Smith Packing Co., Smithville, N.J.

Shipped from Smithville, N.J. (City, town or station) ; To Westville, Ohio (City, town or station)

Final Destination Westville, Ohio (City, town or station) ; Routed via S.J. Trucking c/o Cross Country

Bill of Lading issued by South Jersey Trucking Co.; Date of Bill of Lading May 11, 1972

Paid Freight Bill (Pro) Number 87210 ; Original Car Number and Initial

Name and address of consignee (whom shipped to) Quality Restaurant Supply Co., Westville, Ohio

If shipment reconsigned en route, state particulars:

DETAILED STATEMENT SHOWING HOW AMOUNT CLAIMED IS DETERMINED.
(Number and description of articles, nature and extent of loss or damage, invoice price of articles, amount of claim, etc.)

6 Boxes Catsup - Short at destination	@ $3.23	$19.38
3 Boxes Catsup - Damaged	@ $3.23	$ 9.69

TOTAL AMOUNT CLAIMED

IN ADDITION TO THE INFORMATION GIVEN ABOVE, THE FOLLOWING DOCUMENTS ARE SUBMITTED IN SUPPORT OF THIS CLAIM.*

(X) 1. Original bill of lading, if not previously surrendered to carrier.
(X) 2. Original paid freight (expense) bill.
(X) 3. Original invoice or certified copy.
( ) 4. Other particulars obtainable in proof of loss or damage claimed.

Destination freight bill bearing driver's notation of shortage and damage.

REMARKS: Damaged boxes contained a total of 44 unbroken bottles with labels stained and unmarketable. These were returned to delivering carrier for salvage - receipt attached.

The foregoing statement of facts is hereby certified to as correct:

SMITH PACKING COMPANY

A L Brown Claims Supervisor
(Signature of claimant)

§ Claimant should assign to each claim a number, inserting same in the space provided at the upper right hand corner of this form. Reference should be made thereto in all correspondence pertaining to this claim.
* Claimants will please place check (x) before such of the documents mentioned as have been attached, and explain under "Remarks" the absence of any of the documents called for in connection with this claim. When for any reason it is impossible for claimant to produce original bill of lading, or paid freight bill, claimant should indemnify carrier or carriers against duplicate claim, supported by original documents.

Form No. 1 The Traffic Service Corp., Washington, D.C.

Standard form for presentation of loss and damage claims, illustrating manner of preparation.

CLAIMS

WILLIAM H. RORER, INC. 500 VIRGINIA DRIVE FORT WASHINGTON, PA. 19034 **CLAIM INVOICE** NOTE: ALL PAYMENTS TO BE MAILED TO THE ATTENTION OF MRS. MARY DEVLIN AT THE ABOVE ADDRESS.	DATE March 10, 1980 CLAIM NO 80-106 CARRIER NO. FREIGHT BILL NO. 16-47305-80 FREIGHT BILL DATE Feb 11, 1980

Country Trucking Co
15 Wheatsheaf Lane
Phila, Pa. 19100

SHIPPED BY WILLIAM H. RORER, INC.
☒ 500 Viriginia Dr., Ft. Washington, PA 19034
☐ 2500 165 St., Hammond, IN 46320
☐ 1550 Factor Ave., San Leandro, CA 94577
☐ 4660 Hammermill Rd., Tucker, GA 30084
☐

CONSIGNEE: Smith's Wholesale Drug Co., Columbus, Ohio

SHIPMENT OF DRUGS, MEDICINE, N.O.I.

QUANTITY	DESCRIPTION	WEIGHT	AMOUNT
Eight (8)	cases drugs and medicines Freight Charges	160 lbs	$80.00 6.22

REASON FOR CLAIM:
Bottles broken in each cases causing labels on unbroken bottles to be torn, dirty and illegible
No Salvage

160 LBS. $86.22
RELEASED VALUE   X YES ___ NO

DOCUMENTS SUBMITTED IN SUPPORT OF CLAIM
(X) Bill of Lading (copy)
(X) Paid Freight Bill (copy)
(X) Certified Copy of Invoice
(X) Inspection Report (copy)
( )

INFORMATION PERTINENT TO THIS CLAIM
(Note Statement Checked Below)

(X) We found it necessary, for hygienic reasons, after a thorough examination to completely destroy this merchandise for which claim is filed. There was no salvage.

( ) This shipment moved from stock to distribution point and the price/s used in this claim is/are the due date destination value in the quantity shipped. The amount of claim does not exceed claimant's loss and does not include any prospective profits, brokerage, overhead expense, percent above invoice or other similar items.

F-35 10/78

INDEMNITY AGREEMENT
The claimant agrees to protect the carrier and its connections against any loss resulting from non-surrender of the original bill of lading and/or original freight bill.

The foregoing statement of facts is hereby certified to as correct.
WILLIAM H. RORER, INC.
By _____

The foregoing statement of facts is hereby certified to as correct.
WILLIAM H. RORER, INC.
_____
(Signature of Claimant)

Claim document can be in any form, as shown above, but must contain all essential information.

mat of the claim form can be adjusted to meet the needs of the claimant as show in the "claim invoice" shown on page 195. Note that this form carries the indemnity agreement as well as several other items of particular concern to this shipper.

Concealed Damage

Concealed damages often do not become apparent until many days or weeks after delivery. Or sometimes a large shipment to a distribution center is reshipped later, in the original containers, without examination of contents at the reshipping point.

Where did the damage occur? The carrier may maintain that the freight was damaged when it was moved in the warehouse or in the volume move from the manufacturer to the distribution center.

It is most important that the small shipments, in particular, be examined on arrival for any possible damage. The carrier's liability is no different for concealed losses than for obvious damage, but with the passing of time it becomes difficult to determine cause and to accumulate sufficient evidence to prove the damage or loss took place while the freight was in the carrier's possession.

In an attempt to simplify their loss and damage problems, the carriers in 1969 restricted their liability for losses to fifty percent of the monetary loss. When challenged, the carriers insisted this was simply a guideline. But the ICC, when presented with a formal protest, opened an investigation and its decision, Ex Parte 263, was that these rules were unlawful. The ICC ordered them to be cancelled. Nonetheless, shippers would do well to file claims for concealed damage within five days of delivery.

A procedure for checking the condition of freight at time of delivery should be the basis of any claims program. If shortages or damage are obvious at delivery time, the facts should be clearly stated on the receipt given to the driver. The driver should make a similar notation on the copy of the receipt left with the receiver, sign his name clearly, and date and mark the time of the receipt. Make sure the notation and signature are legible! If there *is* damage to the freight, the receiver should telephone the carrier or his agent and ask for an inspection—and confirm this request in writing.

If there is a loss, the receiver should immediately telephone such information to the carrier. The lost portion of the shipment may turn up elsewhere in the carrier's system.

There are tariff provisions for "shipper's load and count" (SL&C) that have an important effect on a carrier's liability. The notation "SL&C," often written in by the driver on the bill of lading, is commonly taken as an indication that there is no "prima facie" case of carrier liability when loss or damage occurs. The carrier's case here is based on the fact that the shipper loaded the vehicle.

Regarding rail shipments, the courts have held that when the shipper loads, blocks and braces the car, and the consignee unloads it, this relieves the carrier of responsibility. But if improper loading is apparent to the carrier, and he accepts the car, the shipper's load is *not* a defense. The carrier should have drawn the shipper's attention to the improper loading. And if a carrier adds freight on top of the first consignor's, the carrier has broken the "SL&C" defense.

Another type of claim difficulty faces the traffic manager using TOFC (piggyback) rail plans 2½ and 3. Under these plans, the shipper uses leased or local cartage carriers between shipping point and rail ramp and then between destination rail ramp and consignee. Any damage or loss claims are likely to be questioned as to the point in transit where they occured (unless there is a police record of an accident). Since the railroad is not a party to the highway haul, the claim may have to be filed against the local drayage carrier.

Transportation Arbitration Board

Within recent years, a number of shippers have organized themselves into an association called the Shippers National Freight Claim Council Inc. (SNFCC) that publishes a newsletter and other informational material on claims and also conducts seminars on claims handling all over the country. In collaboration with motor carriers, SNFCC created the Transportation Arbitration Board

(TAB). Claimants may present their cases for arbitration to this board with the understanding that its decisions are final and may not be appealed to the courts.

The TAB is sponsored both by the SNFCC and the National Freight Claim Council, which is made up of carriers. The TAB will accept claims on rail, air, motor carriers and freight forwarders (but not water carriers) at a cost of $50 for each claim. The issues are submitted in written briefs only—oral hearings are not permitted. This procedure is less costly and speedier than a full-fledged court hearing. If the board cannot reach a unanimous decision, the claimant is free to take his case to the courts.

Role of the ICC

Even though the ICC cannot adjudicate claims, it does have the responsibility to lay down rules for the handling of claims. It was given this responsibility as a result of Ex Parte 263 and subsequently issued rules for the processing claims that are to be found in 49 CFR 1005. These rules became effective July 1, 1972.

In summary, the commission's ruling says that:

—All claims must be acknowledged by the carrier within 30 days of their receipt.

—The carrier must create a separate file for each claim and assign successive numbers to each claim. The acknowledgement to the claimant must identify that number.

—Claims must be investigated promptly.

—The carrier must either pay, decline, or make a firm offer in settlement within 120 days after receipt of a claim. If the claim is not disposed of at that time, the carrier must report the status of the claim to the claimant every 60 days thereafter.

—Damaged freight rejected by the receiver may, after due notice by the carrier to the claimant, be sold by the carrier as salvage in a manner that will fairly and equally protect the best interests of all parties involved. The carrier must make an itemized record of each item sold. However, the carrier cannot sell the salvage if the owner advises, in writing, that he will not allow it.

Filing freight claims often serves to highlight areas where insurance coverage may be needed. Bills of lading contain provisions to the effect that carriers shall not be liable for articles of extraordinary value unless they are specifically rated and a stipulated value is stated on the bill. Freight rates applicable on some commodities are substantially lower for released valuations. Parcel Post and United Parcel have various limitations on the top liabilities they assume. Insurance coverage is generally available for part, or all, of the various damage possibilities that are not covered by the carriers' liability. A policy commonly used by shippers is for contingent floater transportation insurance. Under this type of policy, the insurer makes up any shipper losses when freight is moving under released value rates and so on. What type of transit insurance a company needs should be worked out by the traffic manager with his company's insurance department.

If a claimant wants information on what cargo insurance cover a carrier has, he can write a letter to the Section of Insurance of the ICC in Washington, D.C. Such a request can be made by letter or by using a form developed by the Section of Insurance. If the form is used it should be forwarded in duplicate with a self-addressed and stamped envelope for a faster reply.

Impact and shock recorders of various types are increasingly used today, both in damage prevention programs and to support claims when shipments have been damaged. The SNFCC advocates their use and publicizes programs offered by manufacturers of the equipment. Impact recorders can be helpful in localizing when and where shipments are receiving rough handling. They can provide a complete handling "history" of the movements of goods from packaging at the factory to the premises of the customer.

Record Keeping

Measuring a carrier's service standards requires keeping records of claims in a form that gives an overall picture of how well he is performing. Carriers must be measured in several areas including time-in-transit, equipment standards and availability, rate negotiations, tracing, "attitude," and of course, claims.

A claim is more than a mere "difficulty" between shipper and

INDUSTRIAL TRAFFIC MANAGEMENT

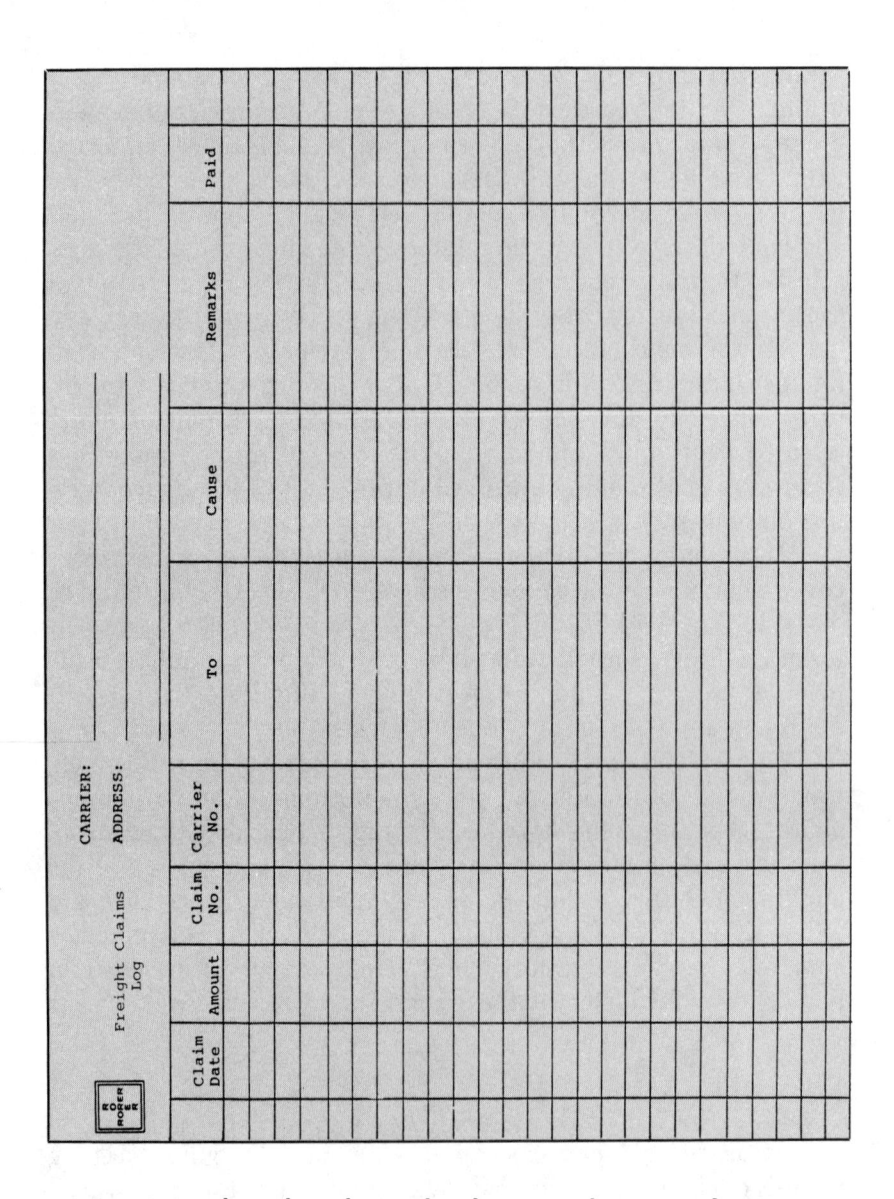

Freight Claims Log

Claim Date	Amount	Claim No.	Carrier No.	To	Cause	Remarks	Paid

CARRIER: _____
ADDRESS: _____

carrier; it involves the relationship between shipper and customer, the loss of a sale by the shipper himself or *his* ultimate customer who does not have a needed item for his retail shelves or the raw material to finish manufacturing a product that brings cash flow and profit.

The customer-service conscious traffic manager should maintain a record of claims against each carrier on a separate "ledger" sheet. If the firm has more than one shipping point, the records should be maintained for each shipping point for each carrier. This sheet should identify date of claim, destination city and state, claim numbers for shipper and carrier, loss or damage, dollar value of claim and payment date or reason for non-payment or variation in settlement. A sample format is shown on page 200.

Such a record, when shown or sent to the carrier, is the signal that corrective action may be in order somewhere if it shows a low or declining service standard. Or it can be used to determine whether a carrier with a slower in-transit record but few claims, might be a better service "trade-off" than a faster carrier with many claims recorded against him. This claims "log" should be in addition to the record of claims and payments developed in a data processing operation. All claims should be part of accounts receivable in the accounting department.

9

DISTRIBUTION & WAREHOUSING

Traditionally, the term "distribution" has been used to describe channels of distribution—i.e., whether the product was sold direct to customers, through dealers, or through jobbers or distributors. This chapter, however, will deal with the actual process of distribution: the physical movement of goods through physical systems consisting of transportation facilities, loading and unloading and storage systems, private and public warehouses. It will also deal with the techniques of planning and control applied in order to maximize the efficiency of distribution systems while assuring that they are operated at the lowest total cost consistent with management's standards of customer service.

Distribution

The National Council of Physical Distribution Management defines "physical distribution" as

". . . the term employed in manufacturing and commerce to describe the broad range of activities concerned with efficient movement of finished products from the end of the production line to the consumer, and in some cases includes the movement of raw materials from the source of supply to the beginning of the production line. These activities include freight transportation, warehousing, material handling, protective packaging, inventory control, plant and warehousing site selection, order processing, market forecasting and customer service."

Note that warehousing is twice mentioned in the definition—once as a management consideration and secondly as a site selection problem. In recent years, the motor carrier industry has advertised itself as the "warehouse on wheels" and, once convinced, consumers (manufacturers and retailers) have reduced their own storage capacities. The argument is that transportation should deliver as needed. This affects in-bound movements to the manufacturer, and in turn, his outbound movements to retailers. Retailers and wholesalers no longer carry heavy inventories. They depend on the manufacturer to carry enough goods close to the consumption point to satisfy all sales needs most of the time.

Thus the need for stock close at hand has promoted the use of warehouses, public and private. The private warehouse is just that—it's available only for the use of the manufacturer or wholesaler or retailer, for his own product line and operations. If a conglomerate operates warehouses for its own companies only, the warehouses are still considered private (even though some companies term them "semi-private") since the services and product lines are limited to those of the conglomerate. Private warehouses can be categorized by functions, into:

— *The security warehouse.* For storing art, documents, precious metals and similar valuable items as securely as possible.

— *The raw materials warehouse.* For storing in-bound raw materials heading for production or for storing materials for peak-production seasonal items or materials bought at a favorable price for use later.

— *The production warehouse* (or the finished goods warehouse). For storing goods coming off the production line awaiting sales or shipping instructions.

Private warehouses today are often combinations of two or even all three types as the needs, economy, size and management philosophy of the company dictate.

Types of public warehouses are more numerous. Generally, these warehouses will perform a whole range of jobs for their customers. The main types are:
- The general merchandise warehouse
- The refrigerated warehouse
- The special commodity warehouse

DISTRIBUTION & WAREHOUSING

- The bonded warehouse
- The field warehouse
- The household goods warehouse

— *The general merchandise warehouse:* These are used for finished products and raw materials by any number of customers. Generally, the products in this type of warehouse do not require any special services other than dry, clean, weather-protected storage. But the warehouseman may have to set off different areas of his warehouse for different products according to their condition, shape and packaging, if finished or raw material, or even because of government regulations. In some states, warehouses must be licensed or bonded.

— *The refrigerated or cold storage warehouse:* These are used primarily for perishable items such as fresh or frozen foods. Space in these warehouses is usually separated by the temperature variations that may be required by customers, from below freezing up to cool. Today's requirements are usually that 80 percent of the space used is for items that must be kept at below freezing while the balance is kept above 32°F.

Some general merchandise warehouses are combined with refrigerated warehouses.

— *The special commodity warehouse:* These are used for agricultural products such as grain, cotton, fertilizers, lumber, etc. Some store only one commodity while others store several, depending on the part of the country in which they are located.

— *The bonded warehouse:* These are warehouses bonded by the U.S. Treasury to secure payment of import duties and taxes due on the commodities stored or handled. The bonds may be written by the U.S. Customs and/or U.S. Internal Revenue. Bonding assures the federal government that the goods will not be released from the warehouse until the duties and taxes have been paid.

Bonded warehouses may also be of several types. Some bonded warehouses are government-owned or leased, and are part of the customs service. There are importers' private bonded warehouses and public bonded warehouses. A bonded warehouse may be a yard or shed for bulky and odd-shapped items or part of a building or yard separated by fence from the rest of the property. It can also be part of a manufacturing operation for export only where imported components or parts are used to finish the manufacturing process.

Foreign trade zones are usually adjacent to a port of entry. That port may be a waterport or an airport. The zone, policed and enclosed, receives imported goods which then undergo some processing and are exported. The foreign trade zone thus promotes foreign trade by allowing importers and exporters to avoid paying duty on each component of a finished item.

— *The field warehouse:* These are warehouses on the property of the owner of the goods, but under the custody of a public warehouse employee. A field warehouse is essentially a form of banking for extending credit. The goods stored become collateral for a loan. Using field warehousing avoids the transportation costs of moving the goods to another warehouse which might be in another community.

Field warehousing, because of its association with banking, is generally not of long duration. Once the owner of the goods has a cash flow, then field warehousing is not needed.

The household goods warehouse: These are used for storing furniture and personal items. Some have vaults for valuables or heirlooms, and some have cold storage for furs.

Role of the Public Warehouse

An early historical usage of warehouses can be found in the Bible when we read of Joseph storing grain and how the grain saved Egypt from starvation during a crop failure. Warehouses today, however, are not primarily used for storage against some future unknown emergency. The warehouse is a buffer between production and the buyer.

Since it is a buffer, it becomes another cost factor and adds to the final cost of the product. Therefore, warehousing should be a short-term expedient. The exception to this is when an inventory must be built up for seasonal items such as Christmas decorations, etc.

DISTRIBUTION & WAREHOUSING

Warehousing's first function, then, is storing goods. But there are many other functions that public warehouses perform today:

Packing: Warehouses will pack and repack goods. Usually this is done for customers who will store unpacked or semi-packed items. Warehouses will also repack items that may need that service because of mishandling in transportation.

Distribution: Warehouses located at strategic marketing points can receive carloads or truckloads of products and distribute them to customers as the shipper directs.

Consolidation: Warehouses serve as a point for receiving many smaller shipments which are then consolidated into carloads or truckloads for the longer hauls.

Sales centers: Through the use of electronic or telephonic communication lines, documents that will start an order filling and shipment "cycle" can be transmitted to warehouses from a company's central office. Customers can be billed from the warehouse at the same time with some of these systems.

In-transit services: Traffic managers can protect through-freight rates by negotiating storage-in-transit rates with carriers, and in-transit rates with warehouses. This will generally reduce total transportation costs.

Fabrication, another service offered by some public warehouses, is seldom used today, but, again, when it is used, manufacturing-in-transit holds down transportation rates. Warehouses can be set up for some manufacturing processes at a point between initial production site and final sale or storage site.

Clerical services: Warehouses may be used as district or regional offices for sales. Some have desk space for sales people or are used as a communication point for special services or sales.

Transportation needs: Depending on the rules of the state within which it's located, a public warehouse may use its own trucks to deliver the goods it stores; if such trucks are certificated by the ICC, they can be used within the authority granted.

These are seven services, besides storage, that public warehouses offer. Some other uses, if the public warehouse has the equipment, are fumigation to meet certain federal regulations, and refrigeration.

All these services have prices attached, of course, but many of the rates are negotiable. All rates, charges and services should be included in an agreement signed by the warehousing company and the customer.

The agreement that was used in public warehousing was approved by the U.S. Department of Commerce in 1926. Since then a form recommended by the American Warehousemen's Association has found greater acceptance and use and is shown in Appendix 1. The warehouse will also issue its non-negotiable warehouse receipt (see page 209) when goods are delivered to the warehouse. Some warehouses will use a "Service Specifications" form such as that illustrated on page 210—for special services required by the customer. The customer can use his own bill of lading for outbound shipments or may elect to use the warehouse's bill of lading (example on page 211).

The liability of a public warehouse for the goods in its keeping differs from that of common carriers. Historically, the warehouseman's liability has remained the same for hundreds of years. The Uniform Commercial Code states that the warehouseman is liable for loss or injury "caused by his failure to exercise such care in regard to goods in his charge as a reasonably careful man would exercise under like circumstances, but unless otherwise agreed he is not liable for damages which could not have been avoided by the exercise of such care."

Under the same code, a warehouseman may not limit his liability. By written agreement a dollar limitation may be set. The limit can be removed, again only in writing, and the owner of the goods may be subject to rate changes, up or down. One who uses public warehouse facilities would do well to become acquainted with Chapter 7, Section 204, Paragraph 2 of the Uniform Commercial Code (UFC).

However, if you rent space from a public warehouse and enter into a lease arrangement, then you have a landlord-tenant relationship. This is subject to an entirely different set of laws than those affecting the warehouseman.

Basically, the public warehouseman's liability for loss or damage to goods in his care must be caused by his failure to exercise

DISTRIBUTION & WAREHOUSING

the reasonable care a prudent man will exercise (UFC 7-204-1). The weight of proving the warehouseman is negligent and did not exercise prudent care falls on the owner of the goods.

Non-negotiable Warehouse Receipt.

INDUSTRIAL TRAFFIC MANAGEMENT

SERVICE SPECIFICATIONS

For customer use in contacting AWA member companies regarding warehousing services.

Full and exact name of your company _____ Date _____

Complete address - Number, Street, State, Zip Code _____ Area Code - Telephone Number

Name of person making inquiry _____ Title _____ Address, if other than above

Commodity, Pack, Type of Container	Overall Dimensions in Inches			Gross Shipping Weight in Lbs.	Value of Package and Contents
	Length	Width	Height		

Estimated Average Stock: _____ Estimated Through-put: _____
INBOUND: Rail ___% Truck ___% Loose ___% Palletized ___% Who Unloads Trucks? ____
  CL ___% LCL ___% TL ___% LTL ___% Pallet Size: L ___ W ___ H ___ (including pallet)
OUTBOUND: CL ___% LCL ___% TL ___% LTL ___% Who Loads Trucks? ____
  Loose ___% Palletized ___% Cartage Needed? ____
Estimated Number of Outbound Orders: _____ Per _____
Outbound Order: Minimum _____ Lbs. Average _____ Lbs.
No. of Different Items or Separations in Account: _____
B/L Used: Warehouse _____ Customer's _____

Check among the following accessorial services those to be required from the warehouseman in addition to the "Handling" and "Storage" services:

____ Handling for immediate distribution. ____ C.O.D. collection on behalf of storer.
____ Reporting marked weights, numbers, etc., on receipt and/or delivery. ____ Furnishing reports in addition to usual notification as to receipts and deliveries.
____ Making out-of-town shipments. ____ Kind and frequency (attach samples).
____ Prepaying freight charges. ____ Weighing on receipt and/or delivery.
____ Filling orders from a credit list ____ Storage-in-transit.
____ Invoicing customers for storer.

Indicate other services required, and give information regarding any special storage characteristics of commodities: (Stacking Limitations, Hazardous, Breaking Cartons, Controlled Temperature, Pick by Package or Serial Number, Odorous or Susceptible to Odors, etc.)

AWA Seal of Security

AMERICAN WAREHOUSEMEN'S ASSOCIATION
222 West Adams Street, Chicago, Illinois 60606

AWA Seal of Security

Service Specifications.

DISTRIBUTION & WAREHOUSING

The Private Warehouse

There is a role for the company-owned and operated warehouse, (nowadays sometimes called a distribution center). But what is it, and when does using a public or private warehouse make the better economic sense?

In the public warehouse, the customer pays for a service only when used and to the extent used. A private warehouse involves a capital expenditure for land and buildings, and has costs for administration, insurance, taxes, utilities, labor and overhead.

The private warehouse's capacity is limited by the size of its

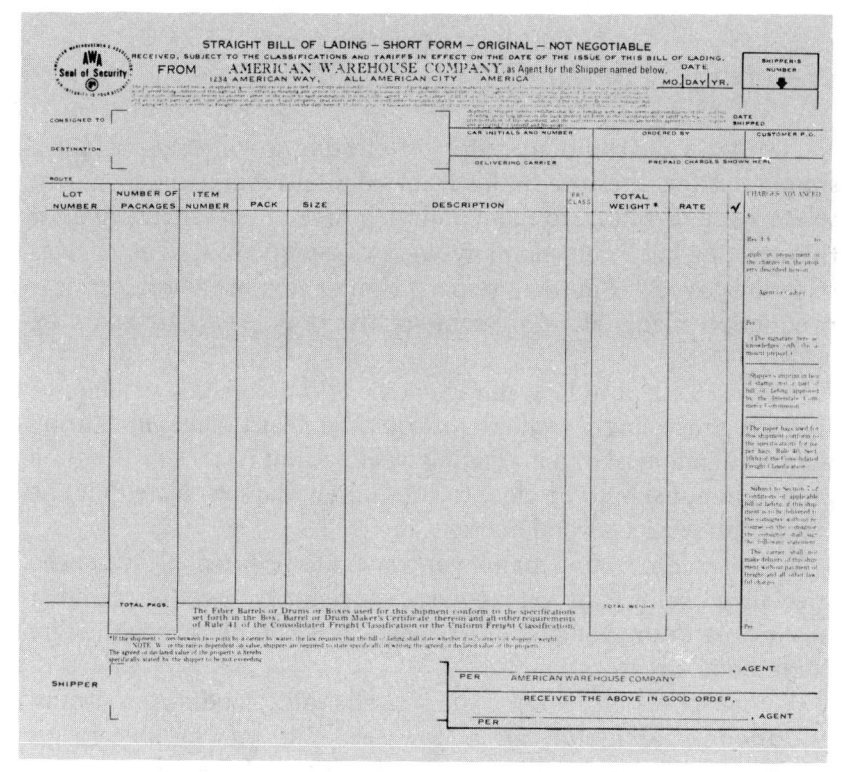

Straight Bill of Lading—Short Form.

211

building. It is "permanent," at least for the duration of the lease or until the building, if owned, is sold.

However, a private warehouse has its advantages, too. Perhaps it can be built where land is cheap, or labor costs low. If the product is valuable, security may be better in a private warehouse.

The size of a company need not be a deciding factor in public vs. private warehousing. Once the need for the warehouse is confirmed, the question of "going public or private" must be resolved regardless of size of the company. A trade-off may be required among a number of factors including the use of private or common carriers to the warehouse, size of inventory, system of inventory control and warehouse operation, labor costs, marketing area to be served and the availability of out-bound transportation services.

Warehouse Location

Siting a distribution center or selecting a warehouse is not a simple matter. Pressure builds from sales and marketing people to select a site favorable to them and their area of operations. Against this must be balanced company-wide values and overall costs. This selection process is in some respects similar to that used for siting a production plant. Here are some of the questions that should be asked:

Purpose. Is it to be a storage point or a processing warehouse? If it is a processing warehouse, it could be near the manufacturing plant raw material or a marketing location, but if it is a distribution point, then it should be close to consuming markets. Note that one center can service several markets.

Labor force. How many workers are to be hired? Do you need specialized help? Can you use part-time help? Is there enough labor in the area or are you going to become very competitive with other commercial organizations?

Transportation. Do you need all modes of transportation—rail, motor, air and water? Will you be able to get specialized services to and from the site, i.e., refrigeration, flatbed equipment, piggy-back, etc.? Is the site on a rail branch line in need of repair? Is the branch line seldom used so that once weekly service is the best you can get? Is the out-bound service efficient?

Community. Are there satisfactory police, fire and sanitation services available? Medical facilities? How is the educational system? Is the water supply satisfactory and constant?

Utilities. Are you going to drain the local power facilities or will the utility companies give you a sufficient supply and not deny others their needs? Are the utility companies upgrading their facilities? How will they deal with a fuel crisis?

Taxes. What are the community tax levies? What are the taxes? What are the county tax levies? What are the state tax levies? What is the impact of all those taxes on your costs and operations? What services are given in exchange for the taxes? Are tax incentives offered to you to move into the community?

The site. Are you allowing for a growth factor over the next ten to twenty-five years? Is the contour of the land correct for the type of building you will construct? Are the roads and road network satisfactory for employee use, for the transportation of freight and for emergency vehicles?

Answers to these questions will enable you to conclude whether you want to locate your distribution center at that particular site. Most of the questions to be answered require a committee approach. They require the evaluation of your company's engineer, tax accountant and personnel manager. The traffic manager should chair that committee.

The Warehouse

The demands made on the distributors of goods become greater as time goes on, requiring a sharper customer service response. The warehouse or distribution center must be ready to meet those demands and the company must impose the system that will satisfy the customer. In this effort, the electronic "mechanization" of office routine and procedures is as much a necessity as the mechanization of the warehouse. Fast order processing, on-line inventory use and replacement computers—these add up to "having the right product at the right place at the right time in the right condition . . . ."

For example, one consumer goods company has a rule that every order must be shipped within twenty-four hours after the traffic department receives it. A goal was established and with each shipping center equipped with its own small computer, the department achieved a 95 percent record of meeting its goal. The proper placement of goods, use of pallets, routing via computer and regular use of same carriers helped achieve that record.

Automation of warehouses is a relatively recent innovation. Some companies have found it feasible to invest in highly-automated warehouses, with order-picking and assembling, inventory control, and shipping document preparation all controlled by computers. In a typical operation, orders are fed into the computer which then groups all orders by destinations, prepares appropriate shipping documents and inventory records, and activates the material handling system so that orders are delivered at the appropriate shipping docks in proper sequence for loading and unloading at destination. In one company, the different orders scheduled for one trailer are all color-coded and delivered to the shipping dock in such a sequence that, as the driver makes each stop, the appropriate order will be in position for unloading and unmistakably color-coded for that destination.

Whether operated by computer, by tape, by card or by push button, the object of such warehouses is to exploit the full potential of mechanical systems through the use of automatic controls. Fast moving items can almost always be automated economically; slow movers, split cartons, etc. are not usually automated because of excessive cost. Some "electronic" warehouses are not economical for large, one-item pallet loads. Where automated, these warehouses use another type of automation with conveyor belts and special lift trucks. Automatic systems are simple or very complex, depending on the problem to be solved. But efficient use of space remains the number one warehousing problem. The heart of any modern warehouse system is live storage!

Remember, planning an automated warehouse should never

be attempted without the guidance of a good engineer and possibly a consulting firm that specializes in the field.

A great deal of the advantage of automated warehousing is lost if transportation scheduling is not coordinated. If shipments stay on the dock without trailers to load them into, or in cars on sidings, the time "bought" by the automated warehouse at a rather high price is quickly lost. For this reason, it is always a good idea to proceed slowly before recommending automated warehousing or sophisticated, high-speed handling equipment. It may work well, but if other parts of the distribution system can not be geared to take advantage of its benefits, the greater part of the investment may be wasted.

Distribution must include traffic management, packaging, material handling, order processing and warehousing. Add to these inventory control and sales forecasting and physical distribution is the result. The advent of the computer has made it possible to measure more accurately than ever before the "tradeoff" of costs between transportation and the other activities we have mentioned. Today's traffic manager should be equipped to use the systems approach, not only in transportation but also in all the other activities that relate to it. He should be prepared to recommend spending more on transportation in cases where he knows it will reduce total cost and improve customer service. He should not hestitate to recommend unit loads and improved packaging where they will produce the same results. He will necessarily be occupied a great deal of the time with rate matters, claims and the like, but he should never let these activities become ends in themselves or obscure the fact that his primary transportation responsibility is part of a total system where it may on occasion be outweighed in importance by other elements such as the cost of inventory.

Whether physical distribution is recognized as such officially by the company, the traffic manager should exercise it to the extent that his influence is felt and traffic management is the major function in the concept.

A good example of what the computer-based systems approach can do in national distribution is a multi-product corporation with 40 plants throughout the United States. Each plant ships all of the items it makes to each of 14 distribution warehouses located close to the corporation's major markets. All customers in each distribution region are served only from the warehouse located in that region. Customers are supplied by regular shipments, daily if desired, of every item sold by the corporation, in mixed shipments from the regional distribution warehouse.

Such a complex is controlled by computer through only one service center. Sales offices throughout the country put their area orders on tape which is sent to the service center by teletype. The service center puts the orders on punched cards which are fed into the computers. Not only do the computers prepare the data for issuing bills-of-lading and loading and routing orders; but they also check the accounts receivable register and determine the customer's credit status and credit limits. The computers also prepare data for invoices and issue daily stock-status reports for inventory control.

Salesmen in the field send their orders to the service center by air mail or teletype. After credit and other checks, orders with full instructions for handling go by teletype to one of the 14 distributing warehouses. There the salesman's order emerges as a loading order, memorandum bill of lading, etc., all set up for the next day's shipping. Next, confirmation of shipment goes back to the service center where stock status inventory control reports are compiled daily for each warehouse. Each regional manager also receives a daily "delivery analysis" report for his area.

Advance scheduling of order handling and carrier equipment means the goods don't have to be "floored"—one of the costliest routines in current warehouse and shipping practice. While the orders are being processed, the computer records the reduction in inventory, makes up the shipping papers and notifies the customer or consignee that the shipment is moving forward.

As regards inventories, control charts and other systems cannot guarantee perfect inbound/outbound balance at all times. There are sure to be seasonal peaks when it may be necessary to use another warehouse temporarily. Care should be exercised to see that old stock is shipped first and that inventory systems include this feature.

The traffic manager who trains himself to think in terms of the total system and keeps up on developments in disciplines that relate to his own should not have too much to worry about from the physical distribution concept. In many ways, he is the logical candidate for the physical distribution "title" by virtue of his training and the scope of the job he is already doing. If he can demonstrate managerial ability and knows how other departments inter-relate to his own within the framework of company objectives, he is well on the way.

While a great deal of physical distribution management appears to have to do with mathematical techniques and computer applications, the traffic manager should bear in mind that the purpose of these is essentially to aid decision-making, and that computers help managers make better use of their skills. They eliminate routine, repetitive decision-making and paperwork so managers can devote more time to major decisions and questions of policy, long-range planning and the like. Naturally, it is in the traffic manager's best interests to know something of what computers can and can't do—and to understand how computer programs are written. It would be impossible, in this book, to go into all the ramifications of physical distribution management. Many of them are so specialized as to apply only to a relatively small group of companies. But the individual traffic manager has a wide range of literature to choose from, and he is limited only by his own ambition. In this respect, he has a distinct advantage over the traffic manager of even a few years ago, who often found that there was no truly helpful literature on the subject.

10

CONTRACT/PRIVATE CARRIAGE

Private and contract carriage have certain features in common when considered, either separately or in combination, as alternatives to for-hire carriage. Both offer a company the advantages of transportation services dedicated to its exclusive use. The improvements in customer service that this almost always brings is a decisive argument in their favor for many firms, especially if there is a good prospect of obtaining such improvements at equivalent or lower costs when compared to what the common carriers are offering.

But in making such comparisons, a company must be thorough and ruthlessly objective if it is to know beforehand what the true costs of having its very own transportation service will be. As with most decisions, there are advantages and drawbacks in trying to decide between private and contract carriage (or some mixture of the two), on the one hand, and for-hire carriage on the other.

Private Carriage

The usual reasons given for "going into" private carriage are improved customer service and the expectation, or hope, of lowered costs. All too often, such factors as the need for return loads if a private carriage operation is to cover its costs, and the legal restrictions that still hedge in private carriers, are forgotten. The fuel shortages of recent times have forced some changes in the ICC's long established policies towards private carriage; these changes are described later in this chapter. But many of the restrictions on the uses to which a private fleet owner may legally put his trucks remain in effect and the subject at this writing is fogged with controversy and pending court decisions.

Despite the attractions that the vision of a vast fleet of trucks bearing the company logo proudly down the nation's highways will have for many corporate executives, a venture into private carriage should "start small." A modest fleet carefully managed in a program of cost comparisons and controls, will serve as a laboratory for testing service levels against what the common carriers can do.

Some companies place the responsibility for managing their private fleets in their transportation or traffic management departments; others have separate departments entirely. Because it requires experience in vehicle maintenance programs, equipment buying or leasing and even labor relations, managing a private truck fleet is more a job for a fleet operations manager than a traffic manager. But whatever the manager's title may be, two important caveats should be noted. The first is that proper staff and help must be provided to run a private trucking operation—it must not be relegated to someone already fully occupied with other duties. The second is that a new private fleet operations should be carefully coordinated with the company's existing for-hire motor carrier operations. A judicious mixture of for-hire and private carriage often gives the best of both worlds and is especially valuable when strikes, fuel shortages or other problems curtail one or the other service.

The information needed to compare the costs of going private with those of for-hire carriage will of course come from the traffic department. To begin with, this might take the form of a simple compilation of figures showing:

—Tonnages outbound to specific market areas or customers

—Tonnages inbound from vendors' plants and frequency of shipments

—What special equipment is needed in shipping to customers or from vendors.

These figures will show the frequency of movement to and from various points, which will give some indication of how well (or badly) private trucks would be utilized, the number and types

CONTRACT/PRIVATE CARRIAGE

of trucks needed, and what effects a private truck fleet would have on service. If this preliminary survey promises an improvement in service, then the next step is to focus on costs.

The costs of buying or leasing trucks, the costs of garaging and maintaining them if the decision is to buy, labor costs, mileage costs, tolls and drivers' expenses (motels, etc.), administration costs (clerical, management, insurance, etc.)—all these have to be determined and put into a form that makes them ultimately comparable to the rates paid to common carriers. The company treasurer must enter the picture here to supply tax costs and a figure for interest on money invested.

If the comparison is *still* in favor of a private fleet, then the question of whether the trucks are to be purchased or leased arises. A buy-or-lease decision needs careful analysis and should be made within the context of a company's own operations. Equipment leases, such as those used for trucks, generally are of two basic types—those in which the trucks alone are leased and those in which the trucks plus maintenance, repair and other services are provided. The first are usually known as "finance" or "net" leases and the second as "service" or "full service" leases. The extent of the services provided in the full service lease are of course spelled out in the lease document itself. A typical truck lease agreement is shown here, together with a "Schedule A" (every lease agreement should have one of these schedules attached; it gives a detailed description of the trucks being leased).

Truck leases are made on varying payment bases such as so much per mile, or per month or some other time period. Some require payments to be made on distance and time combinations. Some leases may include options to purchase the trucks; some have cancellation options, with or without penalties. In some areas, a typical truck lease is for five or six years but in others it may run as long as ten years. But short term leases by the month, week, day or trip are available (concerning "trip leases," and the ICC's rulings on them, see comments later in this chapter). Short term leases are often used by firms wishing to test private truck operations. Most leasing companies print their own forms and are very willing to provide literature promoting the tax and other advantages of leasing.

INDUSTRIAL TRAFFIC MANAGEMENT

RYDER

TRUCK LEASE AND SERVICE AGREEMENT

THIS AGREEMENT is made as of the _____ day of _____, 19___ (hereafter Ryder) and between RYDER TRUCK RENTAL, INC., _____ whose address is _____ (hereafter Customer).

1. **EQUIPMENT COVERED AND TERM:**
   A. Ryder agrees to lease to Customer and Customer agrees to lease from Ryder the vehicles on Schedules A attached hereto and from time to time hereafter executed and made part of this agreement (hereafter Vehicle(s)). Execution of Schedules A shall constitute Customer's authorization to Ryder to acquire the Vehicles. The Agreement shall become effective with respect to each Vehicle on the date the Vehicle is tendered by Ryder to Customer, or 48 hours after the date Ryder notifies Customer that the Vehicle is available for delivery, whichever occurs first, and shall continue for the term specified on Schedule A unless terminated earlier as provided hereinafter.
   B. Acceptance of Vehicles in service constitutes an acknowledgement that the Vehicles comply with Customer's specifications. Customer agrees to pay for any structural alterations (not to be made without Ryder's prior written consent), special equipment, or material alteration in painting, lettering or art work thereafter required by Customer. In the event that, subsequent to the date of execution of this agreement by Ryder, any Federal, state, or local law, ordinance, or regulation shall require the installation of any additional equipment, specifically including but not limited to anti-pollution or safety devices, Customer shall be responsible for the cost thereof, including installation expenses. Ryder agrees to either install same or arrange for the installation of same and Customer agrees to pay Ryder the full cost thereof upon receipt of Ryder's invoice for same.
   C. Where a Vehicle is operated by Customer with a trailer or other equipment not included on Schedules A, or not maintained by Ryder under a separate agreement, Customer warrants that such trailer and/or equipment will be in good operating condition and notwithstanding any other provision of this agreement will indemnify and hold Ryder harmless against any claim or loss or damage resulting from Customer's failure to properly maintain said trailer and/or equipment.
   D. Ryder may finance the Vehicles, or any part thereof, and in that connection may, as security, give the lender an installment sales instrument, mortgage, or security agreement covering such Vehicles or assign amounts due hereunder.

2. **OPERATION OF VEHICLES:**
   A. The Vehicles will be operated by Customer only in the normal and ordinary course of Customer's business, not in violation of any law, rule, regulation, statute or ordinance (including legal weight limitations) and Customer shall indemnify and hold Ryder harmless from and against all fines, forfeitures, seizures, confiscations and penalties arising out of any such violation.
   B. Each Vehicle will be promptly returned by Customer to Ryder's facility specified on Schedule A upon the termination of such lease unless Customer shall have purchased the Vehicle as provided hereinafter.

3. **MAINTENANCE AND REPAIRS:**
   A. Ryder agrees to provide from its facilities: (1) Oil, lubricants, tires, tubes and all other operating supplies and accessories necessary

CONTRACT/PRIVATE CARRIAGE

for the proper and efficient operation of the Vehicles; (2) Maintenance and repairs including all labor and parts which may be required to keep the Vehicles in good operating condition; (3) Painting and lettering, according to Customer's specifications, at the time the Vehicles are placed into service; (4) Washing of the Vehicles; and (5) Road service due to mechanical and tire failure.

B. Customer agrees not to cause or permit any person other than Ryder or persons expressly authorized by Ryder to make repairs or adjustments to Vehicles, governors and other accessories. When repairs are necessary, Customer shall notify Ryder by the speediest means of communication available. Ryder will not be responsible for any repair or service while such Vehicle is away from Ryder's facility, unless expressly authorized by Ryder and unless Customer submits an acceptable voucher for the repairs or services.

C. Customer agrees to return each Vehicle to Ryder for service and maintenance at the facility stated on Schedule A for a minimum of 8 hours each week during Ryder's normal business hours at such scheduled times as agreed by the parties.

4. **FUEL:**

The party designated on Schedules A agrees to provide fuel for the Vehicles.

A. When Ryder is designated:

(1) Fuel, oil and lubricants will be provided from Ryder's facilities or at service stations designated by Ryder. The charges will be based on the Rated Fuel Cost including all fuel taxes and adjustments in the charges will be made as provided on Schedules A.

(2) If Customer purchases fuel from sources other than Ryder's facilities or designated service stations, Ryder will reimburse Customer for such fuel cost upon receipt of an itemized paid invoice. Such reimbursement shall not exceed the Rated Fuel Cost.

(3) Ryder will, where permitted by law, upon request of Customer, apply for fuel tax permits, prepare and file fuel tax returns, and pay the taxes imposed upon the purchase and consumption of fuel by Customer provided: (a) Customer shall provide Ryder weekly with all trip records, fuel tickets or invoices, and other records or documents relating to the use of the Vehicles necessary for the preparation of the fuel tax returns and Customer shall reimburse Ryder the amount of any additional charge, assessment, tax, or penalty, or credit disallowed as a result of untimely or improper furnishing of such documents or information by Customer, and (b) Customer shall reimburse Ryder all such fuel taxes paid on Customer's behalf by Ryder in excess of the fuel taxes which would have been payable had the fuel consumed been purchased in the state of consumption, except to the extent that Customer obtains fuel at places and in amounts specifically designated by Ryder on a Fuel Program signed by Ryder's District Manager.

B. When Customer is designated:

Customer shall indemnify and hold Ryder harmless against any claims or loss resulting from Customer's failure to pay any fuel taxes.

5. **LICENSES:**

A. Ryder agrees to provide or pay for the state motor vehicle license for the licensed weight shown on Schedule A and personal property taxes for each Vehicle in the state of domicile, and federal highway use tax, where applicable, at the rates and method of assessment in effect on the date of execution of each Schedule A.

B. When permitted by law, Ryder will apply for such other vehicle licenses, prorate or state reciprocity plates as Customer may from time to time request at Customer's sole cost and expense.

C. Customer agrees to pay for any special license or pay any taxes required by Customer's business resulting from the operation and use of the Vehicles including mileage taxes, ton mileage taxes, highway or bridge tolls. Ryder shall have the right to pay or discharge any lien or encumbrance asserted against any Vehicle as a result of Customer's failure to pay any claim or assessment for any such taxes and Customer shall promptly reimburse Ryder for such payment.

6. **SUBSTITUTION:**

Ryder agrees to furnish a substitute vehicle at no extra charge for any Vehicle, other than those excepted below, which may be temporarily inoperable because of mechanical failure, the substitute to be as nearly as practicable the same size and appearance as the Vehicle, except that no special painting, lettering or other alterations need be made on the substitute vehicle. The substitute vehicle will be furnished to

223

INDUSTRIAL TRAFFIC MANAGEMENT

Customer whenever possible at the place at which the Vehicle was disabled and shall be returned by Customer to Ryder at the Ryder facility from which it was provided. Ryder shall have no obligation to furnish a substitute vehicle if the inoperable Vehicle is out of service for ordinary maintenance and service time, or is out of service because of damage resulting from collision or upset, or is specialized, or carries a truck body not owned by Ryder, is out of service for repair or maintenance of special equipment or accessories for which Ryder is not responsible, or is of a type Ryder does not have in its rental fleet. Ryder's failure to furnish a substitute vehicle within a reasonable time, where it is obligated hereunder to do so, shall cause the charges applicable to the inoperable Vehicle to abate until the Vehicle is returned to Customer's service or until a substitute is tendered to Customer. Ryder's liability in the event of such a failure shall be limited to the abatement of the charges for the inoperable Vehicle. A substitute vehicle, while in Customer's service, shall be subject to all the terms and conditions of this agreement. While a Vehicle is out of service because of damage resulting from collision or upset, Ryder will, at the request of Customer, rent Customer a replacement vehicle, if available from Ryder's rental fleet, at a rental rate equal to the charges applicable to the inoperable Vehicle. Irrespective of whether or not Customer rents a vehicle from Ryder while a Vehicle is out of service for repair of damage resulting from collision or upset, the charges applicable to the out of service Vehicle shall not abate.

7. DRIVERS:

A. Customer agrees to cause each Vehicle to be operated only by a safe, careful, properly licensed driver, at least 21, who shall be the employee or agent of Customer only, paid by and subject to Customer's exclusive direction and control. Customer agrees to reimburse Ryder in full for loss or damage to Vehicles, including related expenses, if such Vehicles are operated by drivers under 21. Upon receipt of a written complaint from Ryder specifying any reckless, careless or abusive handling of the Vehicle or any other incompetence by or of any driver, and requesting his removal as a driver of Vehicles, Customer will immediately remove such individual as a driver of Vehicles. In the event that Customer shall fail to do so, or shall be prevented from so doing by any agreement with anyone on the driver's behalf: (1) Customer shall, notwithstanding any other remedies of Ryder or provisions of this agreement, reimburse Ryder in full for any loss and expense sustained by Ryder for damage to any Vehicle when being operated by such individual and Customer shall release, indemnify and otherwise hold Ryder completely harmless from and against any claims or causes of action for death or injury to persons or loss or damage to property arising out of the use or operation of any Vehicle when being operated by such individual notwithstanding that Ryder may be designated on applicable schedules A as responsible for furnishing and maintaining liability insurance, and (2) Ryder may at its election, and at any time thereafter upon 30 days prior written notice to Customer, terminate any liability insurance coverage provided by Ryder hereunder, and may, at its election, with respect to each Vehicle, increase the amount of Customer's physical damage responsibility to an amount equal to the agreed value calculated in accordance with Paragraph 11 as of the time of damage or loss.

B. Ryder agrees to assist Customer in developing a driver education and safety program.

C. Customer agrees that the Vehicles will not be operated by a driver in possession of or under the influence of alcohol or any drug which may impair his ability to operate the Vehicle, or in a reckless or abusive manner, or off an improved road, or on a flat tire, or improperly loaded, or loaded beyond the manufacturer's recommended maximum gross weight shown on Schedule A. Notwithstanding any other provision of this agreement, Customer agrees to reimburse Ryder in full for damages, including expenses, resulting from a violation of this provision. Customer will be responsible for all expenses of towing or removal of any mired Vehicle when not in Ryder's possession or on Ryder's premises.

8. CHARGES

A. Customer agrees to pay Ryder the fixed charge for each Vehicle in advance upon receipt of Ryder's invoice for same and to pay all other charges, specifically including, but not limited to, the mileage rate per mile provided for under this agreement within 10 days of the date of Ryder's invoice without deduction or setoff.

B. Mileage shall be determined from odometer readings. If the odometer fails to function, which failure Customer shall immediately report to Ryder in writing, the mileage for the period in which the failure existed may be determined at Ryder's option from (1) Customer's trip records, or (2) from the amount of fuel consumed and the miles per gallon record of Ryder averaged for the previous 30 days.

CONTRACT/PRIVATE CARRIAGE

9. **ADJUSTMENT OF CHARGES:**

A. The parties recognize that the charges provided for in this agreement are based on Ryder's current cost of labor, parts and supplies. The cost of Ryder's operation may fluctuate after the date of execution of this agreement. Customer agrees that for each rise or fall of 1% in the Consumer Price Index (using a 1967 base period, published by the United States Bureau of Labor Statistics), above or below the base index figure on Schedule A, the fixed charge and mileage charge for each Vehicle, including those charges stated in the Mileage Guaranty, if any, shall be adjusted upward or downward as follows:

Vehicles with Fuel provided by Ryder:	1% of one-half of the total charges
Vehicles without Fuel provided by Ryder and Trailers:	1% of three-fourths of the total charges

B. Any and all subsequent adjustments for Vehicles shall be based on the charges stated on Schedule A. Adjustments in charges shall be effective on the first day of each calendar half year period and will be based on the latest index which has been published prior to such effective date. In the event the Consumer Price Index should be discontinued, another mutually agreeable cost adjustment index to adjust charges shall be agreed upon.

C. Customer agrees to pay for (1) any sales or use tax now or hereafter imposed upon the use of the Vehicle or on the rental or other charges accruing hereunder, (2) any increase in license or registration fees, including federal highway use tax, vehicle inspection fees, and personal property tax rates, or (3) any new or additional tax or governmental fees, adopted after the date of the execution of the applicable Schedule A by Customer, upon the fuel provided by Ryder.

10. **INSURANCE:**

A. Liability Insurance Responsibility

(1) Standard policy of automobile liability insurance (hereafter Liability Insurance) with limits specified on each Schedule A shall be furnished and maintained by the party designated on Schedule A at its sole cost, written by a company satisfactory to Ryder covering both Ryder and Customer as insureds for the ownership, maintenance, use or operation of the Vehicles and any vehicle being provided as a substitute therefor. Such policy shall provide that coverage afforded cannot be cancelled or materially altered without 30 days prior written notice to both parties. The party designated shall furnish to the other party certificates to evidence compliance with this provision.

(2) Upon not less than 60 days prior written notice to Customer, Ryder shall have the right to terminate Liability Insurance coverage maintained by Ryder and Customer shall be obligated to procure and maintain Liability Insurance in the limits set forth on Schedule A as of the effective date of termination and the charges will be adjusted accordingly.

(3) If Customer is obligated to procure and maintain Liability Insurance and fails to do so, or fails to furnish Ryder the required evidence of insurance, Customer shall indemnify and hold Ryder harmless from and against any claims or causes of action for death or injury to persons or loss or damage to property arising out of or cause by the ownership, maintenance, use or operation of any Vehicle, and Ryder is authorized but not obligated to procure such Liability Insurance, without prejudice to any other remedy Ryder may have, and Customer shall pay Ryder, as additional rental, the amount of the premium paid by Ryder.

(4) Customer agrees to release, indemnify, and hold Ryder harmless from and against any claims or causes of action for death or injury to persons, or loss or damage to property in excess of the limits of Liability Insurance, whether provided by Ryder or Customer, as indicated on Schedules A, arising out of or caused by the ownership, maintenance, use or operation of any Vehicle leased or furnished hereunder, and any such claims or causes of action which Ryder shall be required to pay as a result of any statutory requirements of insurance and which Ryder would not otherwise, pursuant to the terms hereof, be required to pay.

(5) Ryder will, where required and permitted by law, upon request of Customer, file State and/or Interstate Commerce Commission Certificates of Automobile Liability Insurance covering the Vehicles. Customer agrees to indemnify, defend, and save Ryder

225

INDUSTRIAL TRAFFIC MANAGEMENT

harmless from all claims, causes of action, suits and damages arising out of filing such documents for vehicles other than leased Vehicles.

(6) Customer further agrees to release and hold Ryder harmless for death or injury to Customer, Customer's employees, drivers, passengers or agents, arising out of the ownership, maintenance, use or operation of any Vehicle leased or furnished hereunder.

B. **Physical Damage Responsibility**

The party designated on Schedule A shall pay for loss or damage to any Vehicle subject to the following:

(1) In the event Ryder is designated:

a. Ryder will pay for loss or damage to each Vehicle in excess of the amounts specified on Schedules A EXCEPT (1) any willful damage to the Vehicle arising out of or in connection with any labor dispute; (2) conversion of any Vehicle by an agent or employee of Customer which shall not be considered theft within the terms of this provision; or (3) the loss by theft of tools, tarpaulins, accessories, spare tires and other such appurtenances. Customer shall pay up to the amount specified on Schedule A as deductible, for loss or damage to any Vehicle, including related expenses, from each occurrence and shall pay for all loss or damage to any Vehicle resulting from any perils specifically not assumed by Ryder herein.

b. Upon not less than 60 days prior written notice to Customer, Ryder shall have the right (1) to terminate any physical damage coverage procured and maintained by Ryder and (2) to increase Customer's Physical Damage Responsibility to an amount with respect to each Vehicle equal to the agreed value of such Vehicle computed in accordance with Paragraph 11 as of the time of damage or loss, and each of the charges specified on Schedule A, including those stated in the Mileage Guaranty, if any, shall be decreased accordingly.

c. In the event Ryder terminates physical damage coverage, Customer shall be obligated to procure and maintain physical damage coverage acceptable to Ryder and each of the charges shall be adjusted accordingly. Customer agrees to furnish Ryder certificates necessary to evidence compliance with this paragraph.

d. If Customer is obligated to procure and maintain physical damage coverage and fails to do so, or fails to timely furnish Ryder with evidence of such coverage, Customer agrees to reimburse Ryder all its loss, cost and expense resulting from loss of or damage to the Vehicles or any vehicle being used as a substitute therefor.

(2) In the event Customer is so designated:

a. Customer will pay for all loss or damage to any Vehicle or any vehicle being used as a substitute therefor, including related expenses arising from any cause, but Customer's liability shall not exceed the purchase price of the damaged Vehicle computed according to the provisions of Paragraph 11 at the time of such loss or damage.

b. Customer further agrees to furnish Ryder with a policy of insurance acceptable to Ryder with Ryder as a named insured or endorsed as a loss payee having a deductible amount not to exceed the amount specified on Schedule A, failing in which, Ryder may obtain such insurance and add the cost thereof prorata to the charges for the Vehicles.

C. **Notice of Accident**

Customer agrees to notify Ryder immediately upon the happening of any accident or collision involving the use of a Vehicle by the speediest means of communication available and to cause the driver to make a detailed report in person at Ryder's office as soon as practicable, and to properly render all other assistance to Ryder and the insurer that is requested by either of them in investigation, defense or prosecution of any claims or suits.

D. **Cargo Insurance**

Customer agrees to release and hold Ryder harmless from liability for loss or damage to any goods or other property in or carried on any Vehicle whether such loss or damage occurs in Ryder's facility or elsewhere. Customer shall, at its sole expense, include Ryder as a named insured in any and all cargo or transportation or floater insurance policies covering Customer with respect to any loss or damage to such goods or property. Customer waives any legal right of recovery against Ryder for any such loss or damage. Customer shall reimburse Ryder for loss of

CONTRACT/PRIVATE CARRIAGE

E. Vehicle Theft or Destruction

If a Vehicle is lost or stolen and remains so for 30 days after Ryder has been notified, the lease as to such Vehicle shall then terminate provided all charges for such Vehicle have been paid to that date and provided any amounts due Ryder pursuant to Paragraph 10B have been paid. Ryder shall not be obligated to provide a substitute vehicle during said 30 day period. If a Vehicle is, in Ryder's opinion, damaged beyond repair, Ryder shall notify Customer within 30 days after Ryder has been advised of the loss. Upon receipt of Ryder's notice that the Vehicle has been damaged beyond repair, provided all charges for such Vehicle have been paid to that date and provided any amounts due Ryder pursuant to Paragraph 10B hereof have been paid, the lease as to such Vehicle shall then terminate.

11. TERMINATION:

A. Either party may terminate the lease of any Vehicle prior to expiration of its term on any anniversary date of its delivery date, other than the anniversary date on which the lease term expires, by giving to the other party at least 60 days prior written notice of its intent to do so. If termination is effected by Ryder, Customer shall have the right, but not the obligation, to purchase in accordance with Paragraph 11D all Vehicles with respect to which termination notice has been given on the termination date(s). If termination is effected by Customer, Customer shall at Ryder's option purchase in accordance with Paragraph 11D all Vehicles with respect to which termination notice has been given on the termination date(s).

B. In the event Customer shall become insolvent, file a voluntary petition in bankruptcy, make an assignment for the benefit of creditors, be adjudicated a bankrupt by any court of competent jurisdiction, permit a receiver to be appointed for its business, permit or suffer a material disposition of its assets, the lease of Vehicles shall terminate, at the election of Ryder. Upon written notice thereof sent to Customer, Ryder may at its option demand that Customer purchase within 10 days of termination any one or all of the Vehicles in accordance with Paragraph 11D without prejudice to other remedies Ryder may have under this Agreement and at law.

C. Breach or Default

(1) In the event Customer breaches or is in default of any of the provisions of this Agreement, Ryder may immediately, without notice or demand, take possession of the Vehicles, together with all equipment and accessories thereto, and Ryder shall be entitled to enter upon any premises where said Vehicles may be and remove same, retain or refuse to redeliver the Vehicles to Customer until such breach or default is cured, without any of such actions being deemed an act of termination and without prejudice to the other remedies Ryder may have and Customer shall continue to be liable for all charges accruing during the period the Vehicles are retained by Ryder.

(2) In the event Ryder takes possession of or retains any Vehicle, and there shall, at the time of such taking or retention, be in, upon or attached to such Vehicle any other property, goods or things of value belonging to Customer or in the custody or control of Customer, Ryder is authorized to take possession of such items and either hold the same for Customer or place the same in public storage for Customer at Customer's expense.

(3) If Customer's breach or default continues for 7 days after written notice has been mailed to Customer, Ryder may terminate the lease of Vehicles. Upon termination, Ryder may demand that Customer purchase within 10 days of termination any or all of the Vehicles in accordance with Paragraph 11D without prejudice to other remedies Ryder may have under this Agreement and at law.

(4) Customer shall pay Ryder all Ryder's costs and expense, including reasonable attorney's fees, incurred in collecting amounts due from Customer or in enforcing any rights of Ryder hereunder.

D. In the event Customer (pursuant to Paragraph 11A) shall be required to purchase any of the Vehicles, or should Ryder (pursuant to Paragraph 11B or 11C) demand of Customer that it purchase any of the Vehicles, Customer shall purchase each Vehicle for cash at or within the time aforesaid for its Original Value as shown on Schedule A, less the total depreciation which has accrued for such Vehicle in accordance with Schedule A. Additionally, Customer will, at the time of purchase, pay Ryder for the amount of any unexpired licenses, applicable taxes, including personal property taxes and federal highway use taxes, and other prepaid expenses previously paid by Ryder for the Vehicles prorated

227

INDUSTRIAL TRAFFIC MANAGEMENT

to the date of sale and will be responsible for any sales or use tax arising from the purchase. Customer shall have no obligation or right to purchase any Vehicle as to which the term on Schedule A has expired.

12. **ASSIGNMENT OF LEASE**

This Agreement shall be binding on the parties hereto, their successors, legal representatives and assigns. Customer shall promptly notify Ryder in writing prior to all substantial changes in ownership or any material disposition of the assets of Customer's business. Customer does not have the right to sublease any of the Vehicles, nor to assign this Agreement or any interest therein without the prior written consent of Ryder.

13. **FORCE MAJEURE**

Ryder shall incur no liability to Customer for failure to supply any Vehicle, provide a substitute vehicle, repair any disabled Vehicle, or provide fuel for Vehicles, if prevented by a national emergency, wars, riots, fires, labor disputes, Federal, state, or local laws, rules, regulations, shortages (local or national), or fuel allocation programs, or any other cause beyond Ryder's control whether existing now or hereafter. Notwithstanding Ryder's inability to perform under such conditions, Customer's obligations hereunder shall continue.

14. **NOTICES**

All notices provided for herein shall be in writing and mailed to Ryder and Customer at their respective addresses set forth above or at such other addresses designated in writing by either party.

This Agreement shall not be binding upon Ryder until executed at its Miami Headquarters by a person duly authorized and shall constitute the entire agreement and understanding between the parties concerning the Vehicles, notwithstanding any previous writings or oral undertakings, and its terms shall not hereafter be altered by any oral agreement or informal writing, nor by failure to insist upon performance, or failure to exercise any rights or privilege, but alterations, additions, or changes in this Agreement shall be accomplished only by written endorsement hereon, or amendment hereto, or additional Schedules A made a part hereof duly executed by both parties.

15. **ARTICLE HEADINGS** in bold face type do not constitute any part hereof and shall not be considered in the interpretation hereof.

IN WITNESS WHEREOF, each of the parties hereto has caused these presents to be duly executed the day and year first above written.

RYDER TRUCK RENTAL, INC. (Ryder)   (Customer)

By: _____ By: _____

Name/Title _____ Name/Title _____

Date: _____ Date: _____

Witness: _____ Witness: _____

9-52(4/77)

CONTRACT/PRIVATE CARRIAGE

RYDER TRUCK RENTAL, INC.

SCHEDULE NO. _____
PAGE _____ of _____
(Customer)

SCHEDULE A TO TRUCK LEASE AND SERVICE AGREEMENT
between Ryder Truck Rental, Inc. (Ryder) and _____ Dated _____

Location Name _____

VEHICLE NO. (1)	DATE OF DELIVERY (2)	TERM IN YRS. (3)	YR. & MAKE (4)	VEHICLE DESCRIPTION MODEL & TYPE (5)	SERIAL NUMBER (6)	MFGRS. RECM. MAX. GCM AND/OR GVW (7)	LICENSED WEIGHT (8)	ORIGINAL VALUE (9)	DEPREC. WEEKLY AMOUNT (10)	RATED FUEL COST PER MILE/GAL. (11)	REVISED CPI FOR URBAN WAGE EARNERS & CLERICAL WORKERS INDEX 1967 BASE (12)	FIXED RENTAL CHARGE PER WEEK IN ADVANCE (13)	MILEAGE RATE PER MILE (14)

The Vehicles listed hereon only are leased for the term specified above. As to these Vehicles, either party may still terminate the lease of any Vehicle on any anniversary date of its delivery date, other than the anniversary date on which the lease term expires, by giving to the other party at least 60 days prior written notice of its intent. However, if termination is effected by Ryder, Customer shall have the right, but not the obligation, to purchase in accordance with the Agreement all Vehicles with respect to which termination notice has been given on the termination date(s). If termination is effected by Customer, Customer shall at Ryder's option purchase in accordance with the Agreement all Vehicles with respect to which termination notice has been given on the termination date(s). Customer shall have neither the right nor the obligation to purchase any Vehicle with respect to which the term has expired. All terms and conditions of Paragraphs _____ B. and _____ C. of the Agreement shall remain in full force and effect.

(SIDE ONE)

INDUSTRIAL TRAFFIC MANAGEMENT

Notwithstanding anything in the Agreement to the contrary, it is mutually agreed:

It is agreed that the Original Value, Weekly Depreciation and Fixed Charges per week are based upon manufacturer's quoted price for vehicles as of the date of execution by Customer of this Schedule A. In the event manufacturer's quoted price for such vehicle is increased prior to the Date of Delivery of the vehicles, Customer agrees that for each $50 increase in price (or fraction thereof), the following shall be increased accordingly:

Original Value	Weekly Depreciation	Fixed Charge Per Week
$50.00	$.15	$.30

The amounts in columns 9, 10 and 13 of this Schedule A will be adjusted on Date of Delivery.

1. Liability Insurance
   Responsibility: _____
   (Ryder/Customer)

   Limits: Bodily Injury $ _____ per person
   Bodily Injury $ _____ per accident
   Property Damage $ _____ per accident
   OR Combined Single Limits $ _____ per accident

2. Physical Damage Responsibility by: Ryder ☐ $ _____ deductible   Customer ☐

3. Fuel provided by _____
   (Ryder/Customer)

In the event Ryder's cost for fuel provided for vehicles listed on this Schedule shall be greater or less than that amount set forth in Column 11, Customer shall pay to Ryder or receive a credit from Ryder accordingly. Ryder's fuel cost, shall be developed by Ryder in accordance with standard cost accounting practices, reported and billed to Customer.

4. Domicile of the Vehicles listed on this Schedule (City) _____

5. Service and maintenance location of Vehicles listed on this Schedule _____

6. Vehicle(s) will only be operated in the following states: _____
   Fuel Permits _____

7. Mileage guarantee: Customer guarantees that each of the vehicles described on this Schedule A will be operated a minimum of _____, miles within each _____ month period during which it is under lease ("Guarantee Period").

The payment of the Mileage Rate per Mile shall be made as provided in Paragraph 9 of the Agreement for miles actually operated.

A settlement of the mileage guarantee will be made at the end of each guarantee period as follows: If miles actually operated during the period are less than the miles guaranteed, then Customer will be billed by Ryder and Customer shall promptly pay to Ryder an amount equal to the difference between the number of miles guaranteed and the number of miles operated during the period, multiplied by the Deficiency Rate per Mile specified below. If the number of miles operated during the period exceeds the number of miles guaranteed, then Customer will receive credit for an amount equal to: (1) the difference between the miles actually operated and the miles guaranteed, multiplied by (2) the difference between the Mileage Rate per Mile specified on Side One of this Schedule and the Excess Rate per Mile specified below.

The Deficiency Rate per Mile shall be $ _____ and the Excess Rate per Mile shall be $ _____

Miles operated by vehicles to which the same Mileage Rate per Mile, Deficiency Mileage Rate per Mile, Excess Mileage Rate per Mile, number of miles guaranteed, and guarantee period apply, may be averaged. If the lease of any vehicle is terminated during any guarantee period, the guarantee shall be prorated for that vehicle on the basis of the actual portion of such period it was under lease.

This Schedule A is hereby made a part of that certain Truck Lease And Service Agreement between the parties hereto.

RYDER TRUCK RENTAL, INC.   _____
                                                              (CUSTOMER)

By _____   By _____
Name/Title _____   Name/Title _____
Date _____   Date _____
Witness _____   Witness _____

(SIDE TWO)

CONTRACT/PRIVATE CARRIAGE

An important point to remember in considering the pros and cons of leasing is that ICC regulations do not permit a firm to lease drivers from the same company that leases it the trucks (see the case of Allen vs. the United States, mentioned later in this chapter).

Today a huge number of firms offer leasing agreements for everything from forklifts to jumbo jets. Leasing is a big competitive business and in leasing a truck fleet, there are some pertinent questions that should be asked about both the lease and the lessor. For example:

—Does the leasing firm enjoy a good reputation among its customers?
—What are the arrangements for fueling and lubricating the trucks?
—Preventive maintenance per how many miles or days or weeks?
—Labor, parts and repair?
—Winterization of vehicles?
—Overhaul as needed or at specified intervals?
—Back-up vehicles. Extra charges?
—Service standards?
—Quality of mechanics?
—How many service locations?
—How many fuel stops available?
—Insurance?
—Rates on extra equipment?
—Washing?
—Licenses, taxes and permits?
—Mileage, monthly and annual costs per vehicle (tractors and/or trailers)?
—Reports. What type? Adequate?

Truck leasing companies occupy an important position in transporation and are enjoying fast growth for several reasons. Not the least of these reasons is that their services permit industrial shippers to employ capital more productively elsewhere than on truck purchases. Other reasons include the protection leasing affords against losses due to obsolescence, the assistance it provides in meeting fluctuations in seasonal or cyclical transportation requirements, and the ownership risks borne by the lessor. Further, leasing makes trucks available with no initial investment and resulting cash drain and at times helps improve company financial ratios by which lenders traditionally measure credit worthiness. "Full service" leasing affords a lessee the opportunity to use the more highly specialized maintenance and service facilities, and the trained personnel, of the leasing company. And it is usually much easier to lease a small number of trucks and then expand when starting out in private carriage than it is to plunge unprepared into buying them.

The other side of the coin on owning versus leasing is, most importantly, the extra costs of the latter. Leasing companies can buy equipment and provide service on less expensive bases than shippers. Such companies quite properly, however, must make profits and these as well as costs must be underwritten and reckoned with. Of course, not all of the above benefits are attractive in all cases. Some firms may find long term rent obligations inadvisable, or that special weight should be given the look ahead on inflation. Benefits where depreciation write-offs help taxes may be important.

A company should consider its private fleet as a profit center and operate it in competition to its common carriers. The company fleet should compete for the company's freight and the traffic manager should constantly compare common carrier costs against private fleet costs to determine who gets the shipments.

As we said earlier in this chapter, most industrial companies are in private trucking because of the service benefits and economies it brings to them. Then, too, private trucking is relatively free of restraint and regulation and this appeals to many. Private trucks, by and large, can go anywhere at any time and carry any kind of the owning company's goods. Proprietary motor carriage is somewhat akin in its advantages to a person using his own automobile as against using buses, cabs or trains. More than half of all United States businesses (non-transportation) run their own trucks to haul part or all of their traffic.

Other reasons to go into private trucking may include disenchantment with for-hire carriers, a wish for "complete control" over transportation services, a need to meet the customer service levels competitors are offering and even the use of drivers as salesmen. But generally, a firm goes into, and stays in, private trucking because of customer service problems—late or unreliable deliveries, etc. Once in, getting out of private carriage may be difficult for various reasons—job losses, customers wishing continuances of drivers' help, etc.

Establishing and maintaining an economic private fleet is based on having sufficient tonnage to keep the trucks operating fully at total costs lower than would be charged by other transportation services. Preferably private truck hauls should be loaded both ways to avoid an empty return. Lack of round trip tonnage may not be fatal, but the real criterion is total cost. It is important that the tonnage be available year-round and steady enough for full-truck operation. Running a private truck fleet takes capital which must itself be paid for in finance and interest charges. The result is costs that continue whether freight moves or not. When freight does not move, those costs mount up and red figures appear in the ledger.

On the average, a private fleet should be able to move a company's freight for a lower total cost than a common carrier. All hauling costs should be matched, common carrier vs. private fleet, whether the common carrier rates available for any particular haul are class, exception or commodity. (The comparison must of course be made against the lowest rates available.)

The costs to be used in determining the dollars involved in a company truck operation fall into three categories—equipment costs, operating expenses, and overhead. These may be further categorized as fixed and variable costs, time and distance costs, and so on. They may be lumped into weekly, monthly, or annual totals to compare with total for-hire costs—particularly on peddler, short haul, small shipment, or local operations. A transportation man starting in company hauling should certainly enlist the help of his accounting colleagues in setting up cost figures. Companies differ in their accounting practices and the cost accounting for one operation may not be applicable to another. Sample analysis forms that may be useful in reporting and checking costs for comparing common carrier and private operations are shown on page 235.

A "private carrier of property by motor vehicle" is specified by the I.C. Act, U.S. Code 1970, Title 49, Sec. 303(a)(17), to mean "any person not included in the terms 'common carrier by motor vehicle' or 'contract carrier by motor vehicle,' who or which transports in interstate or foreign commerce by motor vehicle property of which such person is the owner, lessee, or bailee, when such transportation is for the purpose of sale, lease, rent, or bailment, or in furtherance of any commercial enterprise." Industrial companies, by and large, meet these specifications in the proprietary hauling of their goods to their markets. They operate free of requirements for I.C.C. certificates or permits as to commodities and territories. They are also free of rate and service regulations by the I.C.C. Congress recognized shippers' interests in furnishing their own transportation when the Motor Carrier Act was passed. These definitions are important to private carriers and, to operate legally under the Act, their hauling should conform to its various provisions.

While simple proprietary hauling by an industrial company of its own goods to its own markets clearly meets the quoted statutory tests, other situations have arisen which have occasioned court and I.C.C. interpretations. One of prime importance is known as the "primary business" rule. In essence, it says that to be private, the haulage must be performed by a company whose primary business is other than transportation. Two landmark decisions in this regard

				PRIVATE CARRIAGE COST WORKSHEET				
Point of Origin	Point of Destination	Mileage (Highway)	Frequency of Movement	Tonnage Per Year	Commodity Description	Density (Loadability)	Common Carrier Costs	
(1)	(2)	(3)	(4)	(5)	(6)	(7)	Rate	Min. Wt.

A. Total Miles _____
B. Total Tons _____
C. Empty Miles _____
D. Overflow Tons (Tonnage which cannot be handled by own equipment) _____
E. Equipment Needs _____
F. Driver Requirements _____

Work sheet used in connection with private carriage cost analysis.

CONTRACT/PRIVATE CARRIAGE

PRIVATE CARRIAGE COST ANALYSIS

	Company Owned Equipment	Leased Equipment

I. FIXED COSTS

A. Entire Capital Investment .... $ _____ _____
B. Interest on Investment .... _____ _____
C. Finance Charges .... _____ _____
D. License Fees .... _____ _____
E. Insurance (Property Damage, Public Liability, Buildings, Maintenance, Equipment, etc.) .... _____ _____
F. Depreciation on Operating Equipment .... _____ _____
G. Depreciation on Maintenance Equipment and Buildings .. _____ _____
H. Property Taxes .... _____ _____
I. Garage Supervision .... _____ _____
J. Garage Labor (including all insurance, fringe benefits, social security, etc.) .... _____ _____
K. Garage Maintenance Costs (including building repairs, heat, gas, lights, water, telephone, etc.) .... _____ _____
L. Costs to replace broken, worn out tools and equipment . _____ _____
M. Special Equipment Costs (tarpaulins, heaters, refrigerating units, etc.) .... _____ _____
N. Special Garage Equipment (racks, storage bins, etc.) .. _____ _____
O. Fleet Supervisor Salary and Expenses .... _____ _____
P. Administration Expense (includes time management spends on fleet operation problems) .... _____ _____
Q. Costs for maintenance of records (includes checking, typing, posting and payment of bills, postage, and stationery and supplies, other office work) .... _____ _____

FIXED TOTAL COSTS $ _____ _____

II. OPERATING COSTS

A. Primary maintenance of equipment (includes washing, polishing, greasing, painting, etc.) .... $ _____ _____
B. Road Service Costs (resulting from breakdowns, wrecks, or other catastrophies) .... _____ _____
C. Repairs outside company's shop .... _____ _____
D. Gasoline and oil .... _____ _____
E. Tires and Tubes (replacement and repairs) .... _____ _____
F. Replacement parts (Tractor and Trailer) .... _____ _____
G. Anti-Freeze, Tire Chains, Flares, Lamps and other vehicle accessories .... _____ _____
H. Cost of equipment retail (when required due to breakdowns or unusual peak load requirements) ...... _____ _____
I. Drivers' Salary (includes all union benefits, insurance, social security, etc.) .... _____ _____
J. Extra Drivers for vacation periods, sick leaves, etc. .. _____ _____
K. Toll Road Fees .... _____ _____

TOTAL OPERATING COSTS $ _____ _____

	Private	Common
Per Mile	_____	
Per 100 Lbs.	_____	

Miles Operated _____ _____
Cost Per Mile _____ _____
Tons Transported _____ _____
Cost per 100 Pounds _____ _____

TOTAL COST $ _____ _____

Private carriage cost analysis form.

are the Woitishek (43 M.C.C. 193) and the Lenoir Chair (51 M.C.C. 65) cases. Both cases essentially reached the same conclusion.

The Lenoir Chair case involved a furniture manufacturer that sold F.O.B. its factory and transported in its own trucks about one-fifth of its output to its customers. For such transportation the company added separate charges to its invoices comparable to rail or for-hire truck costs. These activities were ruled to not require a change in the furniture company's private carrier status because the primary business of the firm was making furniture and the transportation was subordinate to that activity.

Among other decisions illustrating the primary business test is one by the U.S. Supreme Court in Red Ball Motor Freight vs. Shannon 12 L.ed (2) 341. It was there ruled that the purchase of sugar to provide a back-haul in connection with outbound movements of other commodities was within the scope, and in furtherance of, the primary business of a general merchandise enterprise. The fact that the company was in the business of buying and selling many items including sugar was important in the decision.

Other interpretations have applied what are sometimes called "control" and "substantial burden" tests. The "control" test runs to the effect that, to be private, the shipper must have the right to control, direct, and dominate the performance of the service and must, in fact, do so. The "substantial burden" test requires that the shipper must bear all, or a preponderance, of the normal risks and burdens associated with the transportation enterprise.

These tests are applied often in leasing cases and it has been held in several of them that the lease arrangements were for-hire rather than private transportation. The H.B. Church case (27 M.C.C. 191) went so far as to establish a presumption that a lease of trucks results in for-hire transportation. It went on to point out, however, that the presumption would yield to a showing by the shipper of meeting the "control" test. In a similar vein, Allen vs. United States (187 F. Supp. 625) held that in order to constitute private transportation, where equipment is supplied by a lessor, the shipper must have the exclusive right to direct and control the vehicle as well as the driver. In that case, an individual experienced in

CONTRACT/PRIVATE CARRIAGE

transportation leased trucks to the shipper, supplied drivers, and so on. It was held that this operation was not private.

"Buy-and-sell" transactions, made simply to have transportation provided, are usually regarded by the courts as subterfuges and evasions. Until recently, it was not permissable for one company to cooperate with another and coordinate their private carriage activities through the same fleet of vehicles. This was true even though it permitted the use of trucks both ways that normally would return empty; i.e., loads out for one company and loads back for the other. Even when two affiliates of, say, a conglomerate were involved, such two-way use of vehicles was normally not permitted.

As we noted earlier, the fuel shortages of 1978-80 led to the ICC rethinking its rather hostile traditional policies towards private carriage. The agency has reassessed its views and made a number of substantial changes.

In November 1978, the ICC revised its policy on issuing common or contract operating authority to private carriers when their regulated operations would be incidental to the transportation of their own goods. Under this new policy, designed to eliminate empty backhauls (Ex Parte No. MC-118), a private carrier is eligible to receive either common or contract authority if it meets certain conditions.

First, the applicant must apply for operating authority and meet all the standard public need and fitness requirements that other carriers must meet. Secondly, it has to agree to conduct its private and for-hire motor carrier activities independently—it must keep separate records for its private operations and for its for-hire operations.

In reversing its long-standing policy against granting this kind of authority to private carriers, the Commission considered the question of the most efficient use of limited fuel supplies, which caused it to examine the validity of its prior assumptions. It concluded that private carriers generally seek common or contract authority only to acquire backhaul traffic and that any conflict between private carriers and for-hire carriers would usually occur only on backhauls. In addition, the ICC said that this change could have a favorable impact by reducing deadheading and saving fuel.

Another policy undergoing extensive review by the Commission at the time of writing involves "intercorporate hauling," which is transportation performed for compensation by a parent company for its affiliates or subsidiaries. The ICC has traditionally held that this is for-hire transportation and does not constitute private carriage. Therefore it must be regulated.

The basis for this policy lies in the theory that to promote or encourage private carriage undermines the strength of the regulated for-hire carriers. Following this line of thought, the ICC had tended to assume that such transportation works to the detriment of the public interest and could lead to diversion of traffic from for-hire carriers, less work for owner operators, and substantial losses to the regulated carriers.

In 1978, the Commission conducted a study that found many uncertainties in its assumptions about the effects of intercorporate hauling on the transportation system. Following that study, the Commission issued a policy statement (Ex Parte No. MC-122) which proposed to relax, in large degree, the existing restrictions on intercorporate hauling. Under a new proposal, intercorporate hauling would be permitted if the transportation is performed for divisions or subsidiaries 100% owned by the parent company.

Both issues have caused controversy because of opposition from the common carriers and will finally be resolved in the courts.

Contract Carriage

Like private carriage, contract carriage guarantees that the trucks a company uses to transport its goods are dedicated to its exclusive use. But whereas private carriage is exempt from governmental regulation except for rules concerned with safety and hours of service, a contract carrier requires a permit from the Interstate Commerce Commission.

The ICC defined a "contract carrier" as "essentially an independent contractor whose undertaking is defined and limited by an individual contract which calls for a service specialized to meet the peculiar needs of a particular shipper or a limited number of shippers and operates to make the carrier virtually a part of each shipper's organization." (49 M.C.C. 383, 390, 1949.) Until early 1979, the ICC followed the so called "rule of eight" which limited the number of shipper contracts that a contract carrier could make to a maximum of eight. However, the Commission then abandoned that ruling and contract carriers may now make as many contracts as they wish within the limits of their licenses. Then in late 1979, the Supreme Court ruled that a carrier may hold both common and contract authority, which further obliterated the hitherto carefully drawn distinctions between common and contract carriage. Such dual authority was only granted by the ICC in very rare instances before the Supreme Court ruling.

Contract carriers do not need certificates of public convenience and necessity, but as noted earlier, they do need permits from the ICC. In many areas, contract carriers may have to seek state authority to operate intrastate. In applying for a permit to haul freight for a shipper, the contract carrier will need the support of that shipper in showing a requirement for the carrier's services and how those services are to be tailored to the shipper's specific requirements.

No ICC permits are needed for contract carriers (or, indeed, other kinds of carriers) operating within commercial zones; or for carriers hauling exempt commodities.

As contract carriers haul only for specific companies, their costs can be geared to the particular haulage they perform and their rates can be set on realistic patterns that reflect those costs. It is not uncommon for a contract hauler to work with his shippers in making projections of future transportation requirements and to increase or decrease his fleet to meet those requirements. He may also go over his books periodically with his shippers to establish that his charges are fair and reasonably profitable. Because the contract carrier operates free of rate bureau procedural restraints and Reed-Bulwinkle Section 5a involvements, he can change equipment and rate schedules quickly when this should be done, and concentrate on his operations and on doing a good job for his shippers.

INDUSTRIAL TRAFFIC MANAGEMENT

To make a contract carrier arrangement worthwhile, there must be enough tonnage available to make full use of the trucks—the combined volumes from all of the carrier's shippers should be used to make up tonnage. Also, there should be enough tonnage year-around to provide steady fleet utilization. Back hauls are most helpful and should if possible be arranged, with the purchasing department in the shipper's organization (though it may be necessary for Purchasing to change its purchasing terms).

Some think of contract carriers as highly specialized, such as heavy machinery haulers who combine hauling with rigging operations. While many *are* specialized, contract carriers can in fact haul anything in the way of materials and provide any delivery service that any fleet operator—common carrier or private truck operator—can provide. Where volumes, distribution patterns, and other factors permit, some shippers use all three modes—contract, common and private—together.

The Contract

The usual practice in contract carriage is that the shipper makes a formal contract or memorandum of agreement with the carrier. Such agreements should spell out that the carrier is an independent contractor and is not an agent of the shipper. This can be important in damage and other suits where contentions may be made that the carrier *is* an agent of the shipper and that the shipper is therefore legally liable for the carrier's acts.

Model forms for contract carriage agreements are readily available, but whatever its form, the contract should:

—Identify both shipper and carrier and give their addresses
—Identify the points between which the service will be performed
—List the commodities that will be hauled and the quantities (LTL/TL)
—List and describe the trucks to be furnished
—Ensure that the shipper keeps his right to use common or private carriage (or both) if he need to

—State clearly that all drivers and other persons employed by the carrier are in fact his employees.
—State that the carrier assumes full custody and thus complete liability for any shipment until delivered
—Detail the liability of the trucker for loss or damage and the handling of claims
—Allow advertising of the shipper on the bodies of trailers if he wants it
—Detail insurance coverage for all contingencies—what the insurance will cover, amount of insurance, who will pay for the insurance, renewal factors and submission of proof of insurance
—Cover responsibilities for violations of any municiple, county, state or federal laws
—Detail work to be performed and hours of service
—List services to be performed by the carrier's employees
—State whether freight bills are to be submitted daily, weekly or over what period and when they will be paid
—Cover length of contract, notice of cancellation and/or renewal or extension of contract
—Include a schedule of rates and charges that covers any accessorial services
—Cover replacement of vehicles in maintenance or repair and describe the condition in which the vehicles will be maintained to protect shipments
—Cover the possible need for additional vehicles
—Mention the coordination of shipper and carrier to seek additional authority when warranted.

A contract should be carefully drawn to fit a particular operation and the carrier should never be asked to perform services not covered in it. This tends to blur and break down the contractual relationship. As conditions and requirements change, the contract should be revised accordingly. Try to avoid supplementing a contract by a letter. It's always better to reissue the contract in its entirety or revise it by issuing a formal supplement.

The traffic manager of a company using a contract carrier should continuously examine common carrier rates for comparable services—particularly if his company uses a contract carrier for hauling its own consolidated shipments to a common carrier for break bulk at destination points.

The traffic manager and the contract carrier should work together so that all rates and charges are known and agreed on. Changes necessary for increases in labor or fuel costs are then thoroughly understood and not resented.

For a quick check of whether a contract carrier's rates are reasonable, try to estimate his cash returns per mile or per hour. Such returns are generally calculated by taking the freight payment per load and dividing it by miles run or time taken. The miles run are obtainable from road maps (be sure thay are up to date) and the time taken from estimates of loading and unloading times plus time on the road. This will give a rough estimate of cash return. With experience, an analyst can refine these figures, "weighting" a higher return on short hauls, and so on. If the contract carrier's rates prove to be unreasonable, they should of course be re-negotiated.

The simplest contracts to negotiate are those that use hourly rates only. But hourly rates are not incentive rates for the contract carrier since he is being paid for all the time his men and trucks are away from the garage. Hourly rates also place the burden of efficient dispatching on the shipper.

Tonnage rates or flat-charge rates, on the other hand, are incentive rates because the more efficiently the contract carrier operates, the more profit he makes. This type of rate sets a fixed charge for whatever services are specified—so much per load, per hundred pounds, per gallon or any other unit of measurement that an industry uses. Flat charge rates may be preferable to cover delivery of items that do not lend themselves to tonnage rates. Tonnage rates and flat rates can be more realistically related to a contract carrier's costs than they can to a common carrier's costs.

If hourly rates are to be considered, however, the rates used should reflect the prevailing scale for the locality where the work is to be performed. In setting up a scale of hourly charges, the rate for the vehicle should be considered separately from the rate for the driver. The reason for this is overtime; time-and-one-half or double-time should be applied only to the driver's rate and not to the truck. Where possible, the overhead—office and administrative costs, etc.—should be computed only on the basic labor rate and not on overtime, on the theory that the contractor's overhead is no greater for overtime than for straight time. Computed according to the above, assuming that overhead is 50 percent and using hypothetical figures, the rates for vehicle and driver at $30 per hour, comprising $10 for the vehicle and $20 for the driver would be as follows:

	Regular Time	Time and one-half	Double Time
Vehicle	$10.00	$10.00	$10.00
Labor	$20.00	$30.00	$40.00
Total Charge	$30.00	$40.00	$50.00

Both shipper and carrier share these economies—in lower costs to the shipper and greater profits to the trucker.

If the shipper finds that he has an occasional movement beyond the scope of his contractor's permit, he should contact the Contract Carriers Conference (1616 P St., N.W., Washington, D.C. 20036) for the name of a contract carrier who has appropriate authority to cover the unusual move.

To sum up: The economics in contract carriage are achieved through:

—High utilization of the trucks through complementary front- and backhauls

—Eliminating the terminal operations of the carrier with no loss of efficiency

—Proper ratio of tractors to trailers to allow for loading and unloading of trailers while tractors are on the road

—Loading to full cube on weight allowance

—Elimination of waiting time for loaded trailers.

Contract carriage is not limited to motor carriers. It is also available, in limited applications, on the railroads.

11

EXPEDITING AND TRACING

Expediting in transportation is basically a systematic attempt to hurry a shipment on to its destination so that it arrives faster than it would in the normal course of events. As a transportation technique, it seems to have developed during World War I when there was a need to increase the speed of cars carrying essential war materials and food as they moved through congested rail yards.

Tracing is an after-the-fact attempt to follow the movements of a shipment as it was handled by the carrier or carriers. Usually, tracing is prompted by a complaint that the shipment has not been delivered.

Both expediting and tracing are normally part of the routine operations of a traffic department, although some company managements will allow other departments to attempt their own tracing. This is almost always the most inefficient way to try to find a missing shipment because the traffic department has the reference documents and knows whom to call in the carrier's office for the required information. Then, too, there is the consignee who will periodically maintain that a shipment has not been delivered in order to avoid paying the invoice for it. The traffic department's records should quickly reveal the consignee who is making a habit of losing shipments that tracing shows to have been delivered on time. Constant tracing is costly to both carrier and shipper and is generally an indication that something is going wrong. If a shipper has to assign someone full time to tracing, he would be advised to look at his carrier's service record and consider changing to a carrier that may offer a better performance.

In any event, the traffic department should keep such service records for each carrier it deals with, showing the percentage of on-time deliveries, how often tracing is needed, the time the carrier takes to respond to a tracing inquiry and other useful information. The record should also show the name of the person contacted in the carrier's office when the tracing inquiry was made, and the time and date. The record need not be elaborate—a multi-column pad of the kind available in most stationary stores will do.

A simple but effective method of keeping track of a carrier's on-time delivery record is the postage-paid card. The shipper keeps a supply of preprinted pastage-paid cards with blank spaces for shipment date, quantity and item shipped, and check boxes for the consignee to fill in to show the date the shipment arrived and its condition. The shipper simply fills in the shipment date, quantity, etc., and mails the card at the same time the shipment is dispatched. The responses will provide a useful record of how well the carrier is performing.

The basic information that must be relayed to the carrier when tracing a shipment is the shipper's and consignee's names and addresses, the date of shipment, the commodity shipped, the number of packages and total weight and, of course, a phone number for the reply. This can all be read off from a copy of the bill of lading. Some additional useful items that will speed the search are the freight or waybill numbers, trailer or car numbers, the routing, the rail junctions points and the name of the delivering carrier if the shipment was interlined.

Almost all carriers now use computers to keep their own records so a tracing inquiry should bring a response within a few hours and certainly within a maximum of twenty-four hours. If a shipment cannot be traced, the shipper should quickly file a claim for a lost shipment. Too often, a request that a tracer be put on a shipment brings only a limited search whereas a claim for a lost shipment will bring about a search of the carrier's entire system. This could result in finding the missing shipment in a terminal or rail yard unrelated to the destination shown in its shipping documents.

But as we hinted earlier, the use of tracing can be abused.

EXPEDITING AND TRACING

Someone in sales or purchasing or production may decide to by-pass the traffic department in his anxiety to find out if a particular shipment was delivered to a customer or if long-awaited materials are ready for pick-up. The best way of dealing with this situation is an informal educational program for those other departments in the total costs of transportation. This might well begin with a memorandum pointing out that freight rates reflect not only the costs of transportation operations but administrative costs as well, and that an abuse of tracing increases those administrative costs and hence increases freight rates. The memorandum could go on to say that communications between the company and carriers should go through the traffic department which has the know-how and the records for dealing with the problem.

Whereas tracing is something that happens after the shipment should have been delivered, expediting is something that must be done before the shipment leaves the dock, or at least before it arrives at junction or transfer points. The carrier who has been asked to expedite a shipment can then make his arrangements and pass the word down the line so that personnel and systems are ready.

Some corporations today, particularly in automobile manufacturing, work on such a close scheduling of inbound freight that the transportation system is in fact an extension of the production line. These corporations have expediting departments that track their shipments through the carrier's system (principally railroads) and expedite them by contacting the carrier's operating people at each junction, rail yard or relay terminal. The computer tie-ins that these shippers have with their carriers make the whole process so much easier.

As with tracing, effective expediting requires that the shipper supply the carrier with as much relevant information about the shipment as he can including the date of shipment, name and address of shipper and consignee, commodity, number of packages, weight, car or trailer numbers, freight bill or waybill numbers and the routing.

Expediting, too, may be abused when a carrier is bombarded with unreasonable requests from different departments within a company to rush shipments forward. Too much expediting does more harm than too little. Other shipments may be delayed while the carrier is attempting to handle a shower of requests for expediting. When the carrier is to be asked to expedite a shipment, the request should come from the traffic department which has the shipping documents and the contacts within the carrier's office needed to deal with any problems.

Many companies will assign the junior men in the traffic department to the expediting and tracing jobs. But do not be misled into assuming that anyone can handle expediting without first gaining an understanding of how the various carriers move their freight.

Whether the expediting and tracing job is handled through a centralized traffic department or by local shipping departments in plants or distribution centers will depend on corporate philosophy. As most questions and problems will come up at the local level, handling them on the spot will save time. But records should be centralized in the corporate traffic department so that the carriers' overall performance records can be evaluated.

Whatever tracing and expediting practices a company adopts, it is essential that the people doing the job know the freight handling procedures of railroad, truck, air freight and water carriers and have a clear cut picture of the workings of freight classification yards, local switching systems, pickup and delivery services, piggyback plans, diversions, reconsignments and, just as importantly, of the offices that control the actual moving of freight. The rest of this chapter summarizes the freight handling procedures and practices of the various carriers and offers some suggestions on how to expedite urgent shipments.

Carload Shipments

Carload shipments are loaded by the consignor at his private siding or at a public team track. The car is then sealed, the bill of lading is delivered to the office of the local railroad freight agent.

The car is now ready to be "switched" or moved to a local classification yard. Cars are usually switched at scheduled times. At

EXPEDITING AND TRACING

the classification yard, the car is positioned on a track with other cars going in the same direction so that together they may be made up into a train.

The classification yard in fact consists of a great many such tracks, each designated for a different scheduled train. The classification yards of the major railroads today are electronically controlled. Designed to be operated economically and with maximum automation, they are equipped with television-like scanners that read the "labels" on freight cars and transmit information on the car's contents, destination, etc., to the central control booth and computerized car reporting systems which receive and transmit data on the makeup of trains.

The train also has a symbol reference—a number or name identifying it as a regularly scheduled freight train. When the train is made up, it moves out on its schedule and proceeds to the destination classification yard where it is broken up and each car put through the classification process again or switched to a local yard for eventual spotting at the consignee's siding or at a local public team track. A yardmaster is in charge of each yard. Moving cars to siding or team track is referred to as a "drill."

Ordinarily the first steps in expediting and tracing are taken with the origin carrier at its local office or with its local agent. There is no hard and fast rule as to this, however, and on occasion expediting may involve a call right to the yardmaster's office to make sure a car will move in the first available symbol train. A carload shipment may be traced by getting records of movements from the car service or car record office; these offices are maintained by every railroad on a system-wide basis. In most instances this information is now made available by means of electronic data processing.

At times a special switch for a car may be arranged with the yardmaster. Most symbol trains between large cities go all the way through without any reclassification en route. Most delay is experienced in the local yards. If a car goes "bad order" en route, it is taken out of the through train and sent to the closest "rip track" for light repairs. Cars in need of general repair are usually unloaded and sent to the car shops; when this happens the railroad should advise the shipper. The expediter's job then is to point out the need for quick repairs or transfer to another car, obtain the new car number and ensure that the car is included in the next through train.

Truckload Shipments

Once a truckload of freight has been loaded into a trailer or straight truck, it rarely will be transferred to other trucks or trailers en route. Most truckloads are handled in tractor-trailer units. At times tractors will be changed at division points, but the trailerload will go through from the consignor to the consignee. Even when it is a two-line haul, the original trailer will probably go through to destination (many motor carriers now interchange trailers). If a trailer goes "bad order," the load will be changed to another trailer, if one is available. Otherwise, the movement is delayed until the trailer is repaired. Most established trucking companies maintain repair and reporting stations for service en route, thereby practically eliminating costly delays in transit.

To expedite a truckload, get the trailer and tractor number, driver's name and anticipated arrival time at destination. This information should be wired or telephoned to the consignee. In emergencies it might be well to have the trucking company instruct the driver to telephone the consignee or consignor if any unexpected delay is encountered in transit. The numbers can be included in instructions on the bill of lading. With such information, the shipper or consignee can call the local representatives or agent of the trucking company to correct the difficulty and rush delivery.

Some motor carriers use their sales departments to expedite and trace, while others use their operating departments. Still others have organized customer service departments to handle these problems. The first move when either expediting or tracing is to contact one or the other of these departments of the carrier picking up the shipment for the first line haul of the movement. When local drayage carriers are involved, it is best to go to the line haul carrier which usually has complete information and is thus in a better position to help.

EXPEDITING AND TRACING

In expediting a truckload be sure, by checking with the originating truck terminal, that the load is actually on its way. It's disturbing to find out the next morning that a trailer load is still in town because of the lack of a driver, mechanical trouble, or for some other reason.

Piggyback Shipments

One of the benefits of piggyback services is that generally less expediting and tracing are required. For the most part, piggyback shipments travel in trains that are operating on precise, fast and through schedules. The matters of concern to the expediter are getting the trailer to the piggyback yard in time for loading into a scheduled train, making certain that a tractor is available to haul the trailer from the railroad premises at the end of the rail journey, and that the trailer will be promptly unloaded at destination. On those occasions when it is necessary to expedite or trace, the procedure will depend to a great extent upon the piggyback plan used.

Neither the shipper or consignee is likely to know if his trailerload travels under Plan I, since use of this plan is the motor carrier's prerogative. Such shipments would be expedited or traced under truckload procedures and with the trucking company. Under Plan II, the trailer is in the hands of the railroad from shipper's loading dock to destination; hence the procedure for carloads would be used.

Under Plans II½, III and IV, hauling the trailer from shipper's loading platform to the railroad loading ramp and from the railroad ramp at destination to the consignee's dock depends on the arrangements made by the shipper or consignee, or both, and is therefore under their control at all times. Once loaded on a flat car the number of that car is ascertained and the regular procedure for expediting or tracing carloads is followed as to the rail portion of the journey. Plan V involves the use of motor common carrier at either or both ends of the trip and requires the use of both truckload and carload procedures for the respective portions of the haul.

The above are basic guidelines for expediting and tracing

when shipping piggyback. It is important, in piggyback plans II½, III and IV, to note that local cartage companies are generally used for pickup and/or delivery and their record keeping may not be as detailed and readily available as the records of the railroads and motor carriers. The railroad's records on piggyback will begin with delivery of the trailer at the railroad ramp and end with delivery of the trailer to the cartage agent at the destination railroad ramp. Thus the car number of the flat car used and the identification number of the trailer are important and must be part of the record for tracing or expediting.

Less than Truckload Shipments

The same general procedure outlined for truckloads is to be used for less-than-truckload shipments. Less than truckloads may be transferred en route so it is important to have the carrier's waybill number and the point where the load will be transferred. Then the agent at the transfer point should be contacted to ensure a quick forwarding on the first available outbound trailer load. At times trucking companies will drop lots off en route at locations where they do not maintain a terminal or agent. Be sure that such a lot is loaded last, near the tailgate; otherwise it cannot be dropped off without a great deal of trouble. The lot to be delivered last should be loaded in the nose of the trailer to avoid rehandling and delay.

Nothing should be left to chance or the feeling that the carrier "will do it anyway." If the drop-off en route is part of a stop-off arrangement, the bill of lading should clearly identify what is to be dropped off first, second, third, and so on, and how many packages of what commodity are consigned to each stop and where they are loaded in the vehicle.

The greatest potential delay to less-than-truckload shipments is at motor carrier terminals. A shipment may be picked up promptly but not forwarded the same night. It may get to the destination terminal in good time but delivery is delayed because the trailer has been left standing in the yard instead of being backed up to the platform and unloaded or because the streets round the consignee's

EXPEDITING AND TRACING

dock are congested. Therefore it is necessary to know where the shipment is at all times so that intelligent action can be taken to speed up delivery. Generally, motor carriers provide good service. Many operate with fairly small terminals making it impossible to accumulate much freight on their platforms. They have to "keep it moving."

Freight Forwarder Shipments

Freight forwarders operate in much the same way as shippers, routing their freight in consolidated carloads, piggyback trailers or truckloads via carriers of their selection. Freight forwarders do their own expediting and tracing although, on occasion, an interested shipper or receiver may find it to his advantage to aid in this work.

Usually the practice is to telephone the forwarder's own reference number to the forwarder with a request for expediting or tracing—expected and actual arrival times of the shipment, and so on. The forwarders assign their own reference number in these matters which is known as "bill-of-lading load reference." They also assign a pro or freight bill number to each freight bill. The tonnages they assign and the day-in, day-out routines forwarders establish are such that they can readily trace and expedite. It is rare for forwarders to need help from industrial transportation departments. When they do, special reasons and special situations exist that require particular attention.

Forwarders generally employ a manifold (or equivalent) system of billing, whereby the freight bill, the arrival notice, and the delivery receipts are made up in one billing operation. The forwarding office mails this billing immediately to the receiving office at destination, which means the destination agent knows the contents of a shipment prior to its arrival. Sometimes the billing is sent with the shipment if the shipment is in a through trailer.

Air Freight Shipments

Air freight may be loaded in the belly compartments of passenger planes on regular scheduled flights or it may fly in all-cargo planes that are also operated on a schedule. Shipments may be made airport-to-airport or transported to or from the airport in pickup and delivery service made available by the air carrier. Airlines, too, handle freight through to destination or turn it over to connecting airlines for delivery.

Tracing and expediting air freight must be fitted into this pattern. The chief air freight agent of the carrier at the airport of origin is the prime contact and he should be given full information on the shipment. Air waybills and flight numbers are important and should be provided, or enough information furnished so that they can be ascertained. The agent forwards the air bill and flight numbers to the downline stations to locate the shipment after verifying that it has left his station.

The shipper should give the air carrier as much notice as possible to ensure that space will be available on the plane for a particular flight. If an unusually large or odd shaped piece of equipment is to be shipped, the airline should be told the dimensions and particulars well ahead of time. Since speed is usually the main motive for using air freight, it's a good procedure to phone the flight number, expected time of arrival, etc., to the consignee after the flight has departed. Most air freight still moves "first flight with space available," although some types of package freight are "guaranteed next flight out."

In addition to their regular air freight services, the airlines now offer a variety of express small package (on average less than 50 pounds) and courier type services. Generally the airlines set size, weight and value restrictions on the packages and will not accept dangerous or perishable items. Some services are airport-to-airport, others include pickup and delivery. With speed and premium prices as their outstanding characteristics, such services ought not to have much call for expediting or tracing.

Maritime Shipments

Expediting and tracing maritime shipments is fairly simple. Once a shipment is loaded into the vessel, there is nothing that can be done until the vessel arrives at its destination port. Therefore all efforts should be concentrated on getting the shipment to the forwarding pier in time to be loaded into the ship and then seeing to it that it actually goes aboard. Many a shipment has been left on the dock "shut out" by the steamship line because there is more freight on hand than can be loaded. At destination the job is to get it discharged and placed in an accessible spot for local pickup or loaded into a car, truck or another ship, as the case may be, for its movement onwards.

Each steamship company prepares a ship's manifest in which is listed all the shipments loaded into that ship. Thus there is always available immediate information as to whether the shipment in question is aboard a designated ship.

Parcel Post and Mail

There is no satisfactory way to trace and expedite parcel post and mail. Once a package or a letter is deposited with the post office, the sender loses control or contact until delivery is made at destination. Using air mail services, "special delivery," or "special handling," and posting at the right time and place may help. But if a letter or package is undelivered after a reasonable time has elapsed, there is little that can be done (unless it has been registered or insured). Vocal and persistent complaint to postal officials, however, may result in an improvement in service, if not recovery of the individual item.

The U.S. Postal Service does have one express service for letters and small parcels. It costs extra and the service is available only to or from selected post offices, with guarantees of delivery within specified time limits. If the U.S.P.S. fails to deliver within that time, the shipper is entitled to the return of the postage paid upon application at the origin post office.

Small Package Services

With the demise of REA, the movement of small packages by United Parcel Service and the many regional small shipment companies has increased. These types of carriers rarely use a bill of lading, making do, instead, with a receipt form provided by the carrier.

As with the air carrier package services mentioned earlier, expediting is unnecessary with these companies since they are noted for fast operations and terminals that seldom provide for storing much freight. Tracing, however, may be called for at times, and should begin by contacting the proper individual at the carrier with shipment date, origin and destination information. If the receipt has an identifying number or symbols those, too, should be given.

General—All Services

It is important to keep a good record of persons contacted and date, time, etc., together with the information received, when expediting or tracing. This can best be done by using an expediting and tracing form of your own design. As we suggested earlier, a simple sheet ruled in columns is perfectly adequate. One important precondition in expediting is to find out exactly when the shipment is needed at destination. This should be the final on-the-job "must" date. The traffic department, with its knowledge of the various carriers' transit times, should then be able to specify a route that will provide the desired time of delivery at the lowest possible cost. A substantial amount of money can be saved by adhering to this principle.

In dealing with carrier representatives, try to develop a pleasant telephone manner. Expedite only those shipments that really are urgently needed. Try to work it out so that it is always the same employee in each carrier's office who handles all of your expediting and tracing work. This employee will then become familiar with the character of *your* business and traffic people and will often obtain better over-all results than otherwise. Most carriers want to give their customers good service and are willing to go out of their way to give special handling to urgently-needed shipments; but they have no patience whatsoever with the shippers or receivers who cry "wolf" on almost every shipment. Shippers and receivers should give carriers the same courtesies they expect and look for

EXPEDITING AND TRACING

from the carriers; they should ask for expedited service only on shipments that genuinely must be rushed. And most importantly, they should give the carrier complete and accurate shipping data.

Expediting can be done easily or it can be done strenuously. A good expediter will know when to go all out and when to handle gently. The difference lies in your approach to the carrier and how much you, yourself will do to assure quick delivery.

Sometimes you need only to "watch" the shipment; other times you must do an all-out job. There are degrees of expediting—needed—urgent—emergency. Don't be too hard on other organizations if they fail to give you all the information they should furnish. The traffic department is a service organization and must do the best it can with what it has. If you do not have all the information you need, go out and get it.

Perhaps the simplest and most important part of good expediting is to be sure that the shipments get started right, with correct billing and marking. Inadequate information on the bill of lading as to the consignee's name, address, shipment route, etc., may cause serious delays. If the shipment is a carload, be sure the bill of lading reaches the local agent in time to permit billing, and forwarding on the earliest available train. And in other than carload shipments, check the local shipping platform to be certain that the shipment is not left on the platform but does in fact leave via the right carrier, at the right time, and is loaded in the right manner. This is good claim prevention as well as good expediting procedure.

Expediting is sometimes complicated by strikes, embargoes, or other service stoppages. The challenge, of course, is to find a way to move the freight despite such difficulties. In such instances, having good friends among the carriers pays off in real dividends. They will suggest ways and means. Their help, plus your knowledge and determination will produce results!

Diversion and reconsignment warrant some consideration. If you divert a carload shipment, you change its destination while it is en route. The consignee remains the same. If you reconsign a shipment you change the name of the consignee and if necessary, the destination while it is en route. While the terms "diversion" and "reconsignment" are not actually synonymous, the same rules govern both and usually are published in much the same wording in carrier tariffs.

Of course, a shipment can be diverted or reconsigned only if instructions are issued in time enroute. If the car is already delivered to the original bill of lading destination you cannot reconsign it. Your only recourse then is to reship it on a new bill of lading. Also if you order a diversion from New York to Philadelphia on a car coming from Chicago and the last point at which diversion is possible is Harrisburg, you must see that the carrier issues instructions to Harrisburg before the car leaves there. If you do not, then you are too late either to divert or reconsign. The charge for either diversion or reconsignment is not great, but it varies and is dependent on tariff provisions. Diversions or reconsignment may be undertaken only at the request of the owner of the goods; thus, if you buy FOB destination, the seller must be asked to make the desired change.

Expediting and tracing are made necessary at times because shipping documents are lost or the shipment disappears. These kinds of shipments are often referred to as "no-bills" (the shipment turns up with no waybill to tell the carrier to whom it belongs or should be sent). There are variations on this, of course, but, generally speaking, these problems become "over and short freight" to the carrier, and tracing and expediting involve going into its records and what lading it may have on hand.

Over or short freight is freight, with or without "marks," (including articles in excess of quantity on waybill) that is found in possession of a carrier at any point without a regular revenue waybill. This freight is assigned a "free-astray (F/A) waybill" showing all available details of the shipment, so it can move to a proper destination. The freight charges are controlled by the original revenue waybill, so no charges are assessed on over or short freight (hence the term, "free-astray"). If such freight cannot be identified, it is put into storage, there to await identification, generally through claims or by some other means. In any event, it will not be delivered to the indicated owner without first receiving from him some satisfactory "proof of ownership," as well as a statement from

EXPEDITING AND TRACING

the recipient that credit will be extended to the carrier on any future or previous claim for a shipment with which the material can be identified.

Expediting can be avoided with proper departmental planning. It can be avoided, for example, by Purchasing ordering raw materials in time for normal transportation. It can be avoided by Sales, ordering its sales requirements early enough to meet advertising schedules and special sale dates.

It can also be avoided by purchasing in volume so that the freight moves by the carload or trailerload—a double benefit because this not only usually cuts transit times, but also reduces freight charges.

12

TRANSPORTATION SPECIAL SERVICES

Railroads and motor carriers include in their costs of doing business a factor to cover the expense of furnishing the rolling stock and the truck-trailers that carry the freight. If this equipment is held by shippers or consignees beyond a "free time" considered reasonable for loading or unloading that freight, the carriers' tariffs provide for assessing penalty charges. The railroads call these penalties "demurrage" and the motor carriers "detention charges."

The term demurrage originally meant the detaining of a ship beyond its scheduled time of sailing. Today, it is a penalty charge that railroads impose to discourage the use of freight cars as warehouses. A secondary purpose is to compensate for the "per diem" charges that must be paid to the railroad that owns the car by the railroad that is using it (a freight car may spend months or even years on other railroads' lines before it works its way back to the owning line). If a car is not earning revenue, the per diem charge is an out-of-pocket loss that encourages the railroad on whose line that car is sitting to do something with it. Modern freight cars, equipped with shock absorbing underframes and other special equipment, are very expensive and must be worked hard to show a return on investment.

Demurrage rules and charges are published in one master tariff. It is published by H.J. Positano, 2 Pennsylvania Plaza, New York, N.Y. 10001. The last issue of Positano's tariff (PHJ-6004-L) has been in effect since March 1, 1979 and covers 476 United States railroads participating in all or some of the rules and charges. The tariff is divided into three parts: Part I—"General Car Demurrage

Rules and Charges"; Part 2—"Storage Rules and Charges"; Part 3—"Special Car Demurrage Rules and Charges." Every traffic department that ships or receives by rail should have a copy of this tariff.

Demurrage charges are normally settled under one of two plans. The first is straight demurrage; the second involves an average demurrage agreement between the railroad and the company it serves to keep demurrage charges to a minimum. The whole subject of demurrage rules and charges has been receiving some close scrutiny from the ICC.

Under straight demurrage, a shipper or consignee is allowed two "free" days or 48 hours "free time" to load or unload a car after "active or constructive placement" of the car. "Active" placement is simply positioning the car at the point designated by the shipper or receiver for loading or unloading. "Constructive" placement is defined as what happens when a car cannot "actually be placed because of a condition attributable to the consignor or consignee" (perhaps because other cars are positioned at the dock or some such reason). The car is then held by the railroad, and after the consignee/consignor has been advised, it is considered to have been constructively placed. If the railroad leaves the car on a track serving the consignor/consignee (other than a public delivery track) the car is then considered constructively placed *without notice being given.*

Free time under straight demurrage is computed from the first 7 a.m. after placement of a car on a public delivery track (and after notice is sent or given to the consignor or party entitled to receive it). On cars to be delivered to industrial connections or private sidings, free time begins at the first 7 a.m. after actual or constructive placement. The demurrage calendar included here, courtesy the Chessie System, shows a typical method of calculating free time and charges for both loading and unloading cars. Charges are $20 per day for the first four days after free time has expired, including Saturday and Sunday. However, if free time expires *on* a Saturday, Sunday or holiday, then it is continued through those days and charges do not begin until 7 a.m. the next working day.

After four days at $20 a day, the charges increase to $30 a day

TRANSPORTATION SPECIAL SERVICES

Chessie System
TRANSPORTATION DEPARTMENT
BALTIMORE, MARYLAND
DEMURRAGE AND STORAGE DETENTION CALENDAR
(FREIGHT TARIFF PHJ-6004-L)

TO: PATRONS AND AGENTS

1980

THANKS FOR USING CHESSIE

1980

NOTE: ISSUE NOTICES OF ARRIVAL OF FREIGHT OR NOTICES OF CONSTRUCTIVE PLACEMENT WHEN NECESSARY AND MAINTAIN BASIC RECORDS TO SUPPORT SENDING OR GIVING NOTICES AS REQUIRED BY ITEM 1300 OF THE TARIFF.

CAUTION: In computing time and charges on demurrage and carload hazardous storage, include all Saturdays, Sundays, and holidays after the first chargeable day, except where all days are chargeable without free time, as provided in Item 1225 and Item 1405.

JANUARY
SU	MO	TU	WE	TH	FR	SA
		1	2	3	4	5
6	7	8	9	10	11	12
13	14	15	16	17	18	19
20	21	22	23	24	25	26
27	28	29	30	31		

FEBRUARY
SU	MO	TU	WE	TH	FR	SA
					1	2
3	4	5	6	7	8	9
10	11	12	13	14	15	16
17	18	19	20	21	22	23
24	25	26	27	28	29	

MARCH
SU	MO	TU	WE	TH	FR	SA
						1
2	3	4	5	6	7	8
9	10	11	12	13	14	15
16	17	18	19	20	21	22
23	24	25	26	27	28	29
30	31					

APRIL
SU	MO	TU	WE	TH	FR	SA
		1	2	3	4	5
6	7	8	9	10	11	12
13	14	15	16	17	18	19
20	21	22	23	24	25	26
27	28	29	30			

MAY
SU	MO	TU	WE	TH	FR	SA
				1	2	3
4	5	6	7	8	9	10
11	12	13	14	15	16	17
18	19	20	21	22	23	24
25	26	27	28	29	30	31

JUNE
SU	MO	TU	WE	TH	FR	SA
1	2	3	4	5	6	7
8	9	10	11	12	13	14
15	16	17	18	19	20	21
22	23	24	25	26	27	28
29	30					

JULY
SU	MO	TU	WE	TH	FR	SA
		1	2	3	4	5
6	7	8	9	10	11	12
13	14	15	16	17	18	19
20	21	22	23	24	25	26
27	28	29	30	31		

AUGUST
SU	MO	TU	WE	TH	FR	SA
					1	2
3	4	5	6	7	8	9
10	11	12	13	14	15	16
17	18	19	20	21	22	23
24	25	26	27	28	29	30
31						

SEPTEMBER
SU	MO	TU	WE	TH	FR	SA
	1	2	3	4	5	6
7	8	9	10	11	12	13
14	15	16	17	18	19	20
21	22	23	24	25	26	27
28	29	30				

OCTOBER
SU	MO	TU	WE	TH	FR	SA
			1	2	3	4
5	6	7	8	9	10	11
12	13	14	15	16	17	18
19	20	21	22	23	24	25
26	27	28	29	30	31	

NOVEMBER
SU	MO	TU	WE	TH	FR	SA
						1
2	3	4	5	6	7	8
9	10	11	12	13	14	15
16	17	18	19	20	21	22
23	24	25	26	27	28	29
30						

DECEMBER
SU	MO	TU	WE	TH	FR	SA
	1	2	3	4	5	6
7	8	9	10	11	12	13
14	15	16	17	18	19	20
21	22	23	24	25	26	27
28	29	30	31			

☐ These days are to be excluded in computing detention, except they become chargeable days immediately following the day on which the first chargeable day begins to run. Saturdays, Sundays and holidays, however, are NOT excluded in computing detention on cars for which no free time is allowed (see Item 1225 — empties ordered for loading and not used in transportation service, and Item 1405 — strike interference.)

SAT. / SUN. / *HOL.

(over)

INDUSTRIAL TRAFFIC MANAGEMENT

Part 1 of Tariff PHJ–600A–L
This Chart is Based on Tariff Rules in Effect at Time of Issuance

CHESSIE SYSTEM

Chessie System

NOTE: Chart for Computation of Demurrage on Cars Subject to 24 Hours or 48 Hours Free Time
Under this chart detention is computed on cars actually or constructively placed after 7:00 A.M. In the event cars are tendered at or before 7:00 A.M. then the first free day will begin at 7:00 A.M. of date of tender.

24 HOURS FOR OUTBOUND LOADING; ALSO ON CARS STOPPED IN TRANSIT TO COMPLETE LOADING OR FOR PARTIAL UNLOADING
(SEE NOTE UNDER ITEM 2 BELOW)

(Read Down For Computation of Charges)

	AP	AP	AP	AP
	F	F	F	F
Monday	$20			
Tuesday	20	20		
Wednesday	20	20		
Thursday	20	20	20	AP
Friday	20	20	20	x
Saturday	20	20	20	x
Sunday	30	20	20	x
Monday (Hol.)	30	20	20	F
Tuesday	60	30	20	20
Wednesday	60	60	30	20
Thursday	60	60	60	20
Friday	60	60	60	20
Saturday	60	60	60	30
Sunday	60	60	60	30

KEY: AP = Actual or Constructive Placement
F = Free
X = Excluded
$20 day equals one (1) debit under average agreement.

48 HOURS TO COMPLETE INBOUND UNLOADING

(Read Down For Computation of Charges)

	AP	AP	AP	AP
	F	F	F	F
Monday				
Tuesday				
Wednesday				
Thursday	$20			
Friday	20			
Saturday	20	20		
Sunday	20	20		
Monday (Hol.)	30	20		AP
Tuesday	30	20	20	x
Wednesday	60	30	20	x
Thursday	60	60	20	x
Friday	60	60	20	F
Saturday	60	60	30	20
Sunday	60	60	30	20

1. Under the Average Demurrage Agreement, debit days are $20.00 days. — These can be offset by credits earned on other cars on a one-for-one basis. A credit is earned whenever a car is loaded or unloaded before expiration of first 24 hours of free time. Credits earned on inbound cars cannot be used to offset debits accruing on outbound cars and vice versa. Intraplant cars are only allowed 24 hours free time for loading or unloading. Intraplant cars are excluded from average agreement provisions and must be reported under straight demurrage rules.

   The $30.00 and $60.00 days are arbitrary days and these cannot be offset by credits.

2. Demurrage charges accrue for all Saturdays, Sundays and holidays following the first $20.00 or debit day. (See Part 3, Tariff PHJ–600A–L for Exceptions)
   NOTE: In computing detention on cars stopped in transit to complete loading or for partial unloading, Saturdays, Sundays and holidays are chargeable days, immediately after expiration of the 24-hour free time period.

3. A private car on a private track, when the ownership of the car and the track is the same, is exempt from demurrage. Lease of certain private cars is equivalent to ownership when evidence of lease is shown as stipulated in Item 765. Private cars cannot earn credits.

4. Part 3 of the demurrage tariff outlines rules and charges applicable when specifically referred to in the various incentive freight rate tariffs. These items generally restrict the free time to 24 hours and prohibit inclusion of detention in an average demurrage agreement, and charges for Saturdays, Sundays and holidays apply immediately following the last day of free time or, after expiration of the 24-hour free time period, whichever may be applicable.

This calendar and chart is not a tariff and the information contained herein is subject to change without notice by tariff revision or issuance of Interstate Commerce Commission Service Orders and is applicable only within the United States.

The tariff should be consulted at all times.

(over)

for the next two days and continue thereafter at the rate of $60 a day.

Demurrage is also assessed on cars held for orders, bills of lading, payment of freight charges, reconsignment, diversion, reshipment, inspection or forwarding instructions. Charges for demurrage are made for complete days; there is no pro-rating—fractions of a day are charged as complete days.

Demurrage is not charged on privately-owned or leased cars held on private tracks when both the car and the track belong to the same firm. This exemption includes cars leased to the owner of the track, but the cars must be placarded or stencilled to show that they are exempt.

There are a number of situations in which the shipper or receiver may recover demurrage charges. They include

—Weather so bad that the car cannot be loaded or unloaded safely or without damaging the lading.

—Frozen or congealed lading necessitating thawing before it can be removed from the car.

—The presence of floods, earthquakes, hurricanes or tornadoes, including the damage done to loading or unloading facilities by these natural disasters.

—"Bunching," a result of when cars shipped from the same origin by the same route but on different days are offered for delivery by the railroad in a "bunch" that exceeds the capacity of the unloading dock.

—Delayed or improper notice by the railroad of the arrival of the cars.

—Error by any railroad party to the bill of lading contract that prevents proper tender or delivery of the cars—for example, the railroad, without any instructions from the consignee, makes "actual placement" of "constructively placed" cars, thus muddling up the arrival sequence (known in the trade as "run around").

—Strikes at the shipper's or consignee's premises that prevent the loading or unloading of cars.

In such cases, the regular demurrage charges must first be paid in full and then a claim filed against the railroad, within the various time limits prescribed in the demurrage tariffs, for a refund. In the case of strike claims a special charge of $7 per day per car, without free time allowance, is applicable.

Average Demurrage Agreement

As mentioned earlier, an average demurrage agreement is an agreement between a railroad and a customer company that has advantages for both parties. Such an agreement sets up a scale of incentives for the shipper or receiver to load or unload rail cars quickly, thus minimizing demurrage charges for the company and offering the railroad the prospect of better utilization of its car fleet. Any firm with a steady volume of inbound and outbound carload shipments should consider its benefits.

Basically, an average demurrage agreement establishes a system of debits and credits. Typically, one credit is allowed for each car released to the railroad within the first 24 hours of free time and one debit is charged for each car held beyond the full 48 hours free time. After a car has accumulated four debits—i.e., has been held for four days—there is a charge for the next two days, and then there is a higher charge for all following demurrage days. These charges vary from railroad to railroad and may not be offset by credits. Nor may credits gained on unloading cars be used to offset debits charged on loading cars, or vice versa. At the end of each calendar month the total number of credits is deducted from the total number of debits and a charge is made for each debit remaining. Any surplus credits are cancelled and may not be used to offset debits in another month. A copy of a typical agreement is given here.

A company cannot earn credits on its privately-owned or leased cars when these cars are placed on its own tracks. But debits on private cars constructively placed can be offset by credits on other private cars.

Separate average demurrage agreements must be entered into with each railroad serving a given industry at any one station. Generally, it has been held that no agreement may embrace more

TRANSPORTATION SPECIAL SERVICES

TARIFF PHJ 6004-L

PART 1 - GENERAL CAR DEMURRAGE RULES AND CHARGES - Continued

SECTION 800 - Continued

ITEM 825	**SECURITY**	
	A party who enters into this average agreement may be required to give sufficient security to this railroad for payment of balance against him at the end of each month.	
ITEM 830	**COMBINING PLANTS AND STATIONS**	
	Subject to the requirement of Item 820 debits and credits applying to cars released by one consignor or consignee within the jurisdiction of the same station and served by one and the same railroad must be combined provided that in no case shall debits and credits be combined among two or more customers or are released under the jurisdiction of two or more stations, nor can debits and credits be combined when released on different railroads either at the same or at different stations, except where separate average agreement is requested in writing for each plant of the consignor or consignee in which event each plant will be considered as one consignee or consignor for the purpose of applying this Section. Each station as listed in The Official List of Open and Prepay Stations ICC OPSL 6000, supplements thereto, or successive issues thereof, shall be considered a separate station in the application of this Section.	
ITEM 835	**CARS CONSIGNED TO PUBLIC FIRMS**	
	Cars consigned, reconsigned, or ordered to a public elevator, public warehouse, cotton compress, processing or fabricating plant, serving various parties, shall be combined in one average agreement, and the party signing the agreement as principal shall assume responsibility for all demurrage as assessable thereunder.	
ITEM 840	**CARS CONSIGNED TO GOVERNMENT AGENCIES**	
	Cars consigned to the United States, state or municipal governments, may be included in the account of the construction contractor, the operation of a government agency, or any industrial plant to whom the cars are ordered delivered for unloading.	
ITEM 845	**PRIVATE CARS**	
	Credits cannot be earned on private cars, subject to Item 765, but debits accruing on private cars while held under constructive placement for delivery on private tracks or for other disposition may be offset by credits earned on other cars.	
ITEM 850	**DEMURRAGE AVERAGE AGREEMENT**	
	A demurrage average agreement will be made effective on the first day of the month following receipt of patron's application, except when the credit status of the applicant is not satisfactory to this railroad at that time, the agreement shall be made effective on the first day of the month following approval by this railroad of credit arrangement for the applicant to operate under an average agreement. When the applicant desires to have an agreement made effective prior to the first of the following month, it may be made effective on the date of receipt of the application provided his credit status is then satisfactory; otherwise, on the date this railroad approved credit arrangement; provided, however, that no cars have been released the same month prior to the date the agreement is made effective. The following agreement shall be required for all applicants to operate under the average agreement.	

AGREEMENT

.... Railroad.

Being fully acquainted with the terms, conditions and effect of the average basis for settling for detention of cars, as set forth in .... being the car demurrage rules governing at all stations and sidings on the lines of said railroad, except as shown in said tariff, and being desirous of availing (myself or ourselves) of this alternate method of settlement (I or We) do expressly agree to and with the .... Railroad that with respect to all cars which may, during the continuance of this agreement, be handled for (my or

(Item concluded on next page)

For explanation of reference marks see concluding pages of tariff.

TARIFF PHJ 6004-L

PART 1 - GENERAL CAR DEMURRAGE RULES AND CHARGES - Continued

SECTION 800 - Concluded

ITEM 850 - Concluded

our) account at .... (Station) (I or We) will fully observe and comply with all terms and conditions of said rules as they are now published, or may hereafter be lawfully modified by duly published tariffs and will make prompt payment of all demurrage charges accruing thereunder in accordance with the average basis, as therein established or as hereafter lawfully modified by duly published tariffs.

This agreement to be effective on and after .... day of ...., 19.........., and to continue until termination, by written notice from either party to the other which shall become effective on the first day of the month succeeding that in which it is given, except that for any failure or refusal to pay charges lawfully accruing under this agreement, it may be terminated as of the date of written notice of termination.

Approved and accepted ...., 19 ...., by and on behalf of the above named railroad by ....

ITEM 855   EXCEPTIONS TO SECTION 800

1. An average agreement may not include cars held for:

    (a) purposes other than loading and unloading.

  (34) (b) cars held for loading or unloading in intraplant switch service.

2. Cars subject to the average agreement will not be allowed adjustments provided for in Item 1410 and Paragraphs 1 and 2 of Item 1420, except when bunching has been caused by floods, earthquakes, hurricanes or tornadoes and conditions in the devastated area resulting therefrom, or strikes of railroad's employees, are subsequently delivered to consignee in accumulated numbers.

SECTION 900

ITEM 900   DEMURRAGE CHARGES ON CARS NOT SUBJECT TO AVERAGE AGREEMENT

On cars not subject to average agreement Section 800 and for detention not subject to Item 1405 (Strike Interference), after expiration of free time allowed or without free time allowance, when none is provided, the following charges per car per day, or fraction of a day, will be made until car is released.

(11) (70) $10.00 for each of the first four chargeable days,
(11) (70) $20.00 for each of the next two days,
(11) (70) $30.00 for each subsequent day.

(11) (71) $20.00 for each of the first four chargeable days,
(11) (71) $30.00 for each of the next two days,
(11) (71) $60.00 for each subsequent day.

    (10) $5.00 for each of the first four chargeable days.
    (10) $10.00 for each of the next four days.
    (10) $15.00 for each subsequent day.

(57) The applicable charge will accrue on all Saturdays, Sundays and holidays subsequent to the second chargeable day, including a Saturday, Sunday or holiday immediately following the day on which the second chargeable day begins to run, except as otherwise provided in Item 1225 or Section 1400.

(56) The applicable charge will accrue on all Saturdays, Sundays and holidays subsequent to the first chargeable day, including a Saturday, Sunday or holiday immediately following the day on which the first chargeable day begins to run, except as otherwise provided in Item 1225 or Section 1400.

For explanation of reference marks see concluding pages of tariff.

TRANSPORTATION SPECIAL SERVICES

than one station nor more than one industry but no specific ruling prohibits this last possibility.

Records showing the date and time of receipt and release of all cars should be carefully maintained when operating under an average demurrage agreement. These car records should be kept on a form designed to include all desired information but especially the following:

Car Initial Number	Placement Time Date	Date Released	Credits	Debits (max 4)	$ Days	$ Days	Non-offset Days

This form should have space for summaries and preferably separate forms should be used to list inbound and outbound cars (as the credits on one cannot be used to offset the debits on the other). Some carriers have computerized their demurrage records and in advanced systems the shipper's computer can "talk" to the carrier's computer.

While, by and large, average demurrage agreements represent good deals for industrial shippers and their carriers, changes in distribution patterns may require that an agreement be terminated. Some companies make periodic "call-ups" on these agreements because of this. An experienced demurrage man can look at a statement of demurrage charges and decide whether a more detailed study should be made.

Those who prepare demurrage records must have a clear understanding of the application of the demurrage tariff; otherwise it will be impossible to audit bills received from the railroad. Also, such records should be subjected to constant analysis as an important factor in any program of demurrage control. In any case, traffic managers operating average demurrage agreements should pay particular attention to items 25, 166, 168, 174, 380 to 402, 530, 535, 805, 810, 820, 830, 835, 846, 850, 855, 1415, 1435 of the demurrage rules included in Positano's tariff mentioned earlier.

Demurrage control is essentially the loading and unloading of cars within the free time or offset period so there will be no demurrage bill to pay at the end of any calendar month. This cannot be done unless considerable planning goes into the ordering of empty cars for loading and the scheduling of inbound shipments in a manner that will prevent the arrival of too many cars at the same time. Plans must take into account the car capacity of the sidetrack, the frequency of available switching service and the time or times of day at which the switching can be performed. This may be difficult in a busy plant where there is an urgent need for materials or where shipping and receiving facilities have important requirements and problems. Coordination of all departments is essential, however, and the traffic manager in cooperation with the plant manager can demonstrate his worth by setting up an arrangement that will be workable by all and still minimize demurrage costs.

High bills for demurrage month after month may indicate that unloading facilities are inadequate for the volume of business being done, in which case enlargement or rearrangement of these facilities may be justified ecnomically. For example, if the length of the building permits, a two-car siding might be enlarged to accommodate four cars; or possibly space would permit the extension of the loading platform along an existing siding. Very often the advice and counsel of the traffic manager to plant construction engineers in the initial planning of a plant's shipping and receiving arrangements will avoid costly remodeling of these facilities after the plant is in operation.

Excessive demurrage charges also may arise from the lack or inadequacy of materials handling equipment. Fork lift trucks, portable conveyor systems and palletizing among other things, help to load and unload cars in a fraction of the time formerly necessary. This is particularly true of small-package freight. Savings in demurrage charges may contribute heavily to the justification of improved materials handling systems. Demurrage charges represent a nonproductive and unnecessary expense and indicate help is needed in over-all plant management.

TRANSPORTATION SPECIAL SERVICES

Motor Carrier Detention Charges

Detention is to the motor carrier what demurrage is to the railroad, a penalty for delay in loading or unloading vehicles.

On August 20, 1978 the decision of the ICC in Ex Parte No. MC-88 establishing uniform detention regulations for motor common carriers went into effect. The regulations provide that a motor carrier must publish in its tariff uniform rules for detention of trailers with power units and without power units. Generally exempt from these regulations are carriers of household goods, new furniture, bulk and oversize loads. Contract carriers may be exempt if this is provided for in their contract.

Among other things, the regulations require that the carriers must keep a record of each movement whether or not detention time occurs. The driver of the vehicle must inform the consignor or consignee that the truck is available for work, at which point timing starts.

As with demurrage, free time is allowed and this varies depending on weight of the load. Time is computed during "normal" business hours even though dock-open hours may be much more limited. Thus if a shipper's normal hours are 9 a.m. to 5 p.m. and its docks are closed from 12 noon to 2 p.m. and a truck arrives at 1 p.m., free time is computed from 1 p.m.

Allowable free time is as follows:

Actual Weight	Free Time in Minutes
Less than 10,000 lbs.	120
10,000 but less than 20,000 lbs.	180
20,000 but less than 28,000 lbs.	240
28,000 but less than 36,000 lbs.	300
36,000 but less than 44,000 lbs.	360
44,000 or more	420

Detention charges can only be billed against the party causing the detention and are not subject to prepaid or collect conditions on the bill of lading. The traffic manager should become acquainted with Rules 500, 501 and 503 of the National Motor Freight Classification for guidance in detention matters.

INDUSTRIAL TRAFFIC MANAGEMENT

In piggyback (TOFC) movements, particularly Plan 2½, one must determine who owns the equipment to judge who is liable for detention charges. If the trailer is rail-owned or leased, MC 88 rules do not apply if the railroad does not charge the motor carrier for its use; if the rail carrier does charge the motor carrier for using the trailer, motor carrier detention applies. The same is true for steamship-owned containers moving from the water front to inland destinations. (The National Motor Freight Traffic Association, Inc., 1616 P St., N.W., Washington, D.C. has a synopsis of informal opinions on this subject.)

Care in setting up records and record keeping methods is, as always, important. A useful tip: Put an electric time-punch clock on the dock and keep in and out time cards for all vehicles.

Storage Charges

Storage charges should not be confused with demurrage or detention charges. Storage charges are of various types but generally they apply to the use of carriers' property other than rolling stock by the shipper or consignee while awaiting some form of action. For example, if cars are left on railroad tracks there could be both demurrage charges for the use of the cars and storage charges for use of the tracks. Storage charges are prescribed in the carriers' tariffs.

A motor carrier will assess storage charges when delivery cannot be made and the carrier must take the shipment off the vehicle and put it in his terminal or a warehouse. It's important to remember that the carrier's liability narrows from that of a common carrier to that of a warehouseman at this point.

Switching

Switching—moving cars to and from rail yards to docks or sidings for loading or unloading—is usually limited to a specific terminal area. The cost of these moves is included in the line haul

TRANSPORTATION SPECIAL SERVICES

transportation charges. Various forms of local switching may include moving partially loaded or unloaded cars on an industrial track to accommodate other cars, switching to scales for weighing or to line up cars for the line haul, switching to accommodate an industry that might be served by more than one railroad and so on.

A company having its own tracks and switching locomotives may receive an allowance from a railroad for certain switching services the company does for itself. The allowance and the services must be in a published tariff.

Embargoes

An embargo is an order, issued either by a regulatory agency such as the ICC, or by a carrier to its own employees, that freight is not to be delivered or transported. Generally, an embargo is a recognition that for some reason a carrier cannot continue to perform its job. This may be the result of a natural disaster—floods making a route impassable for example; or it may be the outcome of a man-made breakdown of some sort—a consignee whose docks have become so impossibly congested with accumulated freight that delivery cannot be made. In the latter case, the carrier will declare an embargo against all freight consigned to that receiver.

An industry may request the ICC that an embargo be placed on itself if it has suffered a breakdown in its plant or unloading equipment. But there has to be an excellent reason for the ICC to issue a service order establishing an embargo, which is only issued for a specific period of time or until the conditions that caused it have been corrected. An embargo is a prohibition on the transportation of freight—it has no effect on freight rates or charges.

Transit Privileges: Stopoffs and Other Arrangements

Transit arrangements give the shipper the "privilege" of stopping his shipment en route to add to it or unload some of it, to store it temporarily or to process it in some way, and then to reship it on to its final destination at the through rate that would have applied had the shipment moved directly from origin point to destination. Originally a railroad idea, some types of transit privileges are not offered by motor carriers. There is, of course, usually (though not always) a charge for the privilege but benefits to the shipper in total costs and point-of-sale competitiveness are considerable.

The simplest and probably the most popular of the transit arrangements is a stopoff for a car or trailer to complete loading or partially unload. The car or truck is dispatched with instructions to stop at a specific point en route to finish loading or to partially unload. The shipment is usually destined for one consignee and the freight charges will be computed from the origin point to destination for the highest weight in the vehicle, to which is added a stopoff charge. The stopoff point must be intermediate and all instructions must be clearly noted on the bill of lading.

Milling-in-transit and fabrication-in-transit are two transit privileges that are offered only by the railroads. Milling grain into flour is the oldest and still the most common example of this type of arrangement. Carloads of grain move to a milling point, where they are unloaded, the grain ground into flour and shipped on to destination. The bills of lading and other documents clearly set out that the destination shipment of flour must come from the original shipment of grain. The freight charges are calculated for a shipment of the finished product from origin to destination via the milling point, plus a transit charge. Grain is also stopped off for grading, mixing, cleaning and other processing under similar arrangements.

Fabrication-in-transit is probably most commonly used in shipping iron and steel, which is processed into a more usable form at some intermediate point between mill and customer. Transit privileges of this sort are also granted for logs and lumber. The advantage to the shipper is that the through rate is less than the combination of two separate rates.

Storage-in-transit is an arrangement that is offered by both railroads and motor carriers. Under this practice, a volume shipment moves from origin to a point of storage under a bill of lading that shows the shipment to be a storage-in-transit one, consigned beyond the storage point. The freight charge for the movement to the storage point is the local rate. When the shipment is moved forward to its ultimate destination, for that leg of the movement the freight charge is calculated as though a through rate applied from the original shipping point to ultimate destination, plus a small transit charge (the storage charges are a separate item). This could be cheaper than paying for two entirely separate movements, and has the additional advantages of allowing the shipper to clear his origin-point warehouse and position his products closer to his markets.

Reconsignment and Diversion

A "reconsignment" can be defined as a change in the consignee, destination or route of a shipment that dictates a change in the transit billing. A "diversion" at one time was considered simply a change en route in the destination of a shipment. Today the terms are used interchangeably to mean a change in the destination billing either before or after the shipment reaches the original destination. What gives the privilege its value to the shipper is that the through rate from origin to destination, plus the reconsignment charge, is less than the rate to the diversion or reconsignment point plus the rate from there to destination. Both railroads and motor carriers now offer this service at charges that vary with the carrier and the changes in the billing required.

Reconsignments and diversions can only be made if they are provided for in the tariffs. Note also that if a shipment has reached its original destination *and has been offered to the consignee*, it cannot then be reconsigned but must be reshipped from that point to its new destination, in a new movement, at the local rate.

For shippers of fresh fruits, vegetables and livestock, the reconsignment privilege is often vital, enabling them to start shipments to distant markets and then reconsign them to those markets offering the best prices. Lumber and other products, too, are often sold and resold en route under this type of arrangement.

Side Track Agreements

Railroads actively compete with one another in attracting new industries to the land near their rights of way. They have found that this is an excellent means of obtaining long-haul revenue on the carload business generated by these industries in addition to making productive use of otherwise undeveloped land.

An industrial siding enables a railroad to deliver cars to the premises of an "industry" (company) on its line. It is connected by a switch to the main track of the railroad and the railroad ordinarily furnishes the motive power for placing or removing cars. Sidings vary in size, from single-car to elaborate multi-track layouts with a capacity of several hundred cars. Railroad line-haul rates include one placement on a private siding at origin and destination, but if the arrangement of the siding puts it in the "complicated" category, the railroad is required to make an additional charge for switching.

Except at those plants where the industry uses its own switching locomotives, the industry and the railroad enter into an agreement covering operations over the side track. A so-called "industrial side track agreement" of this type follows a fairly standard pattern. It describes the location and layout of the track and any other structure over which the track is to be built. The industry is called upon to provide, without cost to the railroad, all necessary right-of-way beyond that of the railroad. The agreement confers upon the railroad the right to enter upon the property for the purpose of constructing, maintaining and operating the side track and of removing, upon termination of the agreement, any materials belonging to the railroad. It describes how the materials are to be furnished and generally includes a blueprint of the side track to be constructed.

The agreement also establishes how the cost of construction is to be borne. Usually, the railroad assumes the cost of construction from the switch to the clearance point (where the track leading from the switch clears the main track) and the industry pays the rest. Under certain circumstances, however, the railroad may agree to refund part or all of the industry's outlays at rates per car agreed upon, dependent upon line-haul revenue yielded, for every loaded car originated or terminated (not stopped in transit) on the siding. Cost of maintenance, including removal of snow and ice, is absorbed by the industry, although upon request of the industry, the railroad will furnish the materials and labor and bill the industry for the work. The railroad's right to use the side track for general railroad purposes is established, provided it does not interfere with the use of the track for the industry's traffic, as is the right of the parties to join the track with any other track. Structural clearances in accordance with AAR rules are prescribed which generally mean buildings or objects over the side track less than 21 feet above the top of the rail or alongside the track less than six feet from the nearest rail (more on curves) are prohibited.

A typical liability clause in side track agreements is of sufficient importance to reproduce in full:

> It is understood that the operations of the Railroad Company involve some risk of fire, and the Industry assumes all responsibility for and agrees to indemnify the Railroad Company against loss or damage to property of the Industry or to property upon its premises, regardless of negligence of the Railroad Company, its officers or employees, arising from fire caused by or as a result of the operations of the Railroad Company on said sidetrack, except to the premises of the Railroad Company and to rolling stock belonging to the Railroad Company or to others, and to shipments in the course of transportation.
>
> The Industry also agrees to indemnify and hold harmless the Railroad Company from loss, damage or injury resulting from any act or omission of the Industry, its employees or agents, to the person or property of either of the parties hereto and their employees and to the person or property of any other person or corporation, while on or about said sidetrack unless caused by the sole negligence of the Railroad Company; and if any claim or liability, other than from fire or from the existence of the aforesaid close clearances, shall arise from the joint or concurring negligence of both parties hereto, it shall be borne by them equally.

From this it will be seen that the railroad assumes a minimum of liability for occurrences resulting from its operations on the siding and requires the shipper to hold it harmless as to various features on which liabilities can be substantial. The ICC feels its

INDUSTRIAL TRAFFIC MANAGEMENT

jurisdiction does not include contractual provisions of this sort (61 ICC 120, 25 ICC 352, and 35 ICC 255).

Many industrial concerns obtain insurance coverage for these liabilities. In some cases special policy provisions are necessary, and this should be checked. Additionally, some companies are careful to keep their staff mindful of the liabilities created in these documents.

Some large industrial plants have their own trackage on which they place cars by using their own switching locomotives. In such cases, delivery of cars by and to the railroad takes place on a railroad-owned interchange track. Since the railroad and the industry operate exclusively on their own trackage, no side track agreement is necessary. Industries operating their own trackage may be able to negotiate an agreement with the railroad whereby the railroad will grant an allowance for being relieved of the obligation to perform the switching.

Often the question of responsibility for material in cars standing on private sidings arises. Generally, it is held that the railroad's liability terminates with the placement of the car on inbound shipments. Outbound, the railroad's liability commences when its agent signs the bill of lading, whether or not the car has been removed from the siding, as the bill of lading then becomes an enforceable contract. Unusual circumstances, however, may produce exceptions to this rule.

The carriers, as well as making industrial side track agreements, lease sections of their tracks to industries from time to time, generally on a so-much-per-foot basis. This may be done for a number of reasons: to get cars out of a chock-a-block plant, to bring supplies closer for quicker delivery to consuming points, and so on. As regards demurrage, these arrangements have certain advantages for companies owning or leasing their rail equipment. Such owning or leasing, coupled with the track lease, gives the equipment on the leased track the status of private cars on private tracks—for which there are no demurrage charges. Distribution expenses incident to the use of these leases (no matter whose cars are involved) warrant close and complete checking, however, as such expenses can be quite dear when all cost elements are considered.

Weight Agreements

Freight rates are generally published in weight units of 100 pounds per net ton of 2,000 lbs. or per gross ton of 2,240 lbs. For the railroad to assess proper charges, accurate weights must be shown on the bill of lading even though cars can be sent over scales for verification of weight. Accepting short-weight shipments can result in both shipper and carrier being prosecuted under the Elkins Act.

In the early days of railroading, every shipment was weighed individually. Mass production and the standardization of packages made this procedure cumbersome and time-consuming. Individual railroad agents were quick to recognize that the weight of products shipped repetitively by industries served by their stations did not vary significantly. Soon it was found that by eliminating unnecessary weighing it was possible to move the freight out much more rapidly. Thus the "average" weight agreement evolved which is shown on the next page.

Because agreed weights must be acceptable not only to the originating carrier but to every other railroad in the route, the railroads do not deal directly with shippers in establishing weight agreements. Instead, this function is delegated to weighing and inspection bureaus set up along territorial lines and acting as agents for all railroads in the respective territories. These bureaus have representatives in all major cities who administer the agreements and, in addition, check on bill of lading descriptions and inspect shipments when called upon by the railroads to do so under section 2 of Rule 2 of the Uniform Classification.

An industry desiring to enter into a weight agreement usually notifies the local railroad agent, who refers the matter to the nearest office of the appropriate weighing and inspection bureau. A representative determines the classification descriptions of the products to be covered and the various standard packages of each description. He makes a random selection of at least 10 of each of these packages, weighs them separately and establishes an average weight for each size package of each description. If packages are

INDUSTRIAL TRAFFIC MANAGEMENT

EASTERN WEIGHING AND INSPECTION BUREAU
OFFICE OF MANAGER

2 PENNSYLVANIA PLAZA
NEW YORK, N. Y. 10001

WEIGHT AGREEMENT NO. _____

Effective Date _____ 19____

This AGREEMENT, entered into, by and between the Eastern Weighing and Inspection Bureau, for and in behalf of the Carriers for which the Eastern Weighing and Inspection Bureau is duly authorized to execute this Agreement,

and

of _____ (Town), _____ (State)

WITNESSETH: That, in consideration of the carriers, members of the Eastern Weighing and Inspection Bureau, accepting the weights and descriptions as certified on shipping orders, bills of lading or weight certificates for commodities herein specified as the basis for assessing freight charges, it is hereby agreed:

1. The consignor or consignee, as the case may be, shall report and certify to the carrier correct gross weights and correct descriptions of commodities on shipping orders, bills of lading or weight certificates by placing thereon imprint of certification stamp providing for verification by the carriers, members of the Eastern Weighing and Inspection Bureau. When such weights are obtained on track scales the correct gross, tare and net weights shall be given.

2. When weights of uniform or standard weight articles are based upon averages, the consignor or consignee, as the case may be, shall give prompt notice to the authorized representative of the carrier when any change is made which will affect the weight arrived at by the use of the average, including any change made in package or material used.

3. The consignor or consignee, as the case may be, shall keep in good weighing condition all scales used in determining weights and have track scales tested, maintained and operated in accordance with the Track Scales Specifications and Rules approved by The Association of American Railroads and shall also allow the authorized representative of the carrier to inspect and test them.

4. The consignor or consignee, as the case may be, shall keep his records in such a manner as will permit of correct and complete check, and shall allow the authorized representative of the carrier to inspect the true and original weight sheets, books, invoices and records necessary to verify the weights and descriptions of the commodities certified in the shipping orders, bills of lading or weight certificates.

5. The consignor or consignee, as the case may be, shall promptly pay to the authorized representative of the carrier, bills for all undercharges from original point of shipment to final destination, resulting from certification of incorrect weights or improper description, whether shipment is sold f. o. b. at point of shipment or elsewhere. Overcharges developed from check of consignor's or consignee's records, as the case may be, will be promptly certified by the authorized representative of the carrier in writing for proper adjustment.

6. Shipments made under this agreement will be subject to rates, charges, minimum and estimated weights prescribed by classifications, exceptions thereto, tariffs or rules of the carriers interested.

7. This agreement may be cancelled by ten days' notice in writing to either party; it being understood that the consignor or consignee, as the case may be, shall permit check of business and pay undercharges on all shipments made prior to cancellation.

This agreement applies on

CONSIGNOR—CONSIGNEE

By _____

Title _____

EASTERN WEIGHING AND INSPECTION BUREAU

By _____
          Manager

Per _____

TRANSPORTATION SPECIAL SERVICES

not uniform and an average weight cannot be determined, the shipper's weights may be used if the accuracy of the scales is certified by the bureau representatives. If the shipper uses a track scale, it must be tested, maintained and operated in accordance with the Track Scales Specifications and Rules approved by the Association of American Railroads.

Bills of lading involving shipments which are the subject of weight agreements should be stamped with a rubber stamp as shown here. By signing the bill of lading the shipper is bound by the provisions of the certification. Similar information may be imprinted on the bill of lading forms (instead of using a stamp) provided it carries appropriate "dagger" (†) marks specifying "Shipper's imprint in lieu of stamp; not a part of bill lading approved by the Interstate Commerce Commission."

> The description and weight indicated on this bill of lading are correct.
> Subject to verification by the
> EASTERN WEIGHING & INSP. BUREAU
> According to Agreement.

Shippers using weight agreements must notify the bureau of any change that affects the average weights, keep records in such a manner as to permit checks by bureau representatives, allow inspections of scales and records at any time and promptly pay bills for any undercharges resulting from the use of incorrect weights. Naturally, the weighing and inspection bureau will enter into agreements only with firms of unquestioned integrity.

Only those firms which have a large volume of carload business will be able to justify the expense of installing and maintaining their own track scales, and then only if the major part of their business consists of straight carloads. Bills of lading covering carload shipments weighed on track scales must show the gross weight of the car and its contents, the tare weight of the empty car and the difference between the two, which, less dunnage allowance if any, is the net weight upon which transportation charges are paid. Each car has its empty weight stenciled on its sides and this weight usually is checked at regular intervals, particularly if the car undergoes any modification or rebuilding. Industries with their own track scales, however, frequently make a practice of running their own checks on the weights of empty cars. Each weighing on a track scale should produce a scale ticket which is acceptable under a weight agreement as sufficient evidence of the actual weight of a car or its contents.

Shippers of large numbers of mixed carloads will find it necessary to weigh shipments before they are loaded or use average weights under an agreement. Otherwise, they will find it impossible to benefit from the provisions of Rule 10 of the classification, which permits the use of the carload rate on each different article in the carload. If a track scale weight were to be used on a mixed carload, the entire weight would be charged for at the highest carload rate applicable to any article in the carload.

Certain perishable commodities, mainly vegetables, move from packing sheds close to the fields in which they are grown. Operations at these sheds being of a seasonal nature, the sheds are not equipped with scales, and since the carloads must be moved to destination without delay, the railroads do not track-scale them. Instead, freight charges on shipments of these commodities are based on a scale of estimated weights developed for each type of commodity and published in carriers' tariffs.

The use of estimated weights is a common practice on certain movements of petroleum, molasses, and other liquids in tank cars. These weights are often found in the items naming the rates; i.e., gasoline at 6.6 pounds per gallon and so on. Uniform Classification Rules 52 and 53 provide estimated weights of 4.7 and 5.2 pounds per gallon for liquefied petroleum gas and butadiene. Some motor carriers also use them.

Freight bills, incidentally, on tank car traffic of liquids moving at estimated weights are generally based on the tank shell capacity of the car. Subject to various qualifications, U.F.C. Rule 35 provides for this.

The qualifications referred to in the preceding paragraph include loadings into domes, temperature adjustments, required outages for possible expansion of ladings, and so on. They are quite complicated and numerous. As an example, where a shipper has 10,000 gallons of gasoline to move via tank car between points taking a rate of 30 cents per 100 pounds at an estimated weight of 6.6 pounds per gallon, the freight charges would be $198.00 (10,000 x 6.6 x 30).

This example assumes the liquid moves in a car having a shell capacity of 10,000 gallons. If the shell capacity were 12,000 gallons and only 10,000 gallons were loaded, the shell capacity would control and the freight bill would be $237.60 (12,000 x 6.6 x 30). In matters of this kind "wind weight" payments (here $39.60) require watching and avoidance. Motor carrier tariffs in related fields should be accorded similar treatment.

Whenever possible, consignees should check the weights of all inbound shipments. This is good business because it frequently results in the discovery of figure transpositions in billing (e.g. a weight of 5,870 pounds is transposed to 8,570 pounds). It may also correct weights which have been estimated inaccurately, and it may even pinpoint evidence of leakage or pilferage of the contents of packages. Correcting weights on freight bills or claims may arouse some resistance from carriers, but the reputation of a firm and the proven accuracy of its scales usually will be the deciding factor. A weight agreement in such cases will offer testimony as to the reliability of the scales. More of these agreements are being entered into for inbound shipments, particularly where iron and steel are involved.

Thus far, the subject of weights has dealt with railroad shipments, but it should be pointed out that the practice of accepting shippers' weights also is followed by motor carriers, water carriers, freight forwarders and airlines. For these carriers to weigh each of the millions of shipments they handle would bog down even the most efficient freight terminal and cause no end of delay to the shipments. The motor carriers have their own weighing and inspection bureaus whose representatives conduct continuous sampling checks at carriers' terminals and report inaccurate weighing practices and deviations from correct bill of lading descriptions and packaging provisions. Independently, any carrier may weigh

INDUSTRIAL TRAFFIC MANAGEMENT

shipments, inspect the contents of packages and test shipping containers at any time. Matters generally work out that there is little risk involved in the acceptance of shippers' weights as a basis for the assessment of freight charges.

13

EDP AND TRANSPORTATION

Today, some form of electronic data processing (EDP) is to be found in the traffic management operations of almost every company. Business computers costing as little as $10,000 are on the market, and as the price range and choice of equipment widens, the costs of computerization go down and the variety of systems and programs increases.

The traffic manager's concern is with both the uses to which the equipment will be put and with the equipment itself. In conjunction with the company's EDP professionals, the traffic manager will periodically have to reassess the traffic EDP system by asking some basic questions such as:

—Is the system doing what it was supposed to do—reducing paper handling costs, speeding up procedures, etc.?

—Do the computer people—managers and programers—really understand traffic management problems and operating constraints?

—Is the system providing timely and understandable reports for use in financial planning, freight consolidation programs, equipment-use forecasting, etc.?

—Has the EDP program stayed within its operating budget? Has it justified its original costs?

If the replies are affirmative, another reassessment should be made of the original decision to use "in-house" systems or rent computer time from outside service bureaus or time-sharing companies. That decision will be affected by the data processing services need-

INDUSTRIAL TRAFFIC MANAGEMENT

ed, the size of the traffic manager's company, how much computer time will be used and whether there is anything unique in a firm's requirements.

The basic uses of EDP by traffic departments generally cover
- Order processing
- Freight bill payment and auditing
- Preparing bills of lading with classification, rating and rates
- Outbound and inbound routing
- Extracting and formatting information for management and regulatory agencies, and for use in rate negotiations and budgeting
- Keeping loss and damage statistics for use in prevention programs
- Collating shipments with purchase orders, etc.

Programs to handle all of these basically clerical chores can be made, bought or rented. The management information reports that are (or should be) the valuable by-products of computerization must cover all the activities of the traffic department, summarizing shipments made by warehouse, distribution centers and manufacturing plants by time periods and by any other necessary unit, summarizing expenses by any number of units, revealing exceptions to procedures and highlighting problems. The management information coming from EDP should in fact produce an up-to-the-minute history of the traffic department for long range planning growth. All of this should be as simple as possible in format and language.

Historically, data processing in transportation had its earliest use by carriers preparing statistical reports for the ICC, state commissions and other agencies. Industrial traffic management began to use it much later, usually waiting its turn until accounting and finance were fully computerized. Traffic was given "matching" computer time when it was shown that its procedures and problems (with the possible exception of rates and rating) were particularly susceptible to computerization. Without EDP, many authorities believe that the information required for effective physical distribution management would simply not be obtainable in time to make use of it. The computer may well have saved traffic management from drowning in outdated paper.

The lightning additions, subtractions, and other calculations that the computer can make, its "memories" and ability to recall items, the ways it can highlight and throw out any exceptions when irregularities occur or when mistakes are made, the almost-instant communications it brings by tying in plants and offices miles apart and in doing work simultaneously for two or more departments of a company, have made it an indispensable traffic management tool. How many of these abilities and how much computer power a firm needs (or can afford), must be decided pretty much on a company-by-company basis. Computerized car location, as an example, will be worthwhile for some and not for others. It is important that industrial traffic departments know in general the "state of the art" of EDP and do for their principals what is best for them.

Today, the traffic system of a company will link a complex of plants, warehouses and customers. Traffic is faced with a number of conflicting decisions involving much more than bills of lading or freight bills covering movements between two points. These decisions often require trade-offs to be made between cost, time in transit and consistency or continuity of service. Such conflicts are more easily resolved with the aid of EDP systems that can swiftly supply statistics covering the company's in-transit "experience," shipment loading experience and receiving experience, and then combine all the above in forecasting and planning programs.

Ideally, the traffic department should have, either in its own EDP department or in its time-sharing programs, the capabilities of scheduling shipments, selecting carriers (routing), preparing documents (bills of lading), classifying freight and a selection of freight rates large enough to allow for freight bill auditing. For those firms with a private fleet, the management of that fleet should be "on the computer."

The modern traffic manager, then, should have some knowledge of computer language. In the early days of the computer, such languages as Fortran, Cobol and Algolbo were difficult for any but professional programmers to use. In recent times, experts have come up with simplified languages such as Pascal to give the non-professionals a chance to program computers without weeks of special instructions.

A considerable number of programs specifically written for

traffic and transportation problems can be bought or rented from "software" suppliers, time-sharing libraries, service bureaus, etc. Some of the questions to ask when sizing up these programs are, will it:
- Reduce transportation costs?
- Improve customer service?
- Help establish a proper loading mix?
- Improve equipment utilization?
- Reduce detention and demurrage?
- Reduce overcharges and duplicate payments of freight charges?
- Reduce clerical effort?

One of the earliest uses of the computer in transportation was for reporting the statistics on a uniform basis that the ICC, the state commissions and other bodies require from the common carriers. This led to uniform "codes" or numbers being assigned to certain categories of information. In the case of commodities, the applicable uniform code is the Standard Transportation Commodity Code (STCC). An excerpt from the Uniform Freight Classification listing STCC numbers is shown here.

Let's say a shipment of grain feed (sixteenth line down on the left) is made by a railroad. "Grain feed" is fed into the computer as 20 421 96. All other grain feed shipments take the same number and are totaled at the end of a given period. The STCC number is converted back to "grain feed" in the printout and the total number of shipments goes into the statistics reported. With the other railroads using the STCC number of 20 421 96 for grain feed shipments, the necessary uniformity becomes "automatic." STCC numbers are also used by other carrier modes for such statistical reporting, and today much more sophisticated uses are being made of the STCC in the transportation field and elsewhere.

Another code of particular importance is the Standard Point Location Code (SPLC). As its name implies, it codes origins, destinations, and other location points. Under mainly motor carrier sponsorship it was established about the same time as the STCC. The SPLC employs six digits, the first representing a region, the next two a state, and so on. It is widely used by shippers, motor carriers, railroads, and others. It is reasonably precise, and is also

EDP AND TRANSPORTATION

UNIFORM FREIGHT CLASSIFICATION 13

INDEX TO ARTICLES

STCC No.	Article	Item	STCC No.	Article	Item
35 223 38	Grain drill blades, or points, ot hand	4160	20 421 19	Grain syrup, containing spent grain mash, liquid, feed	37220
35 223 42	" drill shoes ot hand	†4050, 4510	35 516 42	" temperers 59720, †59760, †59810	
35 223 44	" drill tubes, ot hand, rubber	4590	35 514 10	" washer and drier combined	58840
35 223 46	" drill tubes, ot hand, steel, flexible	4600		Grain:	
35 223 14	" drills and fertilizer distributors, SU	3520	14 916 91	Abrasive, natural	2040
35 223 10	" drills, ot hand	3510	32 911 15	Abrasive, synthetic	2040
35 229 82	" dumps	ð	20 419 38	Brewers'	ð
20 418 54	" dust	ð	28 422 37	Poisoned	53130
35 229 12	" elevators, bagging, loading or storing, KD	3550	20 419 94	Roasted, noibn, crushed, ground, granulated or whole	47200
35 229 11	" elevators, bagging, loading, or storing, SU	3550	01 136 90	Sorghum	46890
20 421 96	" feed, noibn	47000	20 823 30	Grains, spent, dried	47040
20 419 92	" flour, noibn	47010	20 823 35	Grains, spent, wet	47050
20 451 20	" flour, self-raising	47020	20 371 30	*Granadillas, fresh, cold-pack (frozen)	40050
35 229 40	" germinating cabinets or racks	19750	01 239 90	*Granadillas, fresh, ot cold-pack	40050
35 516 82	" grinding rollers or rolls. iors	65130	25 314 25	Grand stand chairs, FF	27000
35 516 42	" heater, huller, polisher and temperer combined	59720, †59760, †59810	25 314 30	Grand stand platforms, in flat sections	27010
			25 314 41	Grand stand seat parts	27030
35 516 42	" heaters	59720, †59760, †59810	25 314 40	Grand stand seats, iors and wood, KD	27030
35 229 26	" lifter guards or plates, ot hand	4280	25 314 15	Grand stands, noibn, steel or wood. KD, telescoped or collapsed	26970
20 859 45	" mash, spent, feed	37380	32 811 10	Granite blocks, pieces or slabs, carved, honed, lettered, polished or traced	47640, †47690
35 227 26	" moisture testing machines	62260			
20 418 79	" offal, dry and uncooked	ð	32 811 20	" blocks, pieces or slabs, dressed, sawed or chipped on more than 4 sides, chiseled, hammered	
35 223 14	" planters and fertilizer distributors, SU	3520			
35 223 10	" planters, ot hand	3510		" sand rubbed or steeled	47650, †47690
35 227 40	" probes (samplers)	90640	14 111 10	" blocks, pieces or slabs, rough quarried, or not further finished than chipped, pitched, sawed or scabbled on 4 sides	47660, †47690
20 418 38	" pummies, chopped or ground	ð			
20 418 38	" scourings, chopped or ground	ð			
01 139 30	" screenings, other than ground	47030			
20 418 26	" screenings, ground	47030			
20 933 43	" screenings, oil, liquid or solidified	72490	14 111 20	" bridge stone rough chipped	47680, †47690
			14 219 20	*crushed, not ground nor pulverized	47715
35 227 28	" separating machines, disc	62280	35 597 38	Granulators, clay	60170, 60268
35 516 70	" shovels, power	62270	33 311 65	Granules, copper, hydrometallurgically processed	14825
20 839 35	" skimmings, malthouse	46970			
01 136 90	" sorghums	46890	33 394 10	Granules, nickel-magnesium alloy or nickel-magnesium-silicon alloy	68465
35 223 20	" sowers and fertilizer distributors, ot hand	3580	20 861 40	*Grape beverage	39490
35 223 34	" sowers, broadcast, ot hand	3930	20 341 60	Grape fruit peel, dried, noibn	41720
20 851 36	" *spirits, neutral, in bond	22050, 33800	20 334 70	" juice, concentrated, 60% solid	47225
			20 334 42	" juice, ot frozen, noibn, in tank cars	40220
34 449 32	" spouts, iron, 16 gauge or thicker	30320	20 334 41	" juice, ot frozen, noibn, ot tank cars	40220
34 449 30	" spouts, iron, 17 gauge or thinner	30310	20 334 55	" juice, with sugar or other sweetening and water added, ot frozen	40230
35 229 40	" sprouting cabinets or racks	19750			
34 492 42	" sprouting houses or buildings, wood and sheet steel with equipment	18855	20 352 90	" *leaves, pickled in brine	40560
			20 339 60	" pomace, dry	95460
35 228 16	" stack caps or covers, ot iors	36110	20 339 65	" pomace, wet	95470
35 228 15	" stack caps or covers, iors, flat or nstd	36100	20 933 61	" seed oil	72580
			01 149 65	" seeds	85030
20 421 72	" stalks and heads, sorghum, ground, threshed and shredded, in meal or pellet form	ð	20 841 37	" *spirits, neutral	22050, 33800
			20 999 74	" sugar	ð
01 199 65	" stalks, heads and leaves, sorghum, not threshed, ground, dry or dehydrated	ð	20 871 75	" *syrup	40970
			01 211 10	Grapefruit, fresh	42100
			01 224 10	Grapes, fresh	42030
35 516 42	" steamers	59720, †59760, †59810	32 552 35	Graphite briquettes, iors alloying, not machined	47270
34 492 50	" storage bins or cribs, aluminum and steel, KD	36010	14 919 30	Graphite ore	27310
			40 291 45	Graphite scrap	95480
34 492 51	" storage bins or cribs, steel or wire and steel, KD	36020	32 958 90	Graphite, noibn	47230
24 332 60	" storage bins or cribs, wood and wire	36030	35 799 24	Graphotype machines	59400
			35 314 48	Grapples, dredging, conveying, dumping or hoisting, ot log or tie lifting, iors	60380
24 332 65	" storage bins or cribs, wooden, flat sections	36040	34 239 71	Grapples, log	90740
01 196 20	" straw, threshed	88200	37 428 10	Grapples, tie lifting, steel	62290

For explanation of reference marks, see top of page 15; for abbreviations, see last page of this Classification.

employed in the computer systems of some industries for other-than-transportation purposes, such as state tax computations. Still other codes have come into the transportation picture. These include carrier name codes, carrier equipment codes, patron (user) codes, vehicle identification codes, and so on. Varying degrees of use and recognition have been accorded them.

Often codes and systems developed for collecting statistics proved valuable in inter-plant and even inter-company communications. Thus informational systems today are built not only for traffic, inventory and warehousing control, but for transmitting shipper-carrier, marketing and other kinds of messages. The Transportation Data Coordinating Committee (TDCC) in Washington, D.C., is doing valuable work in coding and computer communications and should be contacted for further information in this field.

Efficient and economical exchange of data is, of course, the reason for developing compatible codes. Compatible systems designs are equally necessary for this kind of data exchange. The logical demands of having to prepare information for the computer often are a bonus in themselves. Domestic and international tariff simplification, for example, has developed because computer systems choke on exceptions. In the meantime, STCC, SPLC, and other transportation oriented codes, and codes used by industry, are being reconciled and made compatible in many cases by what are called "modifiers" or "bridges." But these represent extra steps to be avoided when possible.

A fine example of what can be done when numbers, computers, and motivation mesh is the check-clearing system developed by the banks. By using magnetic-ink, computer-readable code numbers, checks are paid, negotiated, returned to the issuing bank (sometimes even from foreign countries), and finally (with correct calculations, print-outs on bank statements, etc.) returned, duly cancelled, to the party who signed them.

Transportation data processing has developed principally in two fields. One in concerned essentially with operating functions such as service, establishing proper distribution patterns, equipment availability, and so on. The second is concerned essentially with treasury functions, such as auditing freight bills, paying them,

EDP AND TRANSPORTATION

and the like. No hard line can be drawn between operating and treasury functions, and quite often benefits in one ultimately flow to both activities. Joint use by industry and carriers of compatible computer systems is an example. Such joint use is proving of particular value in railroad-industry communications. A company learns of cars enroute to it by car number, location, contents, origin, quantity, departure date, arrival date and so on. The information on outbound cars includes rates and total freight charges.

Railroads, truck lines and air freight carriers are employing EDP to "track" shipments thus enabling them to respond quickly to tracing inquiries. If compatible systems are used by shipper and carrier, the response can be had very quickly.

A substantial group of industries with private car and assigned car fleets now uses computer systems to operate those fleets more efficiently and economically. Under old procedures, operating information was assembled and prepared manually. Some companies now make information from the car location system part of their car fleet administration system. By and large, however, car fleet administration information comes from the data system of the company involved and generally shows by origin, by cars, and by shipment transit times the activity of the private and assigned car fleet. Traffic uses this information to improve turn-around times, loading and unloading schedules and so on.

Another common application of EDP in physical distribution is determining origin points for particular destinations when a choice of several distribution centers is available. Various programs are available to do this but they pretty much boil down to the computer making a selection from the freight rate banks in its storage systems. These, in turn, are applied to values for the freight at the various origins; the computer adds the rates to the values and a "package" comes out showing what are substantially laid-down costs to a territory, destinations, or whatever the case may be, from the various origins.

This freight rate bank information is also at times used to find out how competitors are doing in a particular market. By providing the rates and related data from the competitor's origin to a given

INDUSTRIAL TRAFFIC MANAGEMENT

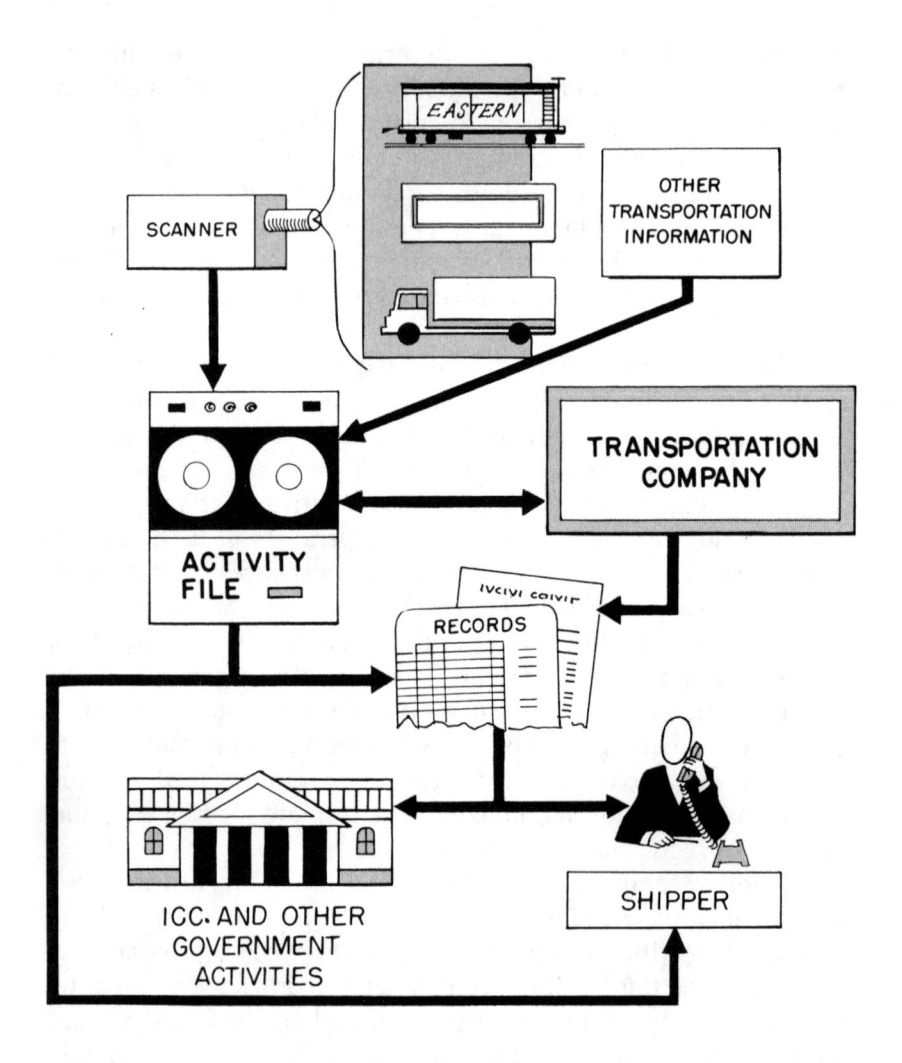

Computerized car location systems such as the one shown above in schematic form are used by the railroads for improving shipment tracing and equipment allocation. The system shown here makes use of an "optical scanner" to read a stripped label fastened in a fixed position on cars, containers and trucks and feed this identifying information to the computer. Other identifying systems use laser beams, microwaves and more advanced technology to "recognize" the vehicle.

EDP AND TRANSPORTATION

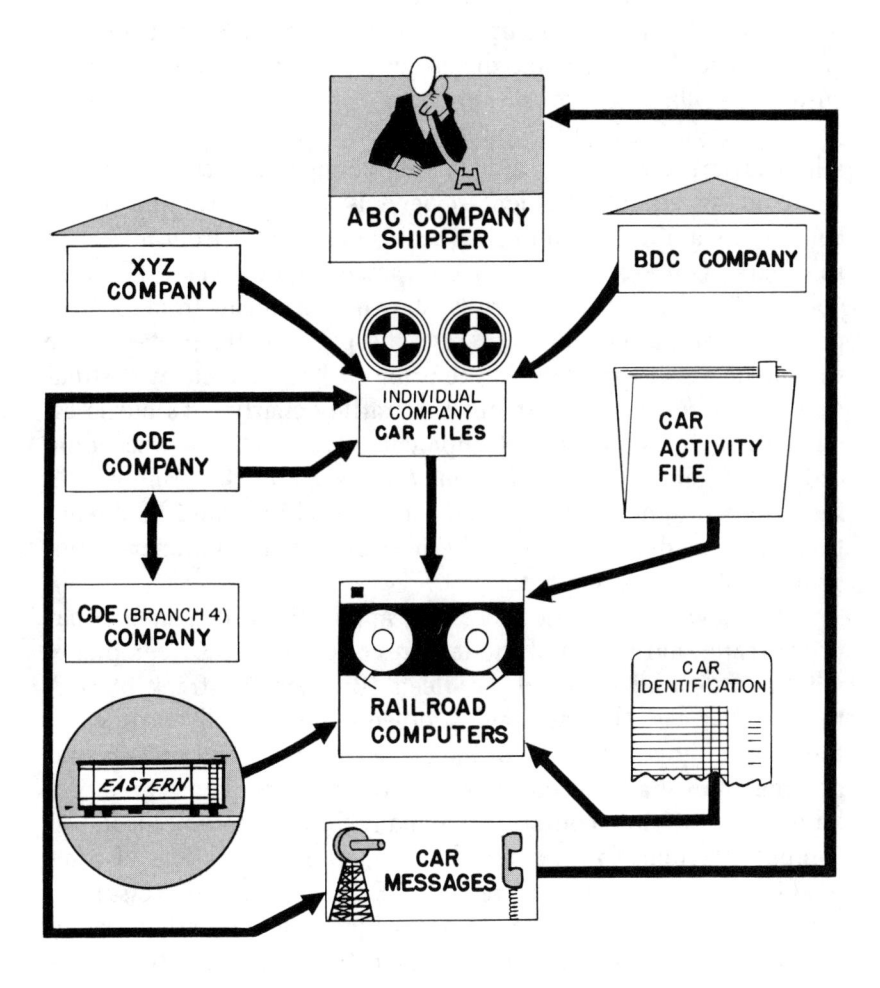

This schematic diagram shows how a railroad's computer system can be linked to a shipper company (ABC Company) and a number of the companies it supplies with the result that car tracing, equipment utilization and the general level of service are all improved.

marketing territory or sales points, the computer can show fairly accurately what it is costing him to serve that market. Cost information of this sort is also used in pricing calculations where freight costs are absorbed in the final price. Often this involves more than simply coming up with a freight rate. Some computer systems take the rate, make the calculations, and even prepare the customer's bill. This type of application, where the computer does almost everything is today quite common, particularly in industries where absorptions are ingrained in marketing practices.

Making actual shipment reviews is another example of the computer's ability to comb significant information from masses of raw data. Quite often these are shipment "spot checks." This involves making periodic "tab runs" showing shipments from (or to) a particular shipping point for, say, the previous thirty days. The shipments are "formatted" in the print-out by product, by destination or whatever may best suit the traffic department's purposes, and can be sorted out so that shipments are arranged in descending order by their volumes and so on. An analysis of the printout will show that shipments may have been made that should be discontinued; rate adjustments may be in order; or any number of improvements may be suggested.

Transportation data processing applications used in international trade share some of the features of those used by companies with private and assigned car fleets. But keeping track of containers overseas, the language problems, intermodal documentation and other peculiarities of the trade require special systems that at times may have special benefits for shippers. For example, a large international container company has its communications-computer systems so arranged that automatic translations of commodity descriptions from French to English (and vice versa) are provided in its shipping papers. Particular codes and other techniques are involved in doing this. Further, the great majority of those using this container company's services do not require negotiable bills of lading. A simple data receipt substitutes for the bill. The information on the receipt (not the receipt itself) is sent by the company's communication system to the port of discharge. This serves to identify the shipment on arrival. No problems with delayed or missing bills of lading with this system!

Freight bills, their payment, and their auditing, are at the heart of most of the paperwork troubles of the transportation business. A great amount of time, error correction, legal problems and other unsatisfactory features result from "normal" freight bill procedures and the field has been a wide-open one for appropriate corrective measures. The computer service firms mentioned earlier in this chapter have come up with various applications for small, medium, and large-sized industrial companies that have been found to be worthwhile.

The banks, too, under their various freight payment plans, have been continuing with their streamlining of debiting and crediting special accounts for freight bills. The banks say that in this they are far ahead of the "checkless society" envisioned for the future. The various transport clearing organizations have been refining their related collecting and remitting operations. An increasing number of firms have "signed up" with these groups and they are performing a great deal of useful work. However, the principal thrust of most bank and clearing house plans is to simplify payments. They tend to bypass the problems of improper ratings, duplicate payments, and the like.

In an effort to take the kinks out of their billing systems, some industrial companies have set up internal procedures for paying the freight bill. This is done selectively with certain kinds of freight by preparing what amounts to a check or draft when the bill of lading is prepared. Applicable freight rates from the tariffs are applied to the weights, and total freight charges due are computed as the freight carrier would compute them in a regular billing operation. The computer then checks the payments and makes any adjustments necessary. More accurate freight payments and fewer duplicate payments are among the benefits.

Freight bill auditing companies, many of whom now use quite sophisticated computer programs, will, for a fee, audit and pay an industrial company's freight bills. Under such an arrangement, a shipper sends his bills of lading to the service company while his carriers send the freight bills covering the same shipmetns. The service company audits the freight bills using the bills of lading and freight rates drawn from its computer freight rate data base. If the audit results are satisfactory, the bills are paid by the service company and it, in turn, is reimbursed by the shipper. If the auditing indicates overcharges, wrong rates, etc., these are handled appropriately. These bill auditing companies also furnish statistical reports according to their client's needs from the transactions processed. This service has many advantages over conventional freight bill handling. It was originally set up for small to medium-sized companies, but large companies now also use bill auditing firms.

14

MANAGING THE TRAFFIC DEPARTMENT

A traffic manager is no longer a clerk with a green eyeshade or the man who sees that the containers are loaded.

Today's traffic manager must be sophisticated in the arts and sciences of distribution, management, communication, administration and business logisitics, although he may have the title of traffic manager. His responsibility often starts with the vendor supplying raw materials, continues through to the delivery of those materials to manufacturing, picks up again at the end of the assembly line and does not end until delivery is made to the customer.

The modern traffic manager is cost conscious and profit minded, but not to the detriment of customer service, being aware of the economics of time utility and place utility, of handling costs, of terms of sale that are fair to his company and to the customer.

The traffic/distribution profession today covers the practices of warehousing, packaging, consolidating, material handling; it deals with the inherent values of each mode of transportation, rate negotiating, using specialized transportation equipment, transportation law and all other activities that affect the movement of goods into or out of a company's plants, no matter how far flung.

The organizational needs and priorities established when the company is small set the pattern of development that is needed as growth begins and continues. The new traffic department in a small company will most likely concentrate on transportation needs, rates, claims prevention and customer service. Growth will determine what additional responsibilities and people are to be added. These considerations will vary, based on company philosophy, industry practices and special needs. There is no standard organization chart that can be made applicable to all, small or big.

It is interesting that certain companies do not know, or even have a reasonable approximation of, what they pay and allow for transportation. They know total costs of wages, of taxes, and so on. Somehow transportation, however, has been lumped with other expenses and has not been isolated as a total figure. Nothing critical should be said about this position in many cases because with some it has simply been practical to treat transportation costs as being in the same category as, say, electric bills—a cost about which little can be done. Such a category should be changed, however; transportation is a real expense factor. Something can be done about such cost items, and total cost figures can be helpful in evaluating what should be done and in doing it.

Five classifications form the nucleus for this discussion of the traffic function. These involve: (1) organization, (2) operations, (3) equipment needs, (4) company relations, and (5) public relations. They are dealt with on flexible lines and they follow in the order named. A particular effort is made to be of help to companies contemplating, or in the process of starting, a traffic operation. Special discussions of aspects they may find worthwhile (e.g., rating and routing guides) are included.

Industrial traffic titles and organization are anything but uniform. The top traffic executive may be designated as general traffic manager, traffic manager, traffic supervisor, manager of transportation, physical distribution manager, traffic director, vice-president in charge of traffic, transportation coordinator, transport control manager, traffic representative, general manager of distribution, corporate transportation manager, director of physical distribution, etc. He may have no particular title or a title in no way descriptive of transportation. Within the company as a whole, the traffic department may be a far-ranging, large group with its head having the status of a vice president or director. The department may be part of a large transportation and supply group, or any other company function. It may be a one-man activity, with that individual also having other duties. The Official Directory of Commercial Traffic Executives lists these by com-

MANAGING THE TRAFFIC DEPARTMENT

panies. A top-flight, highly successful chemical company (not particularly given to titles as is, say, the banking business) operates with simply a "Traffic Department," designates the department's head as "Director of Traffic" and accords him a vice-presidential-plus status in its affairs. That department does a particularly outstanding job for its company.

It is an achievement to become expert in this traffic management field, but it is a greater achievement to harmonize traffic management with the broader needs of company administration, coordinating your own efforts with those of marketing, accounting, production and other functions within the corporation. Here, we deal with the traffic manager, regardless of title, on both a staff and line level and as an operating executive, too, but as an executive in any event.

Sharp increases in freight transportation costs and the realization that the transportation-distribution complex is a major cost element in almost every industry have had a great deal to do with this point of view. Beyond that is the realization that traffic management is tied in closely with important phases of company costs and profits.

Here, then, are the important operating functions of traffic/distribution over which the traffic manager will have direct control or work in an advisory position, dependent on the size and management philosophy of the corporation.

Packaging	Inventory control
Materials handling	Private carriage
Warehousing	Site location
Distribution policy	Vehicle leasing
Data processing	International traffic
Insurance	

These are in addition to his responsibility for:

Shipping	Routing
Receiving	Rates
Classifications	Claims
Consolidations	Freight Auditing
Administration of department	

Functions of an Industrial Traffic Department

Before we review the organizational requirements, let us examine the over-all functions of the transportation/distribution executive's job.

Line Functions

—Establish classification descriptions of all products inbound and outbound and establish packing needs to meet classification claim prevention requirements.

—Establish a consolidation and distribution policy that will improve customer service and lower costs (or increase profits).

—Provide for special needs (i.e., refrigeration, low-bed vehicles, etc.) and availability.

—Provide various transportation accessorial facilities, i.e., stopping-in-transit, containerization, etc.

—Establish alternative modes of transportation, i.e., cooperatives, piggy-back, fishy-back, etc.

—Establish policy for the use of distribution warehouses (company or public).

—Make up routing guides for shipping personnel and include standard weight per package calculations, distinguishing LTL routings from TL and CL routings.

—Establish procedures for weighing of freight, inspection of empty vehicles, containers, etc. prior to loading.

—Establish procedures for receiving freight and inspecting freight before acceptance.

—Establish procedures for auditing and paying freight bills, for settling of claims.

—Establish procedures for reconsignment, diversion, expediting and tracing.

—Manage all the above at all times.

Staff Functions

—Analyze all traffic functions and costs.

—Negotiate for rates, classifications and proper vehicles.

—Establish good relations with carriers through friendly negotiations, through participating in trade organizations and public events affecting transportation.

Cooperative Functions

—Establish rates and routing guide for procurement.

—Advise on site location.

—Advise purchasing on most economical transportation quantities and competitive rates from other vendor locations.

—Advise sales on transportation variations in units of sales.

—Furnish sales with freight rate information from competitive manufacturing or selling points.

—Aid customers in traffic problems.

—Consolidate deliveries to your plant.

—Review terms of purchases to establish transfer of title on all purchase contracts.

—Furnish proof of delivery for accounting where necessary.

—Handle transfer of household effects.

From this rather small list of line functions, staff functions and cooperative functions you gain some idea of the broad requirements of the distribution manager's job, and of the range of other departments—executive, accounting, sales, purchasing and personnel—that have a need for traffic input to work efficiently.

Traffic Organization

Companies starting in business, and small companies generally, do not employ a trained traffic manager. The work is done by (1) a shipping clerk, (2) the office manager or similar executive, and (3) outside sources, such as independent traffic consultants, professional traffic bureaus, shippers' associations and chambers of commerce. These outside agencies maintain professional traffic staffs that perform almost all standard traffic management functions for their members or clients. There are over 300 such associations and consultant services in the United States. The traffic manager of the association or consultant firm is in effect the traffic manager for each of the firms served, and in some cases actually carries a title (with company letter-head) identifying him as such. Some independent traffic management firms represent the shipping interests of as many as 100 or more smaller firms, a situation not unlike that of the corporate traffic manager who controls the transportation activities of as many as 300 or 400 company plants.

Many smaller companies handle routing and carrier selection in accordance with specific instructions furnished them by traffic departments of larger companies to whom they sell. The use of automation for purchase order-writing by such companies has been a factor influencing this, and purchase orders often include pre-programmed shipping instructions. One general traffic department of a major retail chain furnishes such shipping instructions to several thousand of its vendors and has a rule that it will not accept merchandise shipped otherwise.

In a similar vein, these smaller companies will usually permit their inbound shipments to be routed by larger companies from whom they buy because the larger companies employ special skills not available in the smaller companies and do make every effort to give their customers the lowest possible delivered costs.

As traffic expenditures and needs reach a point with a company where the transportation function should be assumed wholly or in part by a company employee, it is preferable that an experienced manager be hired for the job. If this is not in order, the individual assuming the responsibility should, if at all possible, take a traffic course or courses at one of the several good traffic schools. Generally the sessions are after work or at night, and the teachers are experienced and practicing traffic men having responsible, full-time, day employment in distribution. They do good jobs with their students and provide practical instruction related to day-to-day problems. Tuition is not particularly expensive and in many cases is underwritten directly or indirectly by the employer. Where a traffic school is not practical, a good correspondence course can be of real help.

In establishing the traffic organization, one should establish a tariff file, as small as it may be in its beginnings. Tariffs can be kept in ordinary filing cabinets until such time as growth dictates the use of cabinets built for holding tariffs. One should start with the classifications (National Motor or Uniform or both, depending on need), and continue on with exceptions and tariffs of the bureaus affecting the majority of your shipments. Inquire of the carriers being used and those indicating interest in your freight if they publish their own tariffs and ask them to supply you with them.

MANAGING THE TRAFFIC DEPARTMENT

Even before growth requires you to expand the tariff file, you should have the generalized routing guides and city motor carrier directories and Official Guide of the Railways, all of which are acquired by subscription. Rail and road maps are always useful.

There are a number of monthly and bi-monthly traffic publications that are free. One can be overwhelmed by all these publications and the efficient distribution manager develops a sense of discretion as to which he accepts and reads and which is discarded promptly. Two that should be subscribed to promptly are *Traffic World* and *Transport Topics*. (Note the list of publications, as well as other important working tools, in the Appendix.)

The routing guides and motor carrier directories mentioned in a preceding paragraph can be particularly helpful to shippers without much experience in transportation who use less-truckload, truckload, parcel post, United Parcel and similar services. These are compilations from official publications in which information is condensed and simplified so that it is readily available and more easily understood. This type of reference work has been around for some time, is used extensively by many shippers small and large; there are various good ones. Typical of the field are the "guides" published by G.R. Leonard & Co. which are in three basic categories. One includes available service, rate information, and other extensive and detailed data on parcel post, and United Parcel shipments; the second contains motor carrier class rates to the entire country, and the third contains motor carrier routings to all states. Arrangements can be made for "personalizing" shipper information to include particular features (i.e., more specific origins when indicated), and so on. Reissues and loose-leaf arrangements are provided to keep information up to date. The services are leased, rather than sold, and costs vary. The most expensive (made up for particular shippers) is $100 per year. Albrechts specializes in data of a related nature for the West. Similar reference works that may be found helpful include *The Shippers Guide, American Motor Carrier Directory*, and others.

Association memberships are only as valuable as the activities of the organizations offering them, and there are many in the traffic/transportation/distribution arena. Your own industrial

INDUSTRIAL TRAFFIC MANAGEMENT

organization's traffic section can be of great importance. You might also consider organizations such as the National Small Shipment Traffic Conference, National Industrial Traffic League and some specialized ones such as hazardous materials shippers groups, if your products fall into those categories. Depending on your location, joining the transportation committee of your local or state chamber of commerce might be in order and, finally, if you are a registered practitioner before the Interstate Commerce Commission, you should consider the Association of Interstate Commerce Commission Practitioners.

Continuing education for the traffic department staff is important for learning new ideas and new or different practices. Many universities offer continuing education courses and credit in traffic, transportation, distribution and logistics. Seminars are also of value when given by organizations expert in those fields, such as universities, the American Management Association, Penton Learning Systems and others.

Office procedures and the like can take their cues from current practices. By all means make sure good equipment (calculators and the like) are at hand. A possible worth-while step is to arrange for the checking of additions, multiplications, and the like on freight bills over certain amounts. Errors are common here, and recoveries will pay for the equipment and personnel. The checking of rates used in the carriers' bills with the tariffs, should be continuous. A filing system can be important. If it can be melded into an existing system, fine; if not, help can be obtained from business books on filing from local libraries. Filing may be by commodity, territory, mode of carriage, or in other ways or combinations of them. Preferably the system should have a basic foundation such as that provided by a Dewey Decimal type of filing. In most cases the papers should be filed by subject chronologically; special files for tariffs are available. Good arrangements for call-ups should be made; they are sure to be needed. A program for destroying old and unneeded files should also be established and followed.

Starting, expanding, or contracting traffic functions are similar in many ways to doing the same thing for a legal, tax, or other like activity. The legal work of some companies is handled by

MANAGING THE TRAFFIC DEPARTMENT

house attorneys for certain work plus a law firm or firms acting as general counsel on special matters. Similar dealings may be in order for industrial companies in traffic. This can be done by using transportation advisory firms which have been discussed in this chapter. What might be termed a "full responsibility" traffic department, however, can contribute greatly to the success of an industrial concern. See that this function is handled properly, aggressively, and adequately.

The features just mentioned as applying to smaller companies and neophyte companies in transportation also apply to medium-sized companies and larger companies in many ways. The involvements of the latter are on much broader bases; include more carriers and types of carriage, and are concerned with larger freight bills and opportunities to effect greater benefits for their companies. The traffic departments of all companies, whatever their size, have much in common, however, and what follows in this chapter as well as what appears elsewhere in this text applies in many instances to all.

Important fundamentals applicable to all—large and small—are that a traffic-wise company will be competitive with others in the same business, and that freight cost adjustments, when in order, can be of sizeable amounts. With many corporations, very substantial sales increases (if they can even be made) are necessary before profits "flow down" comparable to those that may "flow down" from freight economies. Other fundamentals are that traffic should reflect the management philosophies and over-all directions undertaken by the company and should work to follow and support them. Often this is not easy, because policies change and at times could be more clear and more consistent. The importance of such support, however, is obvious.

Industry does not allocate responsibilities to its traffic organizations on a uniform basis. Some traffic departments have as their responsibilities the obvious traffic functions of auditing freight bills, routing, claims, expediting, tracing and handling rate and classification problems. Other traffic departments have further responsibilities including, in addition to the above, transportation contracts, distribution arrangements, warehousing, receiving and

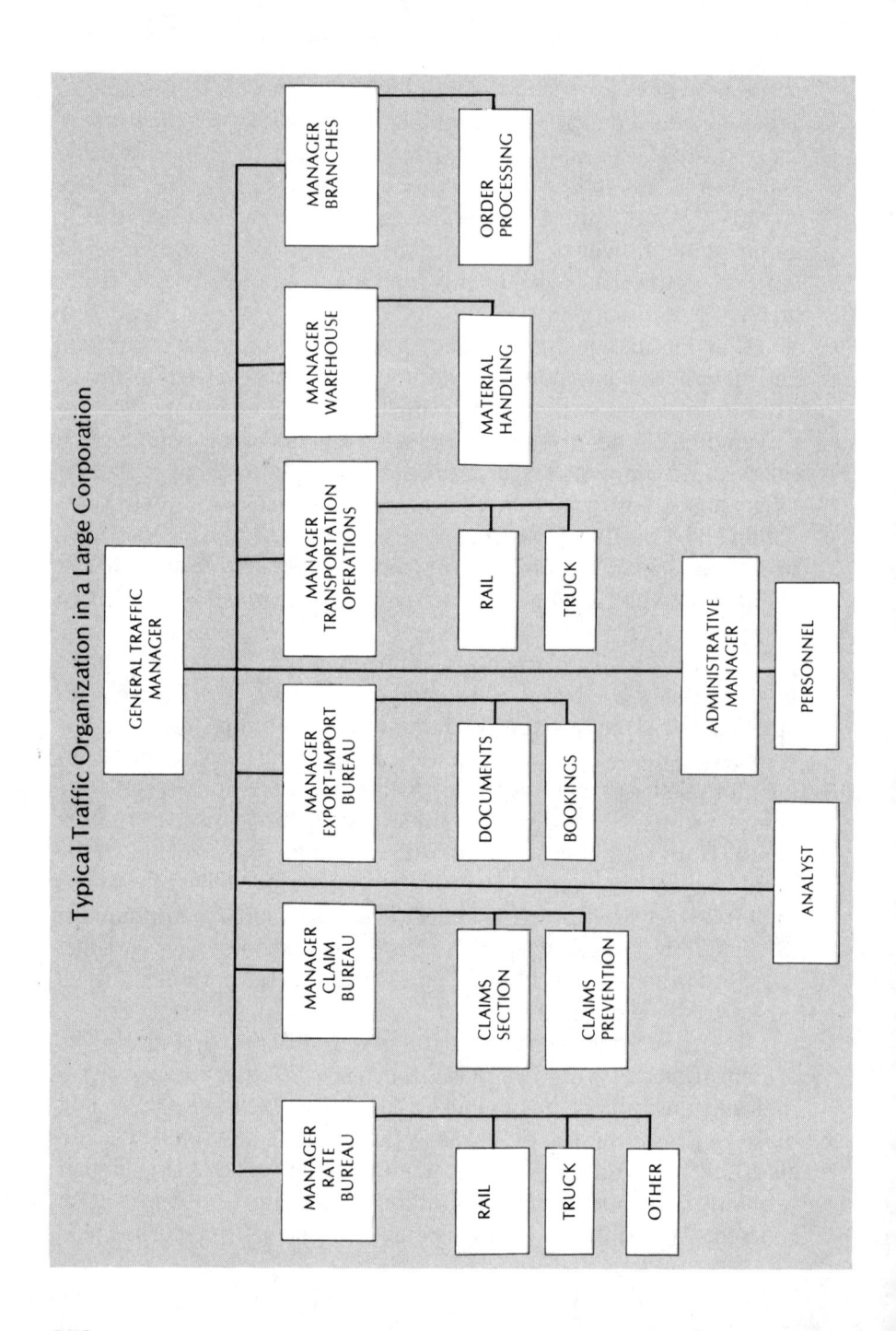

INDUSTRIAL TRAFFIC MANAGEMENT

Typical Traffic Organization in a Large Corporation

- GENERAL TRAFFIC MANAGER
  - MANAGER BRANCHES
    - ORDER PROCESSING
  - MANAGER WAREHOUSE
    - MATERIAL HANDLING
  - MANAGER TRANSPORTATION OPERATIONS
    - RAIL
    - TRUCK
  - MANAGER EXPORT-IMPORT BUREAU
    - DOCUMENTS
    - BOOKINGS
  - MANAGER CLAIM BUREAU
    - CLAIMS SECTION
    - CLAIMS PREVENTION
  - MANAGER RATE BUREAU
    - RAIL
    - TRUCK
    - OTHER
- ADMINISTRATIVE MANAGER
  - PERSONNEL
  - ANALYST

306

shipping departments, packaging, materials handling, local or intercity company-owned truck operations, moving employees' household goods, passenger bookings, rental of drive-yourself passenger cars and trucks, and export and import freight.

Specific organizational chart recommendations are of no value here since each company places its own values on various functions. Diagrams of several traffic departments are offered as generalized suggestions. Insert the specific functions or values that exist in your own company or corporation.

In considering industrial traffic organization charts, don't attempt to list in detail all the functions of each individual employee. This will vary greatly between industries and should be developed by each company according to its needs. Since traffic organizations must be established to meet the needs of each company's type of operation, standard organization charts for uniform use are not practical. Certain generalities, however, can be made as to broad features and other aspects.

Organization charts are helpful in enabling someone to understand how a department works and the functions of its various sections. Many knowledgeable executives, however, use these sparingly for even those purposes. Time taken to maintain them may possibly be spent more profitably elsewhere; often the ramifications of a particular job are of such delicacy that charting presents problems. Under more modern practices "Clerk" designations in job titles are declining. "Analysts," Supervisors" and the like are more favored as being expressive of duties undertaken. Rate clerks are receiving titles of Rate and Tariff Controllers (Comptrollers).

Job assignments can be on territorial, on functional, or on other bases as well as combinations. They should be sufficiently spelled out so that there is a delineation of responsibilities and the staff know what is expected of them. Flexibility should be preserved as much as possible, however, so that when someone sees something that needs doing he is not stopped or discouraged because "it's not my job."

Salaries, benefits, and related aspects of traffic operations (like many other features) are tied to a company's general personnel policies, its profit picture, how much of its gross is paid out for transportation and other items, some of which are of a non-transportation nature. As a part of management, traffic leadership should help in the general framework of these when it can. Personnel practices of other departments similarly situated are to be kept in mind in this framework as well as intra-department personnel dealings. Working within reasonable limits, traffic managers can do a great deal for their companies by taking particular care to reward those who get results, by recognizing increased competence and value through experience and maturity of judgement, and by careful attention to the other elements that are involved in salary administration. Companies with fair and proper salary programs are the most effective and profitable in most cases. As a related item, job promotions should receive particular care, with promotion from within the department or the company being so strong a principle that deviations are made only in exceptional cases. The top executive of one of the country's great companies recently pointed out that industry is expected "to pay fair wages and salaries and treat employees well, providing equal promotion and employment opportunities for everybody."

The following functions can well be treated as traditional responsibilities of the general traffic manager of a smaller company:

- Developing transportation policies and interpreting such policies throughout all company departments.
- Standardizing traffic procedures at all company locations on a uniform basis.
- Filing all complaints with government regulatory bodies.
- Approving all rate and classification proposals and handling them to a conclusion before carrier committees and regulatory bodies.
- Reviewing and approving all trucking and transportation contracts.
- Cooperating with sales, purchasing, manufacturing and other company departments on warehousing, storing, packaging, materials handling, private carriage, routing and distribution problems, plant location studies, inventory con-

MANAGING THE TRAFFIC DEPARTMENT

trol, export/import and electronic data processing distribution studies.

• Conducting continuous studies on all freight movements. Included in this is the development of opportunities to consolidate small lots into pool cars, pool trucks or piggy-back direct to final destination or to strategic points for distribution to final destination.

• Establishing a company-wide tonnage distribution policy for observance in routing freight via all types of carriers.

As companies expand and traffic men are given additional

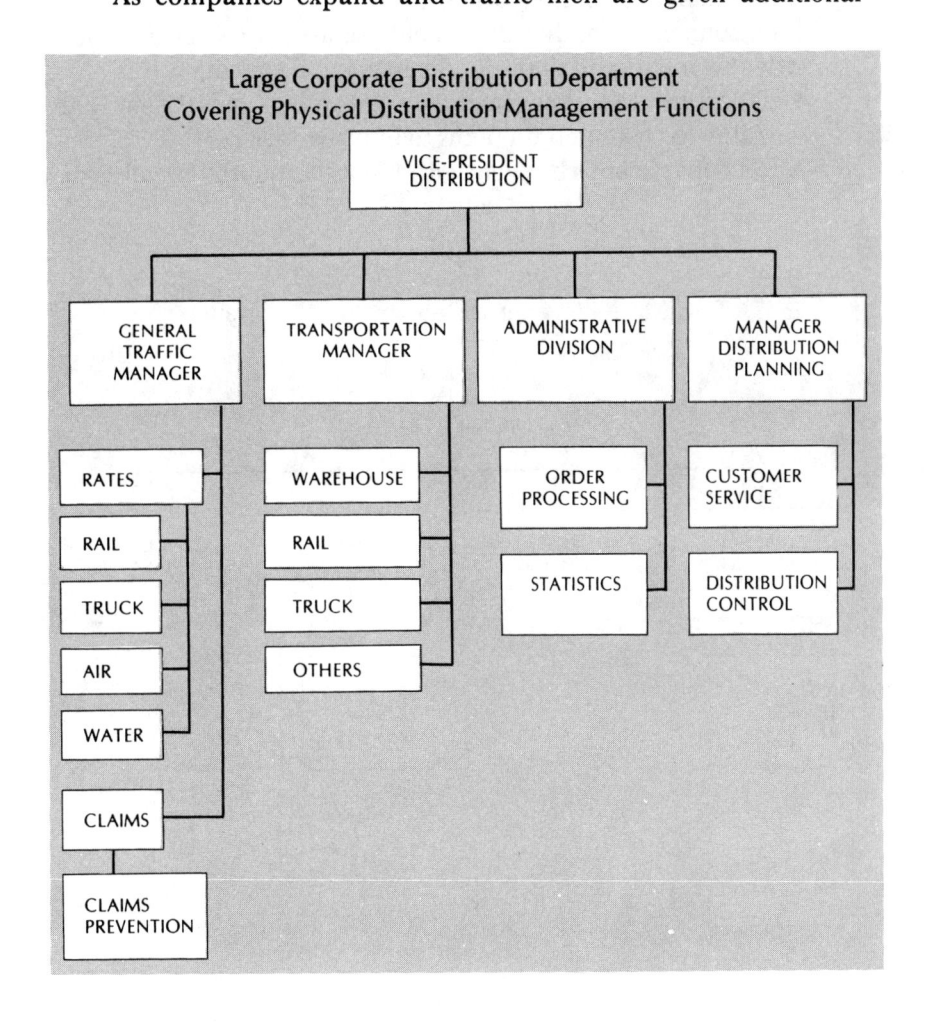

Large Corporate Distribution Department Covering Physical Distribution Management Functions

- VICE-PRESIDENT DISTRIBUTION
  - GENERAL TRAFFIC MANAGER
    - RATES
    - RAIL
    - TRUCK
    - AIR
    - WATER
    - CLAIMS
      - CLAIMS PREVENTION
  - TRANSPORTATION MANAGER
    - WAREHOUSE
    - RAIL
    - TRUCK
    - OTHERS
  - ADMINISTRATIVE DIVISION
    - ORDER PROCESSING
    - STATISTICS
  - MANAGER DISTRIBUTION PLANNING
    - CUSTOMER SERVICE
    - DISTRIBUTION CONTROL

309

INDUSTRIAL TRAFFIC MANAGEMENT

responsibilities, the functions listed above may be broadened and increased. Those listed where traffic is shown as cooperating may well be changed to show traffic as having the direct administrative responsibility.

The modern traffic manager must place his company in the forefront with the customer service it provides. He should have a plan and a list of objectives that he constantly reviews and checks for progress. His list will cover:

- Organization, present and future
- Personnel, plans for placement, salary, titles, evaluations, advancements and plans for departmental education
- Reporting to management, to his own department and possibly to customers on plans to improve service
- His department's budget, with monthly, quarterly and semi-

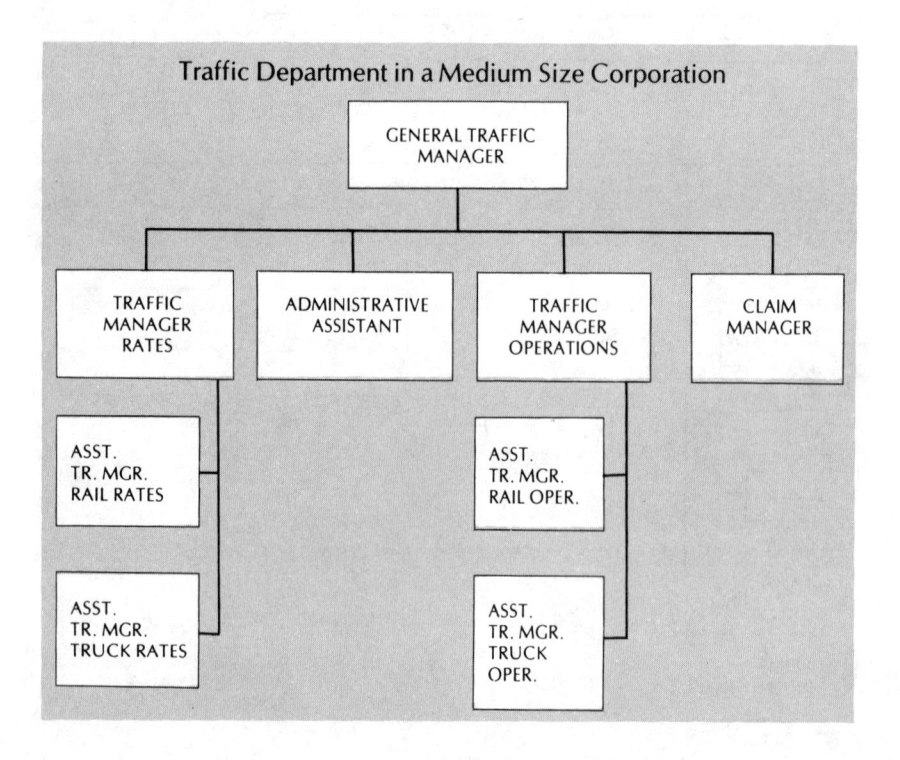

Traffic Department in a Medium Size Corporation

```
                    GENERAL TRAFFIC
                        MANAGER
    ┌───────────────┬──────────────┬───────────────┐
 TRAFFIC       ADMINISTRATIVE    TRAFFIC          CLAIM
 MANAGER         ASSISTANT       MANAGER         MANAGER
  RATES                        OPERATIONS
    │                              │
  ASST.                          ASST.
  TR. MGR.                       TR. MGR.
  RAIL RATES                     RAIL OPER.
    │                              │
  ASST.                          ASST.
  TR. MGR.                       TR. MGR.
  TRUCK RATES                    TRUCK OPER.
```

annual expense reports for comparisons with other departments' budgets
- Analysing methods, legal proceedings, transportation systems, existing facilities and current procedures
- Cooperating with sales, purchasing, finance, marketing, production and personnel
- Participating in outside organizations and judging the value of those organizations to his company and the value to the advancement of his knowledge

The status of transportation in a company's general organization chart is important. The higher it is, the greater is the department's prestige, as a general rule. At times such prestige may be influential in getting things done with carriers and others. Additionally, other benefits flow from such recognition, including encouragement in improving the company's over-all fortunes, more satisfaction and self-fulfillment for the staff, in that their work is recognized as of real value. Obviously there are practical limits to how high this status can be and other features of the company can be as important—often more so. Included in these are worthwhile pay scales and benefits, opportunities for substantial accomplishments, freedom to do quickly and simply what should be done, and actual demonstrated encouragement and appreciation from the rest of the company (including top management) together with fine and responsive help from them when needed.

Traffic should be placed high enough in the organization to assure it the authority and prestige necessary to accomplish the best results. Further, and of equal importance, its situation should be such that top management encourages and supports its traffic executives so that they are assured of consistent company-wide cooperation. Much more than a position on an organization chart is involved. The actual delivery to the company of *bona fide*, worthwhile benefits is basic. Accomplishing these and explaining them in simple, non-technical terms to those affected helps importantly. "Sales" efforts along these lines mean a great deal on many fronts, including, perhaps, assignment of traffic to a higher position on the organization chart.

The executive to whom the traffic manager reports will probably not be familiar with the complexities of traffic management and he will have substantial calls on his time from other company functions. Depending to a great extent on the individual, everything possible should be done, as a general rule, to see that he has an over-all understanding of the department, its strengths and weaknesses, and its short and long term plans. Going over current items with him from time to time is also in order. Efforts in these regards may meet with varying reactions. Many in higher management echelons feel that managers should "manage" their departments pretty much on their own and, if that is the executive's position, it should be respected. If some other tack is indicated, take it. Relations with a top management contact are important and traffic should extend itself to the end that he is cordial, helpful, cooperative, and enthusiastically interested in what traffic does for the company and its advancement. Others in high echelons may be similarly positioned for such particular attention.

Traffic management and the technicalities of transportation are not that complex to traffic personnel, but they are to non-traffic people, just as the activities of lawyers or tax specialists are to the non-specialist. Accordingly, all discussions should be short, to the point and in plain terms whenever possible. The decision you believe should be made should be backed with your reasons. If your top management knows other considerations that make other moves appropriate, they can well modify those you propose. In carrying them out (particularly if they involve unpleasant features) the best effort should be made to "keep the boss off the spot" by explaining the facts behind the action rather than by dumping the responsibility on him. Also, if anything extraordinary involving the department looms on the horizon, it is preferable that this be passed to him right away. Higher echelon colleagues are more effective when they are not put on the spot and are not subjected to "surprises" from "outside" sources.

Providing information that will enable all involved to give better assistance to those needing it is important to traffic's relations with all management. This will also complement other operations within the corporation. In any number of companies "feedbacks" to traffic of impending company actions and developments known at those levels are made to work for the company's benefits. Any number of examples could be cited. One actually occuring recently involved a midwestern manufacturing company that was able to retain a valuable account through a rate adjustment "triggered" by cooperation of the sort here mentioned.

Traffic Operations

Management principles applicable elsewhere in the firm are applicable in transportation, and those that have worked successfully in other connections can well be guides to traffic effectiveness. John D. Rockefeller's concern with the acquisition of "brains" (rather than properties) and Henri Deterding's with "simplification" and "getting the best out of people" as in putting together the fine organizations of Standard Oil and Royal Dutch Shell find particular applications here. A constant looking out for his "money makers" was the principle of Scotsman Alexander Fraser, one of the many excellent Shell Oil presidents who contributed so greatly to its success. Colonel Earl H. (Red) Blaik's winning West Point football teams were brought to stardom on the precept that true success takes more than luck or natural gifts and that "you have to pay the price." More recently, we have the thoughts of Robert Townsend of Avis fame, that business should be "fun" and that the nods for needed executives should go to "anybody who will work full time" and "who doesn't think he's a genius at anything." Paramount of all of these, of course, is the always-to-be-remembered, "do-unto-others . . ." admonition of the Golden Rule.

Management and management leadership techniques are influenced in many ways by those of the company. Due regard is to be given these, and, along with them, such guidance as can be gleaned elsewhere. The American Management Association, (135 W. 50th St., New York, N.Y. 10020, (212) 586-6100) is active in this field and has various texts and publications that may be helpful. Many company personnel departments receive these and they are useful sources of information. A basic management style that appears in varying forms is that of "team management." It has been characterized as the most difficult, the most effective, and the most rewarding of leadership styles. Under it, the executive (i.e., the traffic manager) is concerned for both high production and the needs of his people. The building blocks, however, are not individuals but horizontally and vertically integrated work teams. The leader seeks to involve his subordinates in the work through consultation, participation, and joint problem solving. Tensions are openly admitted and dealt with. Attempts are made to make the work more meaningful rather than hard or easy. While a full application of this philosophy to traffic functions could well be going too far, it has possibilities for doing better jobs in some areas and could be worthwhile on that basis.

High on the list for a traffic department's attention is the matter of cooperating with other departments and securing their cooperation in solving traffic problems. Sincere work is necessary to be sure two-way communications are open and moving freely. Across-the-desk visits with accounting, sales, production, and others with information to exchange, are often of value. Their problems are traffic's when traffic is also involved, and all encouragement possible should be given them to have traffic help when it can. This goes also for the whole organization. When possible, trips can be made to do double duty; when one becomes necessary, including local marketing offices, plants, and other installations can well be arranged. Visits to get up-to-date on what is going on are helpful. In all these contacts, care should be taken that the head man of the other department knows what is going on. The first step on any plant visit might well be a call on the plant manager or, if he is away, to his office. Further, letting the other fellow first do the talking is good psychology in these cases; it's better to listen rather than engage in a conversation in which traffic problems overshadow the others. Traffic's time can well come later.

Another feature on the list for traffic attention is the delineation of plans for the future. No matter how hard a company tries, some of its operations can be improved upon; and some of the conditions with which it will be faced in the future need attention ahead of time. Corporate practices today run more to spelling out goals than in the past. The extent to which they are spelled out varies. Such planning includes timetables within which goals are to be reached as well as the means to be used to reach them. Prospective manpower requirements are a part of those means. A wide range of features may be involved in such plans or they may be comparatively simple. Traffic's timetables should be thought out and spelled out carefully, to the end that all operations needing attention may reasonably be anticipated, recognized and dealt with appropriately. Company budgeting, into which more traffic functions are being drawn, are also to be handled in similar fashion. Take care that time spent in planning, budgeting, and the like, is reasonable and that flexibility is not curtailed because of them.

The personnel picture for traffic has been changing recently. Wider responsibilities, stepped-up activities, and increased complexities in recent years have made this field more critical to companies in many ways. The need for the particular knowledge and abilities that formerly existed is still there, but new techniques of automation and containerization, changing regulations, minority groups, wage-hour considerations and other developments have made added skills necessary. Data processing, engineering, and legal talents have been among those added to traffic staffs in increasing numbers. Coming on stronger also has been a general corporate feeling that management in all brackets (lower to upper) should become more attuned to productivity. Leadership is expected to concentrate more on educating, motivating, overseeing, and determining that jobs are well done. The development of a strong and effective traffic department has always been important and it is even more so today.

At least two management truths are applicable to traffic staff. First, special capabilities are needed to become a good transportation professional; and second, time is also needed. These special capabilities include a "flair" for detail, and being a "digger." Other faculties include being tenacious, not easily discouraged, patient, and a "self starter." As regards time needed to become a good traffic manager, the range can be from two to ten years, the lesser for narrower responsibilities and the higher for the broader ones. This time span is to be hedged with how capable the individual is, exposure received, how successfully the person handles assignments, and so on. Employment activities and plans should include these considerations of getting the best employees and the length of time it takes to train them.

Traffic department heads seeking people "to deliver the goods" in their departments go to a number of places. At one time many industrial traffic organizations were staffed with former carrier employees. Carrier ranks are still used as a source, but traffic trainees may also be sought from other company departments, recruited from universities, trade associations, other industrial companies, and so on. Some, in particular, are sought from a company's field offices, as field experience can be particularly valuable in a main office. Educational specifications, i.e., university attendance and degree requirements, can generally be kept flexible, particularly if the individual involved obviously has a "will to learn" with the job and the drive to progress by getting results for the department. Referrals from others, applicants coming in "cold," are also possibilities, of course.

There is no pat way to tell whether a prospective employee will or will not be outstanding, or, for that matter, whether once hired he will handle competently the increased responsibilities as they are assigned to him. Psychological tests are used by some firms to help in this regard. Others have tried them and discontinued them. Some companies feel its managers are using them as "crutches" and giving them undue weight in hiring, promoting, rewarding and discharging.

Undoubtedly many companies have particular guides that they use in their employment practices. Sometimes these work, sometimes not. They may be helpful. One manager experienced in hiring makes it a point (all other things being equal) to hire applicants that are neatly dressed, and obviously careful of their appearance. The reasoning is that those who take care of their appearance will also take good care of the work they do. Some use school records, searching for clues as to personality, work habits, extra-curricular activities, statistical position in the class and other conditional situations.

Skills for good traffic people can be developed by on-the-job

MANAGING THE TRAFFIC DEPARTMENT

training, which can include problem solving of varying degrees of complexity as the trainee moves along in the department. A program of rotating jobs within the department is often helpful. A continuing education program, including one that leads to the employee taking the examination to practice before the ICC is a good indicator of interest. While the license to practice may not be needed by a particular department, passing the examinations and meeting the research requirements, when brought to the attention of the department head and personnel manager, should result in promotion and increased earnings.

Other educational activities may include gaining membership in the American Society of Traffic and Transportation, and participating in National Council of Physical Distribution Management "Roundtables," attending university courses and seminars. Educational activities should also include courses in the humanities as well as commercial functions.

Company philosophy will determine whether the traffic department will be a centralized or decentralized one. The dialog for or against either system is a long one, but the final concept of management is that of the chief operating officer and it is a concept that will affect many functions in addition to traffic.

Generally, it is difficult to decentralize for one or more modes of transportation and centralize for other modes. The methods may change as top executives come and go and the change over periods can be difficult, and even, depending on personalities, disastrous. Flexibility should, then, be the keynote of the traffic manager's operating philosophy.

A strong central traffic department has many points in its favor. Able rate men are scarce and their talents can best be used for the company as a whole. The development and maintenance of good tariff files are difficult, costly, and time-consuming; good tariff information is best used on a company-wide basis. Experts in specialized traffic functions can perform professional jobs for all plants and divisions of a company better and more economically when working together from a central point. Hiring comparable talents for each individual plant is too expensive and the volume of work does not justify it. By having those specialists in one place and channeling their activities properly, each decentralized unit will receive the full benefit of the best talent that is available.

Centralized traffic organizations do not take any needed prerogatives away from the local departments. Quite the contrary is the case. For example, a centralized tonnage distribution policy directing the distribution of over-all tonnage among the carriers results in maximum benefits throughout the company. One unit may give an individual carrier little tonnage. But all units together may favor that particular carrier substantially. Therefore, if a service, rate or classification problem develops with the carrier in connection with one decentralized unit, the full weight of the total company tonnage is available to assist in working out the local problem. Further, particularly with large companies "everybody's business can become nobody's business" and central traffic organizations help importantly in preventing funds from going down the drain in this regard.

In this manner, we do have a centralized department performing staff duties while the decentralized portion is an operational department.

Traffic help for decentralized departments from a central office is made available in practice in various ways, some or all of which may be used. Reduced to bare essentials, these involve providing the field offices, in uncomplicated form, the rate figures and information they must have; and in working with, and providing expertise to, the field offices so that their activities are within the framework of the law and result in maximum benefits. Experience has shown that the complications of transportation law are such that operating unlawfully without realizing it is considerably easier in decentralized operations.

One program that has worked effectively involves the main office traffic department furnishing to plants and marketing departments in the field (and keeping up-to-date) master shipping instruction forms. These are prepared for locations where repetitive business is involved. As the name implies, data given include the consignee's name and location, the freight rates that may be involved from various shipping points, the routings, and so on. Basically, the plants use these forms to send shipments on their way, and the marketing offices use them when shipments arrive. When a form is needed for a new customer, it is provided by central traffic. When there is an urgent or good reason to deviate from a routing specified, those in the field do so, subject to checking back with the main traffic office for alternates or other handling if indicated. In many cases alternates have already been provided on the forms and their use is automatic. The centralized traffic group also prepares and sends to appropriate units "rate sheets" that give freight rate information on products from numerous origins (including those employed by others) to destinations that may be of interest. These are used in freight studies, sales promotions and other ways.

One aspect of this program is the possibility that some freight bills may have to be checked at a local office, but all freight bills come to the main office for audit and collating with other documents and eventual filing. The same is true for claims. The main office should be the center for collecting and settling claims, keeping track of time limits and following the performance of a carrier serving more than one origin point.

Those doing this work in the field offices are called "operating groups" as a general rule, although "transportation supervisors" is the title used in some instances. Many do other work in addition to traffic. The central traffic office aids the field offices in other ways—for example, by adjusting rates on inbound and outbound shipments, maintaining competitive freight rates, improving carrier service and availability, and generally checking on demurrage practices and other transportation regulations. The main office makes periodic studies to ensure that the company is getting the full benefits of any changes in distribution patterns and the like. No particular point is made as to the individual to whom the office staff reports in the main traffic office; the concentration is on getting a good job done. Functional reporting to traffic is important. Allocations of outlays for main office traffic costs are made, but time spent on them is held to a minimum and details are avoided.

Help and cooperating in this program works both ways. Staff in the field do a top job helping their main office traffic colleagues, and when indicated, they are promoted into the central organization. By being on the spot, they observe and handle promptly many things that the central staff, miles away, cannot deal with nearly as effectively. At the plants, field staff may be particularly concerned with outbound shipments, the scheduling of motor trucks, and making sure sufficient trucks are available for loading. Other field staff concerns may involve shipments inbound to customers, arranging for proper freight rates on them, and so on. In fact, any transportation work that may reasonably be done in the field, is done there.

An important factor in the decentralized operation just described is the local personnel work at the plant site to find good, trainable and promotable people. These may be people who, in their earlier stages of employment, might not want to move to a main office, but with a good training program and with educational opportunities could be ready to move upward.

An industrial traffic executive faced with a decentralized operation can proceed with only one philosophy—to see that the general staff office performs so well that the decentralized offices simply cannot get along without the guidance and help of the central office.

A decentralized traffic organization entails sensitive personal contact with men who have the responsibility for the operations at their local plants or distribution points. There is the matter of making money for the company, and transportation can do a great deal of good in this regard. One approach that has worked well has been to make periodic studies of both the inbound and outward freight movements and present the results, demonstrating the savings made, to the local manager in person so that his department, too, shows up as a "profit center."

The traffic leadership group should be careful not to let itself get bogged down in detail. Trade publication reading (to follow transportation developments) can well be "spread," particulary to younger staff. A "reading" file is useful in many cases. This is made up of extra copies of all correspondence and is passed over the desks of management, section heads, and others who "need to know," thus helping them keep in touch with general developments.

Company passenger travel is traffic's responsibility in many companies. Some use travel agencies, with varying results. Where traffic has the direct responsibility, staffing, again, should be adequate and care should be taken that leadership time is not wasted that could be more profitably spent elsewhere for the company. Most travel is now by air and the guides mentioned later include time tables and fares for checking charges. Air credit cards have the potential for creating problems and should be dealt with carefully and concisely. For those travelling extensively, short-form plane schedules from most important cities are available. These can be particularly helpful in providing leaving and arriving times and in getting places when irregularities in work schedules are normal and travel activities must be kept as flexible as possible.

Equipping the Industrial Traffic Department

How much office space does a transportation department need? There must be recognition of the technical and professional aspects of traffic management, requiring separation of people and things so they may work efficiently. This does not necessarily mean each person in a private office; but sufficient space should be available for tariffs and other working tools at the desks of those who must produce precise information. Distractions should be avoided so that workers can concentrate.

Those whose work requires constant use of telephones should be isolated from those working on tariffs. Employees handling passenger reservations should be placed in a location convenient to visitors calling for tickets or information. Supervisors should be in private offices, since they must interview company associates in other departments, representatives of the various carriers, and employees reporting to them. Typists should be isolated as much as possible because typewriters, too, can be disturbing. Auditors should be near the tariff files. Office lay-out naturally depends on the space available and the floor plan.

Lighting should not be overlooked. Employees working with tariffs must have exceptionally good lighting. The printing in tariffs is often small, resulting in some eyestrain even under the best conditions. Certain non-supervisory employees frequently have callers from outside organizations and should be provided with some facilities for receiving visitors. A chair beside their desks is certainly in order and a separate meeting room or rooms for callers might be desirable.

The passenger group should be well provided with current timetables and information on fares, charges, restrictions, etc. Facilities for checking bills, securing credits, and the like should also be provided where indicated. If possible, this checking should be done as bills are received and should be kept current. Airlines and railroads will often furnish timetables direct to companies upon request. The paramount consideration, however, is to keep colleagues happy as they travel. Proper staffing, reservations, and the like should receive careful attention.

Telephone service is the lifeline of any traffic department. A traffic department is primarily a service organization and it must be able to contact other company departments and outside transportation concerns easily and without delay. No well-managed company will economize by reducing telephones in its traffic department; supervisors, expediting and tracing, reservations, claims, and other activities could hardly function without adequate phone service.

Many companies provide a central reception room where all outside visitors are received and announced to the interested department by telephone. Where this is not the case, special chairs or a bench should be set aside close to a secretary to one of the supervisors who can "double" as a traffic department receptionist.

How should tariffs be filed? Traffic men disagree on this. The bulk of inbound mail coming to traffic consists of tariffs. Large traffic organizations may have 5,000 or more different tariff publications on file. Each tariff has a great many supplements issued to it before it is reissued, and all supplements must be kept on file (even though they are not all effective at any one time) for auditing old freight bills or checking freight rates applied at some previous date. There would probably be an average of ten supplements to each tariff. Even small traffic departments may keep 500 or more active tariffs on file.

There are various kinds of tariff files on the market. Some are

MANAGING THE TRAFFIC DEPARTMENT

made of metal and come in several sections with small drop-front drawers for filing the tariffs. Each section can be as high as desired, depending on the number of banks of drawers used. Usually they are ordered for a workable height of about six feet. Several tariffs can be kept in each drawer, which has a holder for identifying name cards. Some files have a shelf between the top two banks of drawers that can be pulled out and used to hold an open tariff while checking a rate.

You can use regular standard filing cabinets for filing tariffs. Special inserts (holders, frames and pockets) are available which makes them reasonably handy. They are not, however, as satisfactory as the special tariff filing drawers and cabinets. Some companies doing a big freight bill audit job have all their tariffs sitting on open shelves. This may be an inexpensive filing system, but that is its only advantage. It does not look well and the tariffs get torn, dusty, and unusuable.

Here is a suggested order for filing tariffs:
I   Rail
    A. Uniform Freight Classification
    B. Class Rates by Territories or Rate Bureaus
    C. Exceptions by Territories or Bureaus
    D. Commodity Rates
    E. Special Tariffs
        1 — Open and Prepay Station Lists
        2 — Terminal Services
        3 — Accessorial Services
    F. Contracts
II Trucks
    A. National Motor Freight Classification
    B. Class Rates by Bureaus
    C. Exceptions by Bureaus
    D. Commodity Rates
    E. Private Tariffs
    F. Special Tariffs
        1 — Heavy Hauling
        2 — Accessorial Services
    G. Contracts

INDUSTRIAL TRAFFIC MANAGEMENT

   III   Import-Export
         A. Export
         B. Import
         C. Special
   IV   Package
         A. United Parcel Service
         B. Commodity Tariffs
         C. Air Line Tariffs
         D. Specials
             1 — Containers

The department may deviate from this line-up, but it does present a simple system that newly-hired persons will not require weeks of productive time to learn.

A list of the thousands of carrier tariffs currently effective would be very large. Each company must determine what tariff publications it requires for the particular commodities, services and territories within or between which it operates regularly. Carriers will supply any tariffs requested, either free or for a nominal charge (unless a request is for an old issue, the supply of which has been exhausted). Building a tariff file is not easy and takes time. A competent traffic manager, however, will be able to develop a satisfactory tariff file for the average company within a short time. As all tariffs are supplemented and reissued from time to time, and new publications are needed as conditions change, any tariff file, large or small, requires constant attention to keep it currently useful. Supplements and reissues will arrive daily and should be sorted according to application and filed the same day they are received. This responsibility usually is given to a young clerk who thus has an opportunity to learn something about tariffs.

There are quite a few publications that do not fall into the tariff classification but should be obtained by the traffic man because they qualify as being basic tools or serve as a source of needed current general information. In the listings in the appendix, no attempt has been made to judge the relative merits of the publications or to provide any detailed explanation of the subject matter covered. This list is offered primarily as a ready reference and as a reminder to those who may have overlooked certain publications. It is unlikely that any traffic department would need all the publications listed, but most traffic departments could use a substantial percentage of them.

Another suggestion: Why not set up a traffic library where all publications are kept on open shelves and where guide books, reference material, bound data processing reports and other material is kept available to all members of the department? The room should have one table and no more than one or two chairs. Simple open shelves will serve for storage.

Cooperating Functions of the Traffic Department

In many ways a traffic department is a service organization rendering some kind of transportation service to every other department in the company. The traffic department thus has a primary interest in maintaining the best relations with all other departments. This basic fact sometimes should be accentuated more by traffic executives who become preoccupied otherwise in maintaining good public relations with the carriers. Relations with other departments within the company should be promoted with painstaking care.

Many times the personnel in the other departments do not clearly understand what the traffic department does or just how it can help them. If this is the case, some salesmanship is indicated. This can be employed in several ways. One is by personal contact which, when handled diplomatically, pays rich dividends. Another is by issuing memos from time to time on rate adjustments in which the addressee is particularly interested and which can mean a great deal to him. The basic proposition that people like to have attention paid to them finds good application in these instances.

Making periodic, comprehensive reports to management was in the past quite the thing to do in some companies. It is still important now. Top management wants the traffic manager to manage, but still needs reports. It wants them in simple language, short, direct-to-the-point and understandable.

A report form follows that includes a number of items common to traffic operations and an explanation of the treatment given those items. The format will serve as a guide where top management wants such information in periodic bulletins. The information can also be used to supply any "internal" management needs for traffic statistics.

This bulletin is designed to enable an industrial traffic department to report practically all of its activities to management. Those activities that are reported by "number of units" require no explanations, since they are easily understood. Those activities that are reported in dollars saved or recovered may not be so easily understood. In "Special other-than-standard routing arrangements" is reflected the savings that result from the alert use of lower-rated or special services (private carriage is an example) that are not used for the main volume of traffic. These savings can be accumulated by freight bill auditors without too much added effort. Estimating may be used at times, under proper supervisory control.

Next, most concerns have regular scheduled pooling arrangements, and it is proper to show the money saved for each shipment. Non-scheduled pooling for a stop-off is another activity that should be included, and a separate heading is shown to cover it. "Arranging for expedited service in lieu of specified costlier service" means that a shipment has been requested to go by, say, express, but an alert traffic man may suggest an alternate type of service, motor carrier for instance, providing the desired delivery date. If the suggestion is accepted, there is a saving to the company, and it should be reported. "Transportation overcharges recovered" needs no explanation, except to explain that "Charges billed to distributors or vendors" covers those cases where a distributor or vendor does not use good shipping practices—for example, failing to follow lower-rated routing instructions, so a "bill-back" is initiated to cover the amount lost. "Miscellaneous" covers a variety of activities that may be important. This is why all money shown in this column should be explained in a separate sheet headed "Remarks." If a rate is reduced, or a classification rating is adjusted downward, it is proper to show a saving for such activity. It is not feasible to do this on a continuing basis, but it does seem reasonable to estimate the annual savings, and to show this annual figure under "Miscellaneous." Thus the saving is taken only once but

MANAGING THE TRAFFIC DEPARTMENT

Traffic Division: Hqtrs.; Location: Baltimore; For Month of: October MONTHLY DOMESTIC TRAFFIC BULLETIN	Month Issued: October NUMBER OF UNITS
EXPEDITERS	891
TRACERS	320

PASSENGER TRANSPORTATION
  Reservations secured
  Reservations cancelled
  Refunds procured
  Air travel cards issued
  Air travel charges audited
	
	103
	10
	8
	3
	86

HOUSEHOLD GOOD SHIPMENTS HANDLED	15
RATE QUOTATIONS	563
TRANSPORTATION CO. TARIFFS AND SUPS. — Received	121

TRANSPORTATION BILLS AUDITED
  Express
  Freight
  Local Trucking
	
	503
	2010
	600

ADJUSTMENT OF TRANSPORTATION CHARGES
  Bills reduced before payment
  Charges billed to suppliers
  Overcharge claims originated
	
	180
	18
	82

LOSS AND DAMAGE CLAIMS	53
ROUTES ISSUED	601

TRANSPORTATION COST REDUCTION
  Special other-than-standard routing arrangements
    Panama Canal (incl. rail & water)
    Other water lines (incl. rail & water)
	Amount
	
	$ 1,003.00
	4,020.60

Use of company private carriage operations (owned or leased)	$ 2,800.00
Arranging for inclusion of shipments in Scheduled Pool Cars or Trucks (including T.O.F.C. movements, Piggyback Plans II, III, IV or V and use of Shippers' Associations)	$ 4,010.00
Arranging for Consolidation or Stopping in Transit to partially unload or to complete loading	$ 3,100.00
Arranging for Expedited Service in lieu of specified costlier service (reducing use of Premium Transportation)	$10,600.00

TRANSPORTATION OVERCHARGES RECOVERED
  Freight Charges Reduced before Payment
  Overcharge Claims Collected
  Charges Billed Back to Distributors or Vendors
	
	$12,500.00
	6,025.00
	916.00

LOSS AND DAMAGE CLAIMS — COLLECTED Provide separate sheet with details of any claims outstanding six months or more	$ 1,682.00
MISCELLANEOUS. Detail to be included on separate sheet under "Remarks"	$20,016.00
GRAND TOTAL	**$66,672.60**

TOTAL OPEN UNPAID CLAIMS		DEMURRAGE CHARGES PAID		
Number	Amount	Inbound Cars	Outbound Cars	Total Demurrage
15	$4,000.10	$604.00	$38.00	$642.00

Specimen Industrial Traffic Bulletin — Report Form

covers a one-year period. Many other activities such as a contract negotiation for local hauling (where reduced costs result), a plan to reduce demurrage costs, etc., may be shown under "Miscellaneous," either as a lump sum saved by a specific action, or as a sum representing an estimated annual saving. The item for demurrage paid is to acquaint supervision with the amounts so expended. This is particularly helpful on reports received from outlying plant locations.

The traffic department deals with all company organizations on many common functions such as passenger accommodations, movement of employees' household goods, and any and all transportation or allied problems resulting from exporting or importing activities. Additional or more selective traffic activities with other departments are summarized briefly as follows:

Sales

- Quoting freight rates to salesmen
- Suggesting package quantities that make transportation economies and may increase each sale
- Selecting best routing for good customer service
- Advising on normal transportation service and cost and premium cost and service
- Advising when rates change
- Establishing normal transportation patterns to help customers arrange ordering pattern

Manufacturing

- Suggesting improved methods of moving materials intraplant (between buildings)
- Suggesting improved methods of shipping materials intercity (between plant locations)
- Advising as to private carriage operations' ability to haul raw materials
- Advising on package specifications, including unit or pallet loading and/or storage possibilities
- Advising on materials-handling equipment
- Obtaining adequate car and truck supply
- Hiring outside trucks and chauffeurs when required during peak periods

MANAGING THE TRAFFIC DEPARTMENT

- Expediting and tracing raw materials to keep production lines operating continuously

Shipping
- Reissuing information in carriers' tariffs in a simple plant guide to shipping clerks
- Providing classification descriptions for use in preparing bills of lading
- Establishing best bill-of-lading form
- Advising on use of pallets, unit loads or containers
- Obtaining adequate car and truck supply
- Supervising consolidation and pooling of shipments for direct routing or stop-off in transit movement
- Setting up procedures for economical use of piggy-back services

Receiving
- Handling loss and damage claims
- Routing inbound shipments so as to prevent congestion at receiving platforms
- Expediting and tracing urgently needed materials
- Supervising average demurrage agreement operations
- Inspecting inbound shipments subject to damage due to faulty packing or loading
- Auditing freight bills

Insurance
- Discussing and advising on all insurance affecting any form of transportation

Purchasing
- Routing inbound shipments by furnishing routes to be placed on each purchase order
- Advising on quantities to buy and how they should be packaged to protect lowest freight charge
- Quoting competitive freight rates
- Reviewing purchase terms affecting transportation
- Assisting in preparation of contracts
- Arranging transit privileges
- Expediting and tracing urgently needed materials
- Arranging bill-backs to suppliers when failure to use good shipping procedure is detected

Accounting
- Auditing freight bills
- Auditing transportation charges on invoice
- Collecting transportation overcharge claims
- Cooperating on data processing programs
- Advising on general transportation problems
- Setting up credit arrangements with carriers
- Arranging re-audit (secondary) of freight bills by an outside audit bureau
- Paying of freight bills

Legal
- Preparing contracts for transportation of materials or rental of transportation equipment
- Advising on transportation legislation and its effect on company operations
- Cooperating in legal cases regarding claims or other transportation matters
- Preparing transportation data for rate cases
- Seeing, in conjunction with legal, that transportation laws and regulations are observed in all fields of company activities

In connection with the listing under Manufacturing, it is of ever-increasing importance that traffic, the packaging engineer and the materials handling engineer form a sort of triumvirate that works closely on all shipping matters. This group, in turn, should cooperate with sales and other departments in a joint effort to reduce transportation, warehousing and distribution costs through coordination in package, container and materials handling equipment design.

As regards distribution, warehousing deserves careful consideration under any traffic organization discussion because it is part and parcel of the physical movement of goods from factory to customer. Sometimes it involves transportation privileges such as "storage-in-transit" or "processing-in-transit." But usually warehousing is a pause in the through movement and must be arranged for as economically as possible. Thus it is that many companies place the warehousing function, both private and public, directly under the general traffic manager.

The relations of traffic with the legal department are very close in many companies. Traffic's particular expertise carries with it a knowledge of what may be involved when proper freight billings and payments are necessary, when and how operating rights are to be observed, when and how strict demurrage provisions are to be complied with, and so on. These, and other, quite technical aspects of operating within the framework of transportation laws mean that the legal department and the transportation department can do a great deal of good for each other. The possibility of fines, special damage recoveries, adverse publicity, and other unsatisfactory consequences of what may be an unintentional noncompliance with some transportation law are matters of important concern in a company's operations. Both legal and traffic departments are expected to conduct operations so that such difficulties do not arise. Where departmental relations of this type in a particular company can be brought closer together, traffic can well move to see that this is done.

Traffic Department Public Relations

In his various activities the traffic manager must depend not only on excellent relations with other company departments but also on good relations with the carriers and the industrial traffic managers in other companies, particularly those in a similar type of business. So often one hears an experienced traffic manager say that the more one learns about traffic the more one realizes how comparatively little one knows. You can't know it all; the secret is knowing where to go to find out.

"Public relations" in this book refers to the industrial traffic man's activities outside of his company. Such relations may well start with the carriers who handle most of the company's business. By a judicious tonnage distribution policy and friendly relations, a company maintains worthwhile connections with all its carriers. So that its staff requests and gets the finest cooperation from them at all times.

Meeting other industrial traffic professionals would seem to be of at least equal importance. The Drug and Toilet Preparations

Traffic Conference, the Manufacturing Chemists Association, and the American Petroleum Institute traffic committees are but some examples of many instances where companies in an industry help each other. Industry traffic committees are effective in working with the carriers to solve general service and other problems for their particular fields of operations. Through combined efforts they do for all companies similarly situated what one company would find extremely difficult, if not impossible, to do on its own.

There are also industrial traffic groups that embrace all types of industries and work with the carriers on over-all transportation problems, both on a local and national level. Chambers of Commerce and local and regional manufacturers associations often maintain traffic professionals who are also practitioners and offer counsel and help to members. Some groups also include carriers. Information in more detail on some of these follows.

The National Industrial Traffic League (NIT League) is a national organization whose membership is composed exclusively of industrial traffic executives. The League has about 1,800 members representing every type of industry throughout the country. It meets annually and more often when necessary. Most of its work is done in committees that are appointed to cover all transportation issues of national interest. The committee chairmen report to the membership at the annual meetings.

The League supports a full-time executive vice president and staff as well as a general counsel. It appears before the Interstate Commerce Commission and other bodies through its general counsel and committee chairmen. Useful circulars and publications are sent to all members informing them on current transportation happenings. Dues depend upon the size of a company, with maximum at a very reasonable figure. The League is most helpful and does a great deal of good for transportation.

The National Freight Traffic Association is a fine transportation organization composed of both industrial traffic and carrier executives. Its object is to bring together traffic men who are closely involved in business dealings with the freight traffic departments of transportation companies to consider and discuss their common problems and to get to know one another professionally. It

MANAGING THE TRAFFIC DEPARTMENT

meets twice a year and also holds an annual luncheon and dinner meeting in New York City. It has a limited membership, so there is always a waiting list of applicants.

The Transportation Association of America (TAA) is an excellent organization made up of providers and users of, and investors in, transportation. Much of its work is aimed at composing differences between the various modes of transportation to the end that fair and reasonable dealings are accorded them, the basics of private enterprise are preserved, and differences are settled by those directly involved. Work is often done through panels, i.e., rail, air, user, etc. The user panel is made up of top industrial traffic men from all parts of the country, meets regularly, and has a good record of accomplishment. TAA's particular talents are also employed importantly in special situations. Hazardous materials regulations and particular help to certain features of data processing are among the fields in which TAA has done noteworthy work.

The Transportation Data Coordinating Committee (TDCC) has as its important responsibility the coordination of transportation data processing activities between industrial companies, carriers, financial institutions, the federal government, and others. It provides a means whereby common codes and common computer procedures are established and made available so that transportation may benefit to the maximum from the computer. Work is done by full-time staff and task forces. Knowledgeable industrial traffic men with special data-processing training are members of these task forces and contribute importantly there. Periodic meetings and seminars are held which are open to those with an interest. Important support comes from shippers who are presently involved with computers or who expect to be.

The National Committee on International Trade Documentation (NCITD, Suite 1406, 30 E. 42nd St., New York, N.Y. 10017) has as its special field the simplification of documents used in foreign trade. Industry, carriers, banks, insurance companies, the federal government, and others are involved in the activity; cooperating groups with similar aims have been established in foreign countries. Progress has been steady and the committee can show worthwhile accomplishments in the battle against paperwork

INDUSTRIAL TRAFFIC MANAGEMENT

costs. A recent publication, "Paper Work or Profits?," on international documentation catalogues and analyzes practices and procedures in this regard and recommends corrective steps. Support for this activity by industrial companies in world trade is strongly indicated.

The Association of Interstate Commerce Practitioners is a group of attorneys and practitioners who hold certificates permitting them to practice before the Interstate Commerce Commission. Its mission is to maintain high standards of technical procedure and ethics among those who practice or are interested in its field of administrative law. It publishes an informative bi-monthly journal and traffic professionals admitted to practice should consider membership as a part of their activities. (Full-time employees may represent their companies before the Commission without being admitted to practice and a certificate of admission is not normally required. Certificates are excellent to have, however.) There are two classes of ICC practitioners—lawyers and non-lawyers. Admission of lawyers to practice is not as complicated as with non-lawyers. The latter must meet certain educational requirements, undertake certain studies, and pass specific examinations. Details are available from the commission. (See also earlier comments in this chapter concerning the value to a traffic department of staff with ICC admissions behind them.)

Shippers advisory boards, which are sectional and are sponsored and supported by the Association of American Railroads, may also be considered for membership by industrial traffic managers. The boards have both industrial and railroad members and are set up by sections such as the Atlantic States Shippers Advisory Board, the Ohio Valley Shippers Advisory Board, etc. They deal primarily with car supply, car conservation, service, etc. and represent a cooperative way for railroads and shippers to solve mutual problems.

Delta Nu Alpha (DNA) is an organization of those working in transportation who are interested in advancing their usefulness in their chosen field of work by obtaining professional educational and technical training and by helping others to obtain these benefits. It is a group of similar interests and aspirations in the transportation field.

MANAGING THE TRAFFIC DEPARTMENT

Traffic clubs offer a medium for contact with other traffic workers. Almost every city of any size has a local traffic club with a membership representing all forms of transportation as well as industrial traffic personnel. There are traffic clubs for women as well as for men. Most traffic clubs are members of the Associated Traffic Clubs of America which functions along lines similar to the local clubs, but on a broader scale. As well as operating as social organizations, local traffic clubs contribute substantially to the welfare of the traffic profession by sponsoring forums, traffic courses in schools, speakers on transportation subjects and the like. They also help and advise members in employment activities, sponsor public speaking courses for members, and so on. A traffic manager may well join the local traffic club and support it for a number of reasons. In some ways such action is a bet that selecting the traffic profession for a life's work is a smart move.

The American Society of Traffic and Transportation, Inc. (ASTT) is discussed earlier in this chapter in connection with staff development but the Society offers benefits that are much broader than those mentioned in that discussion. Membership, working in, and supporting the organization can contribute to uplifting the general professional status of transportation and physical distribution. The Society's aims and its work contribute importantly to transportation's prestige in American industry.

The National Council of Physical Distribution Management (NCPDM), also referred to earlier, has developed considerable popularity in recent years. Generally speaking its undertakings are in areas closely related to traditional transportation functions. Inventory control and warehousing are examples. Participating in its activities can be of tremendous value.

Various other fine transportation groups also hold interesting possibilities for membership; information on them may be found in trade periodicals and elsewhere.

15

GOVERNMENT TRAFFIC

It would take the combined tonnage of most major U.S. corporations to even approach the amount of freight moved by the United States government. The U.S. Government ships every conceivable commercial commodity and uses almost every conceivable form of transportation. In addition to commercial items, it ships arms, ammunition and military impedimenta in great volume. The scale of the government transportation activities has produced a shipping colossus that not only embraces all normal commercial shipping procedures but often seeks to move commodities where no like commercial traffic exists. Much of what the government ships on commercial transportation moves under provisions of Section 10721 of the Interstate Commerce Act revised and effective October 17, 1978 (P.L. 95-473). Until this revision of the Act, U.S. government domestic traffic was entitled to free or reduced-rate transportation under Section 22 of the unrevised Act; hence government traffic was often known as "Section 22 traffic."

The law now provides for the payment of the full effective rate on government traffic (10721, a, 1) although there are provisions in Section 10721 for the free transportation of certain individuals (customs inspectors and immigration officers) on government service, and some governmental (federal, state or municipal) freight.

Under Section 10722, a common carrier may establish special passenger rates and baggage rules for members of the armed forces when traveling in uniform on "official leave, furlough, or pass or when they have been released from the armed forces not more than 30 days from beginning that transportation. . . ." There are several other categories of passengers that may receive favorable attention; they are listed in Section 10722 (c) and (d). Section 10723 also contains an additional list of persons that can get special consideration while traveling on trains or buses and Section 10724 establishes conditions for carrying certain passengers under emergency conditions.

But it is Section 10721 that governs the movement of government freight. The specific provision for allowing reduced rates is excerpted from 10721(b)(1) and reads:

> "A common carrier providing transportation or service subject to the jurisdiction of the Interstate Commerce Commission under chapter 105 of this title shall provide transportation for the United States Postal Service under chapters 50 and 52 of title 39, and may transport property for the United States Government, a State, or municipal government without charge or at reduced rates."

If the rate quotation or tender of a rate would endanger the national security, the government will advise the carrier and it will not be made public. In all other cases the carrier may file its special rates with the Commission and with the governmental agency that wants to ship the freight.

The sheer volume of government shipping has an effect on most types of carriers.

Government traffic management works under a policy of equitably distributing freight among competing carriers that offer equivalent services, rates and equipment. The last proviso is important because an equitable distribution with those conditions will not necessarily be an equal distribution; rates and the frequency and quality of services will affect the percentages of total government traffic the competing carriers receive. The policy of using commercial carriers whenever possible is based on the sound theory that fostering a healthy transportation system benefits both the economy and national defense.

Government procurement officers, in common with commercial purchasing managers, must consider transportation costs in buying materials. For purchases of any size, it is general policy to request bids both FOB origin and delivered at a named destination. Thus government traffic management will analyze procurement bids with the freight rates available to the government to determine

GOVERNMENT TRAFFIC

how material should be procured. If it is cheaper for the vendor to ship to a named destination than for the government to buy FOB plant and arrange transportation, then the government shipment will be treated as a normal commercial movement. If FOB origin is advantageous, title is taken at the vendor's plant, and then the movement becomes a government movement, subject to normal government procedures and traffic management.

While state and local government do purchase transportation services, the federal government is the major shipper by a tremendous margin. There are basically two principal government traffic management authorities; the Department of Defense and the General Services Administration. The function of the Department of Defense is obvious; the General Services Administration among other things provides traffic management for all civilian departments of the government. In addition, the General Accounting Office establishes basic shipping, accounting or clerical procedures for both military and civilian departments, and the Department of Justice represents the entire government in transportation court cases.

The General Services Administration controls shipments through regional offices, and acts for any civilian department requiring transportation services. A certain amount of traffic management authority is delegated to the various government departments, particularly to the Department of Agriculture and State Department for foreign aid shipments. However, for most domestic commercial shipping, the General Services Administration actually routes government shipments (other than military) and issues the required documentation.

The military is the single largest shipper within the government and its transportation/traffic management organization is complex.

The Military Airlift Command (MAC), under the Secretary of the Air Force, is responsible for overseas air cargo and air personnel movements; the Military Sealift Command (MSC), under the Secretary of the Navy, for ocean cargo and personnel movements (other than inland U.S. waterways), and the Military Traffic Management and Terminal Service (MTMTS), under the Secretary of the Army, for the procurement and use of all modes of transportation in the continental United States. This agency manages surface transportation of military cargo and personnel within the United States, as well as all military ocean terminals with the exception of those used by the Navy in support of fleet activities.

On the subject of government transportation organization, it should be pointed out that the trend is to move away from local traffic management and move towards centralized management on a regional or area basis. This trend is modified to an extent by permitting local government installations to make small or repetitive shipments without reference to the centralized traffic management authority.

In peacetime, economy in military transportation is important; but more important is whether the system can take on larger responsibilities during an emergency or war. Each military service has peculiar problems of its own and there is no easy answer or quick solution that would provide a single system to be used by all services. The military exists for defense of the country. Each service has a job to do and must be capable of rapid expansion. A good military transportation system has but one basic objective—to put men and material where needed and do so as rapidly as possible.

The importance of traffic management to the military is evident in the number of military students studying it at universities and the schools maintained by the three services to teach the subject.

In 1965, domestic military traffic management for all services was merged into one unit. The Defense Traffic Management Service, the Army Terminal Command, the Air Traffic Coordinating offices of the Army, Navy and Air Force and several smaller units were merged into the Military Traffic Management and Terminal Service (MTMTS) for greater efficiency and to eliminate overlapping and duplication. MTMTS reports to the Secretary of the Army but acts as traffic manager in the U.S. for all military services. It controls all domestic land and air transportation for military shipments. It decides whether such shipments go by air, water, rail, motor carrier, or freight forwarder; it decides the port of exit for shipments destined overseas; it reviews all shipment releases for consolidation possibilities and has jurisdiction over all points of origin for all types of cargo regardless of what military agency originates the shipment; it operates military-owned rolling stock in interchange services.

Domestic transportation is arranged almost entirely with commercial carriers, the general policy being that government-owned and operated equipment will not be used in the United States in competition with commercial carriers. The Military Airlift Command operates no scheduled domestic air services, for example, for either passengers or cargo, but does charter planes from the commercial companies to provide a regular air cargo service between key military supply points. Other domestic military air passenger movements are performed largely by the regular air carriers.

Military trucks and buses do not provide point-to-point services in the United States in competition with commercial services, although charter bus and air services rather than common carrier services may be used for specific large movements of passengers.

The industrial traffic manager moving government freight must examine any government contract closely for packaging requirements, which are often different from the normal commercial packaging requirements.

Overseas Traffic

The Military Sealift Command and the General Services Administration are the primary traffic managers for ocean transportation services. In ocean shipping as with domestic, government policy is to foster a strong merchant marine by using America-flag carriers as the primary source of service. However, in the case of MSC, a nucleus of government owned and operated ships is maintained, so that MSC is both a purchaser of commercial services and a carrier for government passengers and freight. The MSC fleet is maintained primarily to insure that the ships involved will be available on short notice to supplement Navy fleet ships in any emergency. The Navy ships in general are used to complement, rather than compete directly with commercial carrier service. The size of the fleet, however, is determined by emergency plan requirements, and it provides some services that could be obtained commercially. Nevertheless an important part of the military dry cargo moves in commercial ships, and the 25 percent that moves in the MSC fleet consists for the most part of cargoes such as Arctic and Antarctic supplies and other types of shipments, such as ammunition, which do not fit in with commercial trade route patterns. Troop lift is provided primarily by government (MSC) troop ships, since no commercial passenger ships are currently available. Originally, the MSC operated a considerable fleet of passenger vessels (transports); but this fleet has been scrapped as use of air service has increased.

Both military and civilian government general-cargo freight movements by scheduled American-flag ships operate on commercial trade routes. Such ships normally carry both government and commercial freight on the same voyage.

After the various military services (called Shipper Services) place their requirements with MSC, the local MSC offices book the cargoes with the commercial operator, using an allocation system based on the number of sailings each steamship line makes over its various trade routes.

This cargo is distributed as equitably as possible among the American lines, with consideration being given to consolidation of cargoes, where possible, to achieve the lowest possible freight rate.

Bulk lifts of government full-cargo lots such as grain or coal are arranged through normal charter procedures. MSC arranges charters for occasional unusual general cargo that cannot be lifted by the MSC nucleus fleet or by available American-flag sailings. To move petroleum products, MSC employs both long-term charters and voyage charters so as to obtain the needed type and number of tankers.

Air Traffic

The Military Airlift Command maintains a sizeable aircraft fleet, to provide airlift of cargoes and troops in the event of military emergencies. This fleet is used almost entirely for movement of personnel and material between the United States and its complex of worldwide military bases and for movements between overseas bases. In recent years there has been an increased emphasis on the use of chartered commercial aircraft to provide some of the required airlift on the regular routes flown by MAC. In addition to MAC, which handles military cargo only, some government shipments to overseas destinations are made with regularly scheduled commercial air lines. Quicktrans and Logair are services set up by MAC through contracts with commercial air lines in the U.S.A. to provide regularly scheduled cargo flights between specific points and bases.

Documentation

Shipments of government property, contracted for on a FOB-origin basis, in domestic and overseas movements via common carriers are almost always documented on government bills of lading. Since government regulations do not permit prepayment of freight charges, the total amount due is collected by the carrier from the government after delivery. The government then normally has title to the goods at the point of shipment. Shippers of goods sold to a government agency FOB origin who substitute their own commercial bill of lading for the government bill of lading (GBL) will be billed by the carrier; the government will not pay the freight charges unless the shipment moves under a GBL. If this happens, the shipper should endeavor to substitute a GBL for the commercial form as quickly as possible. On the other hand, a consignee of a GBL shipment will be relieved of paying the freight charges by filling in the "certificate of receipt of goods" on the GBL and turning it over to the carrier.

Regulations concerning the use of government bills of lading are issued to all government departments by the General Accounting Office in its "Policy and Procedures Manual for the Guidance of Federal Agencies, Transportation." Instructions to carriers and non-government shippers or consignees are contained on the reverse side of each original GBL. Information to be shown on government bills of lading is identical to that required for commercial bills, insofar as description of articles in accordance with the governing freight classification or tariff, and the identification of shipper and consignee are concerned. Certain additional accounting data and shipment control data for the government's internal use may also appear on GBLs.

It should be emphasized that the GBL is not used to ship or pay freight under special government contracts, but is used for shipment by all modes of transportation world-wide, where the payment of freight is based on the rates, terms and conditions of published commercial tariffs or specifically negotiated rates (the former Section 22 rates).

Government shipments made under special contracts or charter arrangements, and shipments moving in government-owned trucks, aircraft or ships, are documented on special forms developed for the purpose, rather than on GBLs. Payment to contract carriers for such shipments is made in accordance with the terms of the contract or charter party, and the form used to document the freight may or may not be the basis for payment to the carrier. As an example, the Military Sealift Command has shipping agreements, container agreements, and some shipping contracts with American-flag ocean carriers for the movement of Department of Defense cargo. Each line handling cargo under these arrangements receives a copy of the military manifest listing all cargo on a particular ship. MSC then provides a detailed shipping order for each ship sailing. In these instances no specific bill of lading of any type is issued. Charges are billed by each line against the shipping order copy.

The Interstate Commerce Commission only enters into any of the above in the acceptance and filing of special rates for government traffic. However, to ensure enough commercial motor carrier capacity for government freight and to open the door to minorities interested in trucking, the ICC decided in January 1980 (in Ex Parte MC-107 "Transportation of Government Traffic"), to simplify the licensing procedures for motor carriers.

Under the new rules, the commission will issue a "master certificate" authorizing operations by all qualified motor carriers that

United States Government Bill of Lading.

want to compete for the federal government's transportation business. In doing so, the Commission has eliminated the former requirement that individual trucking companies file applications and pay filing fees. Thus a motor carrier can now simply request to participate in the "master certificate" and the Commission will confine itself to examining the trucking company's financial and operations fitness to conduct transportation services.

16

IMPORT-EXPORT TRAFFIC

Although international trade is a relatively small percentage of the total tonnage of goods produced and transported in the United States, it is an essential percentage. Without it, the international balance of payments system would break down and the free-trading nations would probably retreat into economic isolationism. And with the transforming of so many U.S. companies into multinationals in recent years, the scope of industrial traffic management now more than ever embraces exporting, importing and international shipping.

Despite the considerable progress made since the end of World War II in simplifying procedures and standardizing documents, international traffic management remains a complex and document-burdened business, with its own specialized requirements in bills of lading, dock receipts, export declarations, consular invoices, export and import licenses, packaging, etc. How are exports to be paid for—open account, letter of credit, or some other means? What are the terms of sale for the export shipment? Is the company prepared to price its goods so they can be landed at a foreign destination and sold at a price that meets or beats the local competition, which has none of these expenses or problems? The international traffic department must have the answers to these questions in its policies and practices.

How the international traffic department itself is organized will vary from company to company, often depending on non-transportation factors such as taxes, domestic and foreign government requirements, the overseas sales organization and so on. In

INDUSTRIAL TRAFFIC MANAGEMENT

many companies, international sales are handled by separate but home-based corporations that are subsidiaries of the parent company; and that pays fees to its traffic department for transportation services rendered. Other companies set up affiliates overseas with varying degrees of autonomy covering traffic management and other activities. Whenever possible, however, international *and* domestic traffic operations should be controlled by one department, thus bringing to bear all the company's traffic expertise on any problems arising.

Brokers are widely used in international transportation, even by companies with large international operations of their own. The brokers essentially fall into three categories—independent ocean freight forwarders, customs brokers and ship brokers and charterers. Ocean freight forwarders differ considerably in the range of services they offer from the domestic forwarding companies that simply consolidate and forward shipments. Some of the large European forwarding companies combine all three categories and in fact provide a veritable department store of transportation services at all prices.

Traffic moving into and out of Canada and Mexico is of course "international trade," too, but for U.S. companies it is relatively uncomplicated. Most of the actual movements are by rail or truck across the borders, with an increasing use of air freight today. Customs and paperwork requirements are reasonably straightforward and understandable. In the case of Mexico, many U.S.-based companies make it a point to maintain broker connections at border crossings to deal with any customs or other problems on inbound and outbound movements.

The export-import traffic manager does need a much wider vocabulary of sales terms than his domestic counterpart. The familiar F.O.B. (origin or destination) occurs in such variants as F.O.B. (named inland carrier or named inland point of departure); F.O.B. (named inland carrier at named inland point of exportation); F.O.B. Vessel (named port of shipment); F.O.B. (named inland point in country of importation). Other terms of sale frequently used include F.A.S. (free along side); C.&F. (cost and freight); C.I.F. (cost, insurance and freight); and Ex Dock (named port of importation). A complete list of terms of sale abbreviations, and the obligations they entail for seller and buyer, is given in Appendix 3. Basically, terms of sale define the point to which seller or buyer must pay for transportation, insurance etc., in addition to the price of the goods. A useful reference is the "Revised American Foreign Trade Definitions" published by the U.S. Chamber of Commerce.

With all the complications that are seemingly inescapable in foreign trade even in this age of computerized communications, the international traffic department nonetheless has a great many sources of help at hand. Federal and state government agencies, the Federal Maritime Commission, the Department of Commerce, port authorities, banks, industry associations are all eager to promote foreign trade.

Ocean Carriers

Until recently, "breakbulk" ships had dominated the ocean trading routes for hundreds of years. The term "breakbulk" is derived from the method of loading and unloading the traditional cargo ship, which in fact entails breaking the bulk of the cargo into lots small enough to be loaded and unloaded through the ship's hatches. The advent of containerization and other methods of "unit load" load handling has greatly changed the ocean shipping scene. But breakbulk ships continue to ply most shipping routes.

Ocean carriers fall into two broad types—liner services and tramp services. Liner services are confined to particular trade routes and fixed regular sailing schedules; tramp services go where cargo is available, following no fixed routes or regular schedules.

Liner companies generally operate as members of "steamship conferences." These are voluntary associations of competing carriers, with ships flying the flags of many different maritime nations. Conferences normally serve a particular trade route or area—there is a "Far East Conference," for example, and an "American West African Freight Conference." The Federal Maritime Commission (FMC) publishes the "Approved Conference, Rate, Interconference Pooling and Joint Service Agreements and Selective Major

Cooperative Working Arrangements of Steamship Lines in the Foreign Commerce of the United States" (available from the U.S. Government Printing Office in Washington, D.C. in 1980 for $5.50). This publication lists 108 conferences, together with their addresses and other information.

In many ways, steamship conferences resemble domestic rate bureaus, although they operate much more informally. Meetings of member lines are held more often as a general rule and lines may join or withdraw from conferences as they choose. The conferences operate under anti-trust exemptions similar to those accorded to domestic rail and truck rate bureaus and conference agreements and are in return subject to FMC approval. A line will often be a member of several conferences. The rationale of the conference system is that it stabilizes rates and prevents predatory rate wars and the other evils that its advocates say would result from unrestricted competition.

A number of U.S. ship lines receive operating subsidies from the federal government in return for agreeing to provide a minimum and maximum number of sailings on one or more of a number of specified essential trade routes. Under these agreements, the ship operator may not deviate from specified trade routes unless special permission is obtained in each instance from the Maritime Administration.

Cargo preference laws are another important feature of maritime transportation economics. In the years since the end of World War II, Congress has passed a number of these laws giving preference whenever possible to American flag operators for shipping "government" cargo—goods and equipment shipped under foreign aid programs, shipments to U.S. bases, etc. U.S. government sponsored cargoes comprise a substantial percentage of the revenues of the U.S.-flag ships. (To qualify as an American-flag ship, a ship must be built in the United States to specifications of the American Bureau of Shipping, be registered with the U.S. Customs, be owned and operated by U.S. citizens with a U.S. crew and be governed by U.S. safety regulations. Most U.S. large cargo vessels today are built with the aid of federal construction subsidies.)

Both U.S. ship builders and operators *and* the U.S. maritime unions have recently intensified their long standing campaign of urging American importers and exporters to use American-flag ships. The rapid development of a huge Soviet merchant marine operating free of the "capitalist" constraints of profit and loss and able to cut rates below costs could drive American flag ships from the sea lanes, say these groups. Congress has responded to this perceived threat with a number of programs designed to promote the long-term health of the American Merchant Marine.

Ocean Freight Rates

Ocean freight rates, rules and regulations are published both by the steamship conferences and by the individual lines. The rate tariffs must be filed with, and are subject to the jurisdiction of, the Federal Maritime Commission, and by law must be made available to the public. There are watching services, generally based in Washington, D.C., that will report to subscribers as rates are filed with the FMC. But traffic departments heavily involved in foreign trade will want to get the rate tariffs directly from the publishers as they are issued. Subscribing to a conference's rate publishing service will keep tariffs up to date with rate and other changes. Arrangements for obtaining the individual lines' tariffs must be made directly with the carriers involved. (Considerable volumes of cargo move under individual lines' tariffs.) The format and the information shown on an ocean freight rate tariff have much in common with domestic rail and truck tariffs. The cover and a page from a typical conference tariff is shown on the next page.

Ocean freight rate tariffs cover most of the world's trade routes. They include contract and non-contract types of rates. Ocean carriers generally think in terms of commodity rates rather than class rates (although some tonnage does move under class rates). The reason for this is that distance is not very important as a controlling factor in establishing a schedule of port-to-port freight charges. The value of the commodity, or the price at which it will be offered on the international market generally receives more attention in the rate-making process. Many ocean freight rates are published with expiration dates.

INDUSTRIAL TRAFFIC MANAGEMENT

Far East Conference

Freight Tariff No. 25 • **F.M.C. No. 5**

(CANCELS FREIGHT TARIFF NO. 24 AND CORRECTIONS THERETO)

2ND REVISED TITLE PAGE	
CANCELS: 1ST REV. TITLE PAGE	
EFFECTIVE: January 12, 1972	
Correction No. 2656	Cancels Corr. No. 2550

NAMING CONTRACT AND NON-CONTRACT COMMODITY RATES

FROM: UNITED STATES ATLANTIC and GULF PORTS.

TO: JAPAN, OKINAWA, KOREA, TAIWAN (Formosa), HONG KONG, PHILIPPINE ISLANDS, VIET NAM, CAMBODIA, LAOS, and MAINLAND CHINA.

VIA: DIRECT CALL OR TRANSHIPMENT.

FOR RATES TO THE FOLLOWING GROUP PORTS REFER TO THE COMMODITY RATE PAGES PUBLISHED HEREIN

Group 1 Ports: NAGOYA, YOKOHAMA, KOBE, OSAKA, MANILA and HONG KONG

Group 2 Ports: CEBU and ILOILO

Group 3 Ports: TAKAO (Kaohsiung) and KEELUNG (Chilung)

Group 4 Port : SAIGON

REFER TO PAGES NOS. 1 and 2 HEREIN FOR THE NAMES OF THE PARTICIPATING CARRIERS.

FOR RATES TO OUTPORTS AND THEIR APPLICATION REFER TO THE OUTPORT SECTION OF THIS TARIFF.

RATES PUBLISHED HEREIN ARE SUBJECT TO THE RULES AND REGULATIONS OF THIS TARIFF UNLESS OTHERWISE INDICATED IN THE INDIVIDUAL COMMODITY ITEM.

Issued: January 15, 1969   Effective: January 15, 1969

GERALD J. FLYNN, Chairman
11 Broadway
New York, N.Y. 10004

I: New Matter effective January 12, 1972.

Illustration: Typical Conference Ocean Freight Title Page Showing Trade Areas, Commodity Rates Involved, and Other Information.

IMPORT-EXPORT TRAFFIC

Far East Conference
Tariff No. 25 **F. M. C. No. 5**

ORIG REV	PAGE
17th Revised	237
CANCELS	PAGE
16th Revised	237
EFFECTIVE DATE	
June 1, 1972	
CORRECTION NO	3086
CANCELS CORR NO	2845

From: UNITED STATES ATLANTIC and GULF PORTS.
To: JAPAN, OKINAWA, KOREA, TAIWAN, HONG KONG, PHILIPPINE ISLANDS, VIET NAM, CAMBODIA, LAOS and MAINLAND CHINA

EXCEPT AS OTHERWISE PROVIDED HEREIN, RATES APPLY PER TON OF 2000 LBS., OR 40 CUBIC FEET, WHICHEVER PRODUCES THE GREATER REVENUE

"C" denotes "CONTRACT" rates.   "NC" denotes "NON-CONTRACT" rates.

Commodity Code	Commodity	Rate Basis	1 Nagoya Yokohama Kobe Osaka Manila Hong Kong $	2 Cebu Iloilo $	3 Tokao Keelung $	4 Saigon $	Item No
	Iron and Steel Articles Plain or Galvanized (Not including Stainless Steel) Viz: (cont'd)						
	Bale Tires, Wire	C 2240 A:	54.00	57.50	61.00	56.00	
	Billets	NC lbs.	62.10	66.10	70.15	64.40	1353
	Blooms	or 40					
	Bolt Stock, Steel	cu.ft.					
	Bolts, Mine, Roof						
	Bolts, Railway		Apply the following rates as special rates on				
	Bolts, N.O.S.		STEEL BILLETS TO PHILIPPINE SAFE PORTS,				
	Clevices		HONG KONG (Rule 1(f)(7) not applicable):				
	Conduit Fittings, (not Cast Iron)		Rate declared Open subject to tariff rules				
	Cuttings, N.O.S.		and regulations and also subject to minimum				
	Cuttings, Plate		quantity of 1000 revenue tons per vessel per				
	Cuttings, Sheet		port of loading and port of discharge.				
	Expansion Shells		() Effective through September 30, 1972.				
	Forgings, not machined						
	Forgings, not machined beyond deburring, with or without lifting lugs.		PIPE FITTINGS, NOT CAST IRON - TO TAKAO: Effective through September 30, 1972, the following rates will apply:				
	Hoop Steel		A: C - $58.75 - NC - $67.55 per 2240 lbs. or				
	Hoops		40 cu.ft., Berth Terms, or at Ship's Option.				
	Ingots		A: C - $52.50 - NC - $60.35 per 2240 lbs. or				
	Nails, Common Wire		40 cu.ft., Free In Stowed - 50% discount in				
	Nails, Leadhead		Heavy Lift Charges.				
	Nuts						
	Piling, N.O.S.						
	Pipe Fittings, not Cast Iron						
	Rivets						
	Screw Stock						
	Screws						
	Screws, Brassplated, Bronze plated, Cadmium plated or Copper plated						
	Shafting, not forged						
	Shafting, not machined						
	Sheet Bar						
	Skelp						
	Slabs						
	Spikes, Boat						
	Spikes, N.O.S.						
	Spikes, Railway						
	Strips, N.O.S.						
	Strips, Copper Coated, N.O.S.						
	Studs						
	Thread Protectors, Pipe						
	Tire Bead Wire		(Continued on next Page)				

A: Advance in rates effective June 1, 1972.
() Expiration date extended from May 31, 1972.

Illustration: Ocean Freight Tariff Page Naming Rates and Other Information on Specific Commodities.

INDUSTRIAL TRAFFIC MANAGEMENT

The FMC can review, approve or conditionally disapprove ocean freight rates and compel carriers to publish their tariffs (Public Law 87-346, amending Section 15 of the Shipping Act of 1916). Increases in rates must be filed at least 30 days in advance of the effective date; reductions in rates become effective immediately on filing. Some tariffs have cargo booking provisions that may extend the advance notice required for rate increases.

Ocean freight carriers do not publish the freight classifications that are a familiar feature of domestic land transportation; nor must they observe specific rules as to the format and content of their tariffs. Consequently, there are a number of "rate terms," some of which are peculiar to ocean freight rate tariffs. The following define some of these familiar and not-so-familiar terms in their maritime context:

A commodity rate is a rate on a specific commodity; it normally takes precedence over class rates.

A class rate applies to a group or class of articles.

Minimum rates are those below which a carrier may not quote or charge.

An arbitrary rate includes a charge over the basic rate.

Heavy lift rates (also extra charges for long or large one-piece shipments) are added to basic rates (or quoted separately) if the items are abnormally large or heavy.

General rates apply when there are no commodity or class rates.

Charter rates are charged for the charter of an entire ship by a single shipper or group of shippers.

Optional rates offer an option as to the port of call for discharging the cargo.

Ad valorem rates are charged for items of a value greater than the limits of liability shown on the bill of lading.

Deck cargo rates (generally lower than hold rates) apply to cargo loaded on deck rather than in the ship's holds.

Refrigerated cargo rates include extra charges for the refrigeration service.

Dangerous or hazardous cargo rates must be specially negotiated with the carrier.

An open rate is a rate that members of a conference cannot agree

IMPORT-EXPORT TRAFFIC

on; the rate is then declared open and each conference member is allowed to charge what it can get.

Ocean freight rates may be assessed on either a weight or measurement basis, but are usually quoted on a "per tonM/W" basis, meaning per ton weight or measurement, whichever produces the greater revenue for the shipping line (see the note at the top of the tariff page taken from the Far East Conference tariff). Rates weight basis are generally quoted per net ton of 2,000 pounds, although long or gross tons of 2,240 pounds and kilo or metric tonnes 2,204.6 pounds are sometimes specified, as are rates in cents per 100 pounds.

When measurement bases are used, each shipment is normally measured at dockside in accordance with the rules and regulations of the governing conference or tariff publishing authority. Basically, the cube of a package is determined by measuring its length, height and width, using the outside or largest dimensions, and multiplying the three measurements together.

When the tariff provisions specify that the rates are on a weight-measurement "ship's option" basis, the freight charges are generally computed on the gross weight or on the overall measurement of the individual pieces. On regularly moving commodities, or where specific packaging standards have been set up for particular industries, the conferences and their shippers often arrange weight-measurement agreements similar to those used in rail and truck tariffs for assessing rates. Thus rates on lumber are quoted "per 1,000 board feet," and rates for oil are, of course, "per barrel." Other rates may be quoted per case, per bag, per bale, per linear foot and so on. Some conferences use inspection bureaus to handle cargo weight-measurement problems, while others have standardized on commercially-available cubic measurement tables.

While, as we noted earlier, there are no specific rules covering the format and content of an ocean freight tariff, the international traffic manager would do well to check on the following items in any tariff he is examining:

—Do commodity rates take precedence over class rates?
—Are exceptions to the "Freight-must-be-prepaid" rule listed?
—Are rates subject to change without notice?

—Are participating carriers listed?
—Are mixed goods in the same container subject to the highest rate for any item included?
—Are consular and export control rules mentioned?
—Are hazardous cargo regulations, packing and marking rules, cargo measurement regulations referred to?
—Are "To Order" bills of lading acceptable?

Remember, too, that ocean freight rates do not normally include marine insurance, that items exceeding a specific weight or length are subject to additional charges, that on-deck cargo is accepted at "owner's risk," and that there is usually a "minimum freight" rule below which a shipment will not be accepted for transportation.

The usual procedure when a shipper wants to change a freight rate or a tariff rule is to send a letter to the chairman of the steamship conference. This is better than making a request through one of its members. Some shippers may prefer to call in person on the chairman, a conference committee, or appear before the entire conference membership. In the case of conferences headquartered abroad, shippers may have no choice but to send a letter or relay the request through a member carrier who will present the proposal to the conference. Letters should include all possible data such as a complete description of the commodity, origins, destinations, value per pound, weight per cubic foot, damage factor, volume expected to move, F.A.S. selling price and freight rate from foreign country (if the request is to meet foreign competition), how packed including package dimensions, comparative rates on similar commodities in the same or other trades, and anything else that will assist the conference in making its decision. Most conferences use traffic analysis forms for this purpose on the lines of the one shown here. Some firms send a copy of their conference transmittal letter to all conference member steamship lines with whom they route or might route shipments. A rate change sought from an individual steamship company is, of course, to be handled with its representatives.

Some steamship conferences will offer "contract" rates up to 15 percent lower than standard rates if the shipper will agree to route

IMPORT-EXPORT TRAFFIC

ASSOCIATED LATIN AMERICAN FREIGHT CONFERENCES TRAFFIC ANALYSIS FORM

1. Name of article and its trade name, if any: Clay Tile
2. Description of article: Clay Tile including ceramic, wall and floor and quarry floor tile
3. What it is used for: Decorative floor and wall covering
4. State if hazardous or inflammable: Neither
5. State what label is required for ocean shipment: None
6. State if liquid, paste, flake, powdered, granulated or solid: Solid
7. State how packed (box, barrel, bale, carton, crate, etc.): Crates
8. Package Dimensions: Length 17" Width 10" Height 4½" Cu. ft. per package: 0.5
9. Package gross weight is 40 lbs. 24 Cu. ft. per 2,000 lbs.:
10. From U. S. port of: New Orleans To foreign port of: Rio Haina, Dominican Republic
11. Present rate: $79. per 2,000 lbs. Rate proposed by applicant: $50. per 2,000 lbs.
12. Place where material is made or produced: Various points in U. S.
13. How article is described for railroad movement: Clay Tile
14. How described in Shipper's Export Declaration: Clay Tile
15. Schedule B Commodity No.: D662.4610
16. Tonnage of present movement to foreign port named in Item 10: Nil
17. Extra tonnage expected to move at proposed rate: 240 tons per annum
18. State if movement is continuous, seasonal or sporadic: Continuous
19. F.A.S. value per pound or package (state which): $6.00 per package
20. Name of competitive or substitute article: Tile from other countries
21. F.A.S. value per pound of competitive article: Unknown - C.I.F. about $350. per ton
    (If this information is not obtainable, but the delivered competitive price is the basis of this application, it must be indicated at least approximately)
22. Reason for proposed change:
    Shipments to other areas, competing with other supplying countries, are on the increase. Therefore, feel if comparable freight obtainable here, business would increase.
23. Application submitted by: Slippery Floor Co., Inc.
                         Per: George Mudd, T. M.
    Date: Address: 678 Oak Street, Catalpa, Alaska

Note.-Applicant to keep one completed copy of this form. The remainder to be mailed to C. D. Marshall, Conference Chairman, Room 2100, 11 Broadway, New York 4. N. Y.

Steamship Conference Traffic Analysis Form.

all his freight via conference members. This agreement is in fact a formal written contract, and such contracts are available to all shippers, regardless of the volume of their shipments.

Each conference has its own contract forms and the contracts are subject to FMC jurisdiction and to federal legislation. Some features of these contracts may cause problems, particularly with regard to the routing of freight and the extent to which company affiliates are bound by contracts they did not directly sign. Company affiliates have differing lines of corporate authority and manage their traffic in so many different ways that just what constitutes an affiliate for contract rate purposes may be hard to determine.

As regards the shipper's obligation to route in these rate contracts, it "covers only those goods" for which he has "the legal right at the time of shipment to select the carrier," according to the basic statute. Further, he may not divest himself (or permit himself to be divested) of the "legal right to select the carrier" before the time of shipment "with the intent to avoid his obligation under the contract." The FMC and the conferences endeavor to interpret these laws realistically. Important considerations in routing ocean freight are the arrangements between the seller and the buyer; when title passes; the kind of title involved (i.e., whether title is retained simply for security purposes); who is shown as shipper on the transportation documents; the extent of the shipper's participation in the making of the transportation arrangements, and so on.

A shipper is not obligated to refuse a sale simply because a customer insists on making his own transportation arrangements and, if the customer does insist on doing so, his instructions may be followed without a contract violation. The FMC has taken a firm position to this effect, and clauses so providing are to be found in the various contracts. Terms of sale may be important in determining when title passes and an obligation to route occurs, but such terms are often so loosely used that they are simply a factor for consideration. As an example, F.O.B can be "free on board" a vessel or dock anywhere, and further specifics would be needed to make the term truly binding. A conference may, at its option, provide in its contract that where a shipper participates in the transportation arrangements, or where his name appears as shipper on the bill of lading or export declaration, there is a presumption (which is rebuttable) that he had the right to route.

Conference rate contracts usually have a cancellation clause that gives each party 90 days to cancel. There may also be a provision releasing the shipper of his contractual obligation if the conference cannot provide suitable ship space.

Not all conferences have a dual system of contract and noncontract rates, but many do. Nonetheless, the independent ship operator should not be overlooked. The independents are giving the conference carriers increasing competition these days and it's always worthwhile to check their rates.

Insurance and Claims

Steamship companies are liable for loss or damage due to their negligence, fault or failure in loading, storage, unloading, carriage or care of cargo. They are not liable for losses due to perils of the seas, acts of God, acts of the public enemy, inherent defects of cargo or packages, or to negligence attributable to the shipper. However, steamship companies can risk cargo and deviate from their intended voyage in order to save life or property at sea. There are many opportunities for misunderstandings and differences of opinions as to responsibility for loss or damage to overseas shipments, and companies engaged in international trade must be sure they have the best possible insurance protection. Careful selection of insurance carriers is also important in handling any claims that might arise against international carriers.

There are three major types of marine insurance, two of which are of little concern to the traffic manager unless a ship charter is to be used. The latter two types are freight insurance used by ship operators to cover any loss of earnings if they are unable to collect freight charges and hull insurance to protect the ship's owners against damage to, or loss of, the ship.

The type of insurance most commonly used by shippers is cargo insurance. This is bought by the owner of the merchandise to protect himself because of the limited liability of the vessel owners. Incidentally, the shipper who suffers a loss at sea must understand that while he is entitled to collect from his insurance company, he is obligated, under the terms of a marine policy, to assist that insurance company in any action it takes to collect from the steamship company.

Specific insurance—insurance placed for each specific shipment—is usually handled by a special risks department of an insurance broker. However, by far the greater volume of ocean marine insurance is now written under what are known as open policies. These are insurance contracts that remain in force until cancelled, providing continuous insurance coverage for all the shipments the insured shipper may make.

The shipper gains many advantages from the use of an open policy. In the first place, he has automatic protection (up to the maximum limits stated in the policy) from the time shipments leave the warehouse at the place named in the policy until they are delivered to the consignee. The policyholder warrants that shipments will be "declared" to the insurance company as soon as practicable, but unintentional failure to declare will not void the insurance since the goods are "held covered," subject to policy conditions. In effect, this is coverage for errors and omissions, and it insures against the possibility that, because of the press of business, a shipment may be dispatched without being insured. Thus, the open policy relieves the shipper from the necessity of arranging individual placings of insurance for each shipment.

Under an open policy, too, the shipper knows beforehand the premium that will be charged. This in turn makes it much easier to quote a landed sales price.

Finally, the use of the open policy creates a sound, long-term business relationship between insurer and insured—(in one company an open cargo policy has been in effect for more than 75 years). This permits the insurer to learn the special requirements of its insureds and so to provide them with individualized protection, tailor-made to fit specific situations. This may be an important factor in loss adjustments at out-of-the-way ports around the world, or in overcoming problems peculiar to shipping a given commodity.

IMPORT-EXPORT TRAFFIC

There are two methods of declaring or reporting shipments to the insurer under open policies. For imports, a short form of declaration will usually be sufficient, since there is no need to furnish evidence of insurance to third parties. This form calls for the name of vessel and sailing date, points of shipment and destination, nature of commodity, description of units comprising the shipment, the amount of insurance desired, and the number of the open policy under which the declaration is made. The declaration forms are prepared by the shipper and are forwarded daily, weekly, or as shipments are made. When the full value of the shipment is not known at the time a declaration is made, a provisional report may be sent in. The "provisional" is closed when value is finally known. The premium is billed monthly in accordance with the schedule of rates provided by the policy.

The same short form of declaration may be used at times for exports, if claims are to be returned by foreign consignees for adjustment in this country, in which case no special marine policy is issued. However, in most cases the exporter must furnish evidence of insurance to his customer, to banks, or to other third parties in order to collect claims abroad. This calls for a special marine policy, occasionally referred to as a certificate.

The special marine policy contains information similar to that shown on the declaration. In addition, it calls for the marks and numbers of the shipment, the name of the party to whom loss shall be payable (usually the assured, "or order," thus making the policy a "negotiable instrument" upon endorsement by the assured), and the applicable policy provisions. Some of these provisions are standard clauses and are incorporated by reference only, while others are specific and apply to the individual shipment in question.

The special marine policy is prepared in four or more copies. The original and duplicate are negotiable and are forwarded with the shipping documents to the consignee. The remaining two serve as office copies for the assured and for the insurance company.

A special marine policy may be prepared by the shipper, by the freight forwarder, by the insurance agent or broker, or by the insurance company. It is important that it be made out with care. The shipment should be described in sufficient detail to make identification clear, especially if more than one shipment is going forward by the same vessel. The origin-to-destination transit should be explicitly stated, as "New Castle, Pa., by rail to Philadelphia, SS. John Doe to Bahia, Brazil." If the ultimate destination is an inland city, this should be made clear.

The terms "special marine policy" and "certificate" are both used to describe the policy document. The practical effect of the two is the same. In former years, the use of "certificate" was customary. This, as the name implies, certifies that a shipment has been insured under a given open policy, and that the certificate represents and takes the place of such open policy, the provisions of which are controlling.

Because of the objections that an "instrument" of this kind did not constitute a "policy" within the requirements of letters of credit, it has become the practice to use a special marine policy. This makes no reference to the open policy and stands on its own feet as an obligation of the underwriting company.

In some cases, exporters who use freight forwarders also insure their shipments through them. While this has the merit of simplicity, there are definite advantages in having one's own policy. The shipper gets conditions of insurance suited to his own requirements, rather than coverage designed for general application over a wide diversity of commodities and trades. Rates will be based upon the known hazards and experience of the shipper's business, rather than upon a average of which his business is only a part. Finally, the shipper will be able to call upon the services of his local insurance agent or broker, for loss prevention service and packing advice, and in making claims.

The use of his own policy need not necessarily require the shipper who uses forwarders to prepare special policies. Arrangements can be made allowing the freight forwarder to provide these special policies through the facilities of the shipper's insurance company.

In order to satisfy the terms of a letter of credit or the request of a buyer or seller abroad, an exporter or importer must have more than a passing acquaintance with the various terms and conditions of the insurance coverage he may be called upon to supply.

Coverage given by the "free of particular average" (F.P.A.)

policy is the narrowest in common use. This claim restricts coverage generally to total losses. In addition, it provides that partial losses are recoverable only in the event that the vessel or interest insured has been stranded, sunk, burnt, on fire, or in a collision. There are two forms of F.P.A. coverage, "under deck" and "on deck." The under-deck coverage is customarily written according to English conditions (F.P.A.-E.C.) which are:

> Free of particular average (unless general) or unless the vessel or craft be stranded, sunk, burnt, on fire, or in collision with another vessel.

The on-deck coverage is customarily written according to American conditions (F.P.A.-A.C.) which are:

> Free of particular average (unless general) or unless caused by stranding, sinking, burning, or in collision with another vessel.

In the English form, partial losses resulting from sea perils are recoverable, provided one of the stated accidents has taken place, without requiring that the damage actually be caused by the peril. In the American form, partial losses are not recoverable, unless the damage is caused by the stranding, sinking, burning or collision of the carrying vessel. Under both English and American forms, both general average and salvage charges are recoverable.

"With average" (W.A.) coverage is a broadening of the basic F.P.A. form and may be stated as follows:

> Free of average under three percent unless general or the vessel and/or the interest hereby insured be stranded, sunk, burnt, on fire, or in collision with any substance (ice included) other than water, each package separately insured or on the whole.

In simpler terms, this clause means that if the vessel is stranded, sunk, burnt or in collision, and the partial loss is due to another sea peril such as salt water or unusually heavy weather, the loss is paid in full if it amounts to three percent or more. Insured losses not amounting to the three percent are not collectible.

The all-risk clause which is widely used and is the most complete of the coverages available reads:

> Against all risks of physical loss or damage from any external

cause (excepting risks excluded by the F.C.&S. (free of capture & seizure) and strikes, riots and civil commotions warranties unless covered elsewhere herein), irrespective of percentage.

In all the forms of insurance coverage previously mentioned, certain perils are excluded. The risks of war, strikes, riots and civil commotions are especially excluded, but can be included by special endorsement, or by a separate war policy. Claims for loss of market and for loss, damage or deterioration arising from delay are excluded. These are claims of consequential and indirect damages and not ones of physical loss or damage caused by an external force.

Another exclusion in every policy is that of "inherent vice." Damage that can be foreseen, or is due to the internal structure or nature of the commodity, is known as inherent vice. In other words, the claims that are allowable under the marine contract are in most cases caused by an external force. "Trade" losses such as normal loss of weight of certain types of foodstuffs, chemicals, and organic substances are not recoverable. Allowances for such weight discrepancies are taken into consideration by traders when negotiating for the merchandise.

When determining the type of coverage needed, an exporter or importer should obtain the most complete coverage practical to suit not only his buyer or seller but also the goods he is shipping. He must determine with the help of his broker whether or not F.P.A., W.A., or all-risks conditions are warranted, taking into consideration, of course, the cost of the insurance and the susceptibility of the cargo to damage.

Marine insurance rates are based on the premium and loss experience of the individual shipper together with the over-all experience of certain types of commodities shipped to and from various places in the world. In determining rates, careful consideration is given to commodity, packaging, destination, ocean carriers and connecting conveyances involved, climatic conditions encountered during the voyage, and whether there is any additional storage and trans-shipment throughout the venture. An exporter or importer, by cooperating with his broker and underwriter, may often improve his loss experience by careful attention to these

IMPORT-EXPORT TRAFFIC

various factors and by being acquainted with the reliability of the foreign customers with whom he is dealing.

General average (G.A.) is a marine insurance term referring to assessment against all shippers or interests to cover loss sustained when some cargo is jettisoned to save the vessel and the remaining part of the cargo. General average has been authoritatively described as "that which has been destroyed for all shall be replaced by the contributions of all. A general average loss is the result of a sacrifice voluntarily made under fortuitous circumstances of a portion of either ship or cargo or a voluntary expense for the sole purpose of saving the common interest from total destruction."

The expenses, losses, and sacrifices involved in a sea disaster are borne by all those with an interest in the goods on the ship and also the ship itself. The cargo and ship owners must contribute together and recompense those who have suffered loss or damage to their goods.

The method of arriving at the contribution of each is handled by general average adjusters. Some cases are solved and settled quickly, but if there are many interests involved, the process of the general average adjustment is a complicated one and may take months to settle.

In order for there to be a general average assessment, four conditions must be present: First, there must be an immediate impending physical peril, common to vessel and cargo; second, there must be a voluntary sacrifice or extraordinary expenditure to avert the peril; third, there must be a successful result; and fourth, there must be no fault on the part of those claiming contribution.

If the shipper is insured against general average losses, his underwriter will usually furnish the guarantee for the general average bonds or security. The deposit receipt for such a bond is a negotiable instrument and should be held in readiness to be presented to the adjusters upon completion of the adjustment for refund if there should be an excess payment over and above the general average contribution.

There is a popular misconception among shippers that they are required to pay the costs of damage to the vessel sustained during a peril. This is not true. Only sacrifice and extraordinary expenses incurred after the peril or to avoid the peril are charged into general average.

In all general average losses, the shipper is called upon to do the following:
— Post security in order to release cargo at port of destination.
— Establish documentary evidence of the value of his cargo.
— Pay the general average contribution when assessed by the adjusters.

The marine insurance policy is a "valued" policy; that is, when the value of the shipment is declared a certain policy value is agreed upon. In case of total loss, the total amount insured is paid; if a partial loss, a percentage of the total insured value is recoverable, based upon the extent of the damage to the goods.

An importer or exporter wishes to receive in return for damage to his merchandise a sum of money that will indemnify him fully. The various factors that go into making up the insured value include invoice value, freight charges, packing, consular charges, other fees, and insurance premiums. Anticipated profit of the transaction is also taken care of by adding a percentage of "advance" to the total value of the various factors above. While usually ten percent, an advance may be varied as an assured deems desirable to protect his position. It is most important that a trader in commodities such as sugar, coffee, burlap, rubber, cotton, etc., which are subject to price fluctuations in the open market, make certain that his valuation clauses are so drawn as to give him protection not only for his original costs but also for market changes.

Insurance Claims

When a shipper suffers a loss under a marine policy, he must, to the extent that he can, act with speed to keep any loss at a minimum. Expenses to prevent any further loss—further than that which is obvious—are usually recoverable if reasonable. But loss or damage should be reported immediately to both the carrier and the insurance company's agent even if the "report" is only a notice of intent to file a claim—that notice of intent then acts as a notice of loss.

Shippers should note clauses in the insurance policy or ocean bill of lading setting any time limits for notifying losses. Under the U.S. "Carriage of Goods by Sea Act" of 1936, notification of loss or damage must be submitted to the carrier at the port of discharge before or at the time of removal of goods into the custody of the party entitled to receive them. And this should be followed, at the same time, with notice to the party named in the insurance policy as the claim agent.

The following are the steps to take in filing a claim:
—Report the loss to insurance agency in written form
—Report the loss to the shipping line advising it that you hold it responsible. A copy of this letter should also be sent to the insurance agency.
—Send to the insurance agency:
(a) Original or copy of bill of lading
(b) Original or copy of insurance certificate
(c) Original or copy of commercial invoice
(d) Packing list and any other statement evidencing condition of cargo at time of shipment including dock receipt
(e) Inspection (survey) report covering proof of loss or damage
(f) Your statement of loss or damage in detail

Documents

In domestic transportation, the traffic manager is essentially concerned with only four documents—the bill of lading, delivery receipt, freight bill and an invoice that establishes the financial relationship between buyer and seller.

International transportation requires a great deal more paperwork. The basic documents of concern are the purchase order with packing and marking instructions, dock permits, dock receipts, ocean bill of lading, letters of credit, bankdrafts, insurance certificates, customs declaration, government export license, consular invoice and certificates of origin. Of course, if the shipment is by air

INDUSTRIAL TRAFFIC MANAGEMENT

then a dock permit and a dock receipt will not be required and the air waybill will take the place of the ocean bill of lading.

A typical procedure for handling export shipments begins when the plant advises the traffic department the cargo is ready for packaging. After ensuring that the packaging is in accordance with the buyer's specifications, insurance requirements, the carrier's rules and the destination government rules, the packages are weighted and measured, the figures recorded on a "packing slip," and this information phoned to the pier where the vessel is to dock. The shipper is then given a date to deliver the cargo to the dock. This is the dock permit, also known as the delivery permit.

After it receives the shipment, the shipping line prepares the dock receipt in quadruplicate (see sample on page 369; the shipper should make sure it is signed by the carrier's receiving clerk). This document identifies shipper and consignee, describes the cargo, its weights and measurements and number of packages, gives the name of vessel, and may identify owner of the cargo. The dock receipt also shows the forwarding agent's name, if a forwarder is used. The original signed copy is returned to the shipper or his agent, and this serves as documentary proof of satisfying a sales contract. It may also give other information concerning the handling of the cargo on the vessel or on the dock.

The ocean bill of lading has much in common with its domestic counterpart. It identifies the terms and condition of carriage of the goods and is the contract of transportation between the shipper and the steamship company. It is also a signed receipt from the ocean carrier for the goods and will normally identify the owner of the goods (see sample on page 371).

The number of copies needed will vary from carrier to carrier. The shipper will take the signed original and the next two or three copies, also signed. The steamship company will take from three to eight copies, and the foreign consul of the destination country will want from two to four copies.

All shipping lines today use essentially the same format for their bills of lading. "Short" form ocean bills of lading similar to the "short" form domestic bill are common. Some variants on the standard bill are:

IMPORT-EXPORT TRAFFIC

DOCK RECEIPT

SHIPPER/EXPORTER MONSANTO INTERNATIONAL SALES COMPANY ST.LOUIS, MO. 63166 ID NO. 43-0975900-00	**DOCUMENT NO.** **EXPORT REFERENCES** DI-29234 REF:38062 CC: R.HINES
CONSIGNEE ORDER OF: MONSANTO FAR EAST LTD/ MALAYSIA BRANCH/K 116 JALAN SEMANGAT PETALING JAYA SELANGOR, MALAYSIA	**FORWARDING AGENT - REFERENCES** H.A.GOGARTY, INC. FMC#464 325 CHESTNUT ST. PHILA., PA. 19106 **POINT AND COUNTRY OF ORIGIN**
NOTIFY PARTY SAME AS ABOVE	**DOMESTIC ROUTING/EXPORT INSTRUCTIONS** LTO #9 79 SEP -4 A10 26
PIER OR TIOGA MARINE TERMINAL	
EXPORTING CARRIER ADRIAN MAERSK	**PORT OF LOADING** PHILADELPHIA	**ONWARD INLAND ROUTING**
AIR/SEA PORT OF DISCHARGE PETALING JAYA VIA:	**FOR TRANSSHIPMENT TO** PORT KELANG	

PARTICULARS FURNISHED BY SHIPPER

MARKS AND NUMBERS	NO. OF PKGS.	DESCRIPTION OF PACKAGES AND GOODS	GROSS WEIGHT	MEASUREMENT
HOFEL PETALING JAYA VIA PORT KELANG 36/79 MADE IN USA	1	CRATE: ELECTRICAL INSTRUMENTS FREIGHT PREPAID	360# 163 KILOS	45CFT

UNITED STATES LAW PROHIBITS DISPOSITION OF THESE COMMODITIES TO THE PEOPLE'S REPUBLIC OF CHINA, NORTH KOREA, NORTH VIETNAM, SOUTH VIETNAM, CAMBODIA, SOUTHERN RHODESIA, UNLESS OTHERWISE AUTHORIZED BY THE UNITED STATES.

DELIVERED BY: LIGHTER TRUCK ARRIVED– DATE .... TIME .... UNLOADED–DATE .... TIME .... CHECKED BY.......... DR PLACED IN SHIP LOCATION 0037	RECEIVED THE ABOVE DESCRIBED GOODS OR PACKAGES SUBJECT TO ALL THE TERMS OF THE UNDERSIGNED'S REGULAR FORM OF DOCK RECEIPT AND BILL OF LADING WHICH SHALL CONSTITUTE THE CONTRACT UNDER WHICH THE GOODS ARE RECEIVED, COPIES OF WHICH ARE AVAILABLE FROM THE CARRIER ON REQUEST AND MAY BE INSPECTED AT ANY OF ITS OFFICES. RECEIVING CLERK: IF SHIPMENT IS NOT IN PROPER SHIPPING CONDITION, NOTIFY US AND WE WILL ARRANGE FOR THE NECESSARY COOPERAGE & INSPECTION. PHONE: N.Y. - 212-2694040 - PHIL. PA. 215-923-5154 H. A. GOGARTY, INC. FOR THE MASTER BY .... RECEIVING CLERK DATE 9-12-79 **ONLY CLEAN DOCK RECEIPT ACCEPTED.**

Printed and sold by THE MARTIN PRESS, Inc., Form No. 133-A

Dock Receipt Used to Deliver Deliver Cargo to Pier.

The straight bill of lading, which is a non-negotiable document. It shows the shipment consigned straight or directly to the consignee rather than "to the order of . . . ." The shipper thus has to find some way to protect the payment due him before the shipment is delivered to the consignee. Hence the much more widely used. . . .

The Order bill of lading which is a negotiable document. It requires the carrier to notify the consignee that his cargo has arrived and he must produce his "clear" copy of the order bill of lading before the cargo will be released to him.

The through export bill of lading combines the domestic rail and ocean bill of lading. It is generally used for carload shipments from Western inland points to Pacific ports.

The through export bill, because it covers the shipment from the U.S. inland point right through to its foreign destination port, contains all the rules and regulations of the domestic bill of lading *and* the ocean bill. It names the steamship line, port of exit, destination port, shipment package numbers and markings, etc. The shipper pays the railroad both the rail and ocean charges, and the railroad then pays the ocean freight charges to the steamship line.

Through export bills are not used universally, despite such advantages to the shipper as the near elimination of handling, wharfage and other port accessorial charges (which are absorbed all or in part by the rail and steamship carriers) and reduced rates. Foreign importers may insist on an "on-board" ocean bill of lading which is absolute evidence that the shipment is en route on a specified ship; and because ocean freight charges must be prepaid, the steamship company must agree to a rate before it receives the cargo without the option of using the standard ocean weight-or-measurement rate base.

The through export bill of lading has found its greatest use where it originated, among Western rail carriers serving Pacific ports. Many shipments to Hawaii and Far East points move under through export bills of lading, and freight forwarders and household goods carriers in particular find it useful.

The through ocean bill of lading, not to be confused with the through export bill of lading, is used when the cargo is to be transhipped to another vessel in two-carrier service.

IMPORT-EXPORT TRAFFIC

Ocean Bill of Lading Issued by Steamship Company.

INDUSTRIAL TRAFFIC MANAGEMENT

The steamship company normally uses an "arrival notice" (see example on next page) to tell the consignee that his shipment has arrived. This shows the name of the shipper, cargo, charges if any, the name of vessel and such other information as may be important to the consignee.

Export Documents

Documents required by foreign governments will vary from country to country. Some countries will insist on a copy of the shipper's commercial invoice as well as a consular invoice which usually must be prepared in the language of the destination country; others may also require a certificate of origin (see example on page 374). Most of the other documents demanded by foreign customs will be supplied by the overseas buyer.

The U.S. government, however, requires a number of documents from U.S. exporters. The first of these is an export license, which must be obtained from the U.S. Department of Commerce. The government issues both general and validated licenses. A general license authorizes an exporter to ship certain goods and items, and the technical data necessary to use those items, without any further governmental documents. A validated license, on the other hand, authorizes the shipment of goods and items only within the limits described in that license. A validated license may for example impose restrictions on price, resale, "end use" of electronic equipment, etc.

Shipments to U.S. possessions do not require licenses, nor do most shipments to Canada. Some types of exports to Canada may need a validated license, however, and shippers in doubt should check the Commerce Department's Export Administration Regulations.

The U.S. shipper must also file a "Shippers Export Declaration" for an export shipment (see example on page 375). This must be cleared by the U.S. Customs before the cargo can be accepted and bills of lading (or air way bills) issued by the carriers. (Exceptions to the "export dec" filing requirements can be found in Sub-

IMPORT-EXPORT TRAFFIC

part D of the Census Bureau's "Foreign Trade Statistics Regulations.") The export declaration form is prepared in triplicate (quadruplicate if the shipment is subject to export licensing) and must describe the commodities, and give the number and kind of

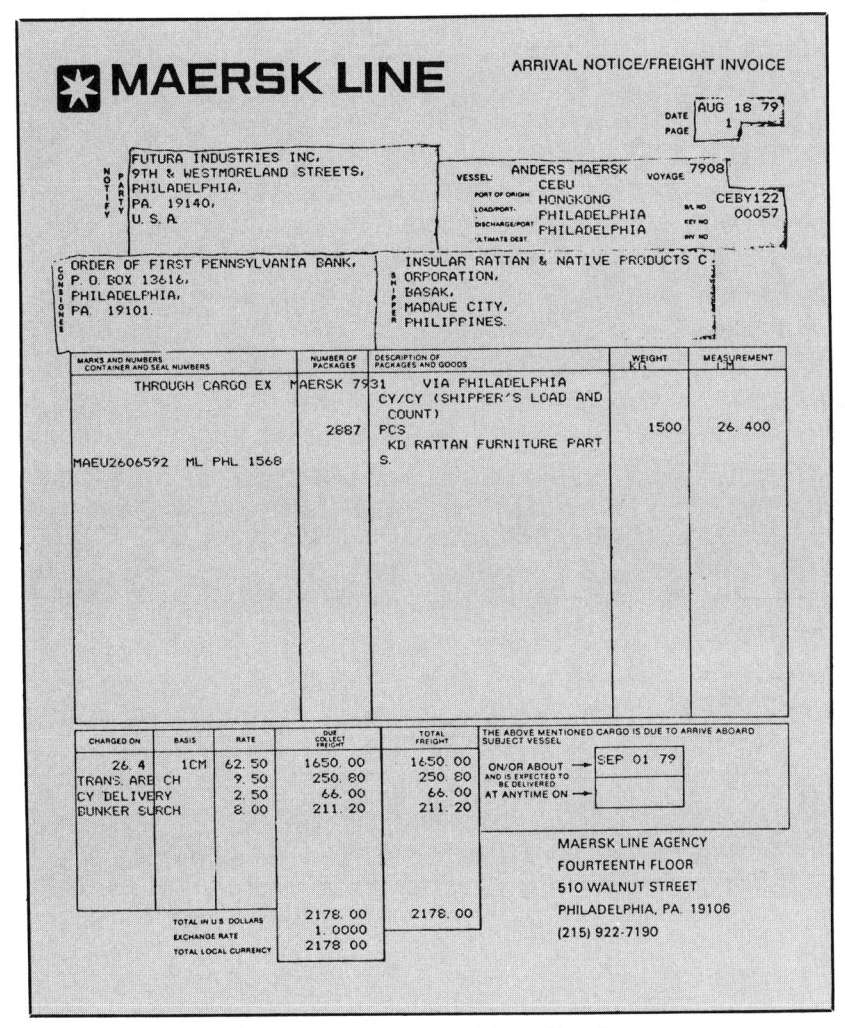

Arrival Notice From Steamship to Consignee.

INDUSTRIAL TRAFFIC MANAGEMENT

CERTIFICATE OF ORIGIN
FOR GENERAL USE AND FOR THE FOLLOWING COUNTRIES
AUSTRIA BRAZIL BRITISH MALAYA (Singapore) and FEDERATION OF MALAYA COLOMBIA EGYPT ERITREA FINLAND GERMANY (Western)
GREECE INDIA IRAN ITALY LEBANON REPUBLIC OF KOREA NETHERLANDS SAUDI ARABIA VIET-NAM

The undersigned _____
(Owner or Agent, or &c)

for _____ declares
(Name and Address of Shipper)

that the following mentioned goods shipped on _____ S _____
(Name of Ship)

on the date _____ consigned to _____

_____ are the product of the United States of America.

MARKS AND NUMBERS	NO. OF PKGS. BOXES OR CASES	WEIGHT IN KILOS GROSS	NET	DESCRIPTION

Sworn to before me

Dated at _____ on the _____ day of _____ 19 ___
this _____ day of _____ 19 ___

_____
(Signature of Owner or Agent)

The _____
a recognized Chamber of Commerce under the laws of the State of _____, has examined the manufacturer's invoice or shipper's affidavit concerning the origin of the merchandise and, according to the best of its knowledge and belief, finds that the products named originated in the United States of North America.

Secretary _____

APPERSON BUSINESS FORMS, INC., L.A. - N.Y.
FORM X101 REV. 1-58

Certificate of Origin Required by Some Governments.

374

Shippers Export Declaration.

packages, export license number, value and weight of the shipment and other information listed on the reverse of the form.

The use of proper commodity descriptions is important in preparing these papers. Such descriptions should conform to those on the export license if one is used. They also should conform to the descriptions in the U.S. Commerce Department "Schedule B" commodity list, including the commodity code numbers given there. These are seven-digit numbers and are used in the Department's data processing system for collecting statistics. "Schedule B's" may be obtained from most Commerce Department offices. Additionally, the descriptions should correspond as closely as possible (though they need not be exactly the same) to the description on the ocean bill of lading. Further refinements may be necessary for hazardous materials, where import requirements may make the use of certain phraseology advisable.

The declaration must be signed by an officer of the exporting company, although power of attorney may be issued by an officer to permit signing by another employee or a foreign freight forwarder. The original and one copy are retained by the custom house; a third copy is given to the steamship company at the time of delivery; and a fourth copy, bearing the custom house number, is retained on file by the freight forwarder or the shipper. The steamship company is required to attach a copy of each declaration to its ship's manifest, which is filed with and validated by the Collector of Customs, before sailing clearance papers can be issued.

Shipper's export declaration forms are used as export control documents as well as for statistics. Government licensing of the export may be by one of the formal documents described above or by endorsement on the export declaration.

Methods of Paying for Exports

Foreign financing carries with it added paper work and it cannot be emphasized too strongly that absolute accuracy in preparing and executing documents is essential! Errors or discrepancies can be costly and, if serious enough, may result in delayed payment or no payment at all. Take care that the information on the various forms is properly reconciled. Variations in such information are as bad as errors. Finance, shipping, billing, and other papers go out as "packages" in many cases, and the need for all the papers to "tell the same story" is important. One of the leading foreign trade banks reports discrepancies in more than 30 percent of the documents involved with letters of credit.

U.S.-based companies exporting to other countries will, of course, wish to satisfy themselves of the buyer's ability to pay, usually by checking with banks engaged in international finance and with commercial credit agencies. In addition to these familiar steps, however, an exporter must try to apply his credit principles and knowledge to his buyer's country. Exchange regulations, both domestic and foreign, must be examined and the political and economic climate of the importing nation should be kept in mind, too.

As we mentioned earlier, the U.S. government requires export licenses for some products, and in many foreign countries exchange payments are controlled through import licenses. Both exporter and importer should know how to meet such requirements. The U.S. Department of Commerce and commercial banks engaged in international trade, most of which publish reports containing this type of information, are useful sources of help.

The methods of paying for exports are:
- cash upon receipt of order
- open account
- drafts
- letter of credit
- extended terms

In all cases the terms of sale will determine the method of payment. For consumer goods, a draft or letter of credit are the most popular methods of payment.

Cash upon receipt of order—in effect paying in advance—is a very limited method. The importing country generally will not allow its currency to be sent out of the country before shipment has been made from the exporting country. Also, considerable time elapses between the date of payment and the date the goods are delivered and this is a disadvantage to many consignees.

Open account terms are used most often in the settlement of shipments between a parent and its foreign subsidiary, or to very reliable customers, agents or representatives. Such terms are granted only when complete confidence exists between the buyer and seller because the exporter normally makes shipment, awaits payment, and bears the brunt of all of the financing. This method also is little used.

Settlement by draft is the most prevalent way of paying in international trade with letters of credit a close second. This is particularly the case with short term financing. In many transactions both drafts and letters of credit are used and at times they may be used interchangeably. Drafts are also known as bills of exchange and are "negotiable instruments." Some drafts are essentially the same as a check that a company would use in paying for domestic merchandise. Others have added features similar to those of promissory notes in that they are to be paid in 30, 60, 90, or 180 days or some other period.

A letter of credit is what its name implies. It is a written representation by a bank that funds are available and will be paid for goods sent to an importer on presentation of bills of lading and other documents that will permit him to take delivery of those goods. Such letters, generally speaking, mean that both the bank and the importer are liable when the terms of the sale are fulfilled by the exporter. Bank acceptances and acceptance credits are also used in these agreements to pay the exporter promptly for the goods and to allow the importer time (usually a short period) in which to pay for them.

Drafts, letters of credit, acceptances, and the like are essentially forms of bank financing and many banks are geared to handle them on a routine basis. These banks also are generally closer to, and better equipped to handle, exchange matters, currency availability, and related problems, than are most industrial companies. Further, when banks are involved, importers and exporters take greater care to pay on time because of possible reflections on their credit standings. Also, in certain situations, U.S. industries

IMPORT-EXPORT TRAFFIC

must first go to their banks for financing before going to the federal government. Good banking connections are important and often indispensable to companies in foreign trade.

Three types of drafts are in general use in foreign trade—sight, time and date. With a sight draft or documents-against-payment arrangement, the exporter has the draft, bill of lading, and other papers forwarded through his bank to its branch or correspondent in the importer's city. There the importer pays the draft and takes delivery of the documents that permit him to obtain the merchandise from the carrier. The branch or correspondent in turn remits the payment to the exporter's bank for credit to the exporter's account.

While such traditional handling survives it is outdated in actual practice today. Air mail generally brings the draft "package" to the importer quickly and there may be a substantial time lag between its receipt and the arrival of the merchandise. Payment in such cases may be burdensome to the importer. Also, where dollar payments are required, there may be times when dollars are not available. Further, some countries forbid payment for merchandise before customs are cleared. This eventuality is generally dealt with by provisional deposits of local currency and agreements to adjust in the event of changes in exchange rates; but other potential problems exist in such cases.

Drafts and Other "Instruments"

A draft, also commonly known as a bill of exchange, is simply an unconditional order directing the buyer to pay a fixed sum of money to the bearer on a set or determinable date. The most common terms are those normally included in a time draft, which require payment within 30, 60, 90 or more days after sight. After exporter and buyer arrive at mutually satisfactory terms and shipment has been arranged, the exporter presents the draft and related documents to his bank for endorsement. A full set of export documents may include the following: A draft, in original and duplicate ("first and second of exchange"); bills of lading; commercial and consular invoices; an insurance policy or certificate; certificates of origin; inspection certificates; weight or packing lists; and any other documents required by the destination country, the nature of the shipment, or both. There are two types of drafts—sight and time—distinguished from one another by the "tenor," or time element, which is negotiated beforehand between the buyer and seller.

A sight draft instructs the purchaser to make payment when the draft is presented or "sighted." The draft is usually accompanied by various shipping documents and is directed to the exporter's local bank for collection. The bank in turn forwards both the draft and the appropriate documents to its correspondent bank overseas, very often located in the buyer's city. When these have been received, the correspondent bank informs the purchaser that the documents are in its hands and will be released to him when he "sights" the instruments, approves them and pays for the merchandise described in them. This money is then remitted through regular banking channels until the exporter's account is credited in his home bank. Many American exporters prefer to sell on a sight draft basis in order that their funds might be returned more quickly.

A time draft is handled in substantially the same manner as a sight draft, with the difference that the documents are usually released to the buyer when he accepts or "sights" the draft, while the draft itself is either retained by the correspondent bank abroad

Bank Draft.

or returned to the seller. At maturity 30, 60, 90 or more days after its sighting, as previously agreed between the buyer and the seller, the draft is again presented to the buyer-importer for payment, and the funds are transmitted via the banks back to the exporter's account in the United States.

But there are some things to watch for in using drafts. Even if the importer is willing and able to pay his bills in dollars according to terms, the dollar exchange situation in his country may be so tight as to preclude prompt payment. And when exporting on a draft basis, particularly by means of a time draft, the seller must be thoroughly satisfied that the buyer is a good credit risk. Negotiating for the payment of past-due bills with a customer on the other side of the earth can be a frustrating and often unrewarding experience.

The commercial letter of credit is applied to almost every type of foreign trade transaction. There are two broad types of letters of credit—revocable and irrevocable. A revocable letter of credit is the least satisfactory of the two in that is is merely a confirmation of an order. It does not guarantee payment because it can be cancelled without notification. Revocable letters of credit are therefore rare and generally unacceptable. On the other hand an irrevocable letter of credit, short of obtaining full payment in advance of shipment, offers the seller the most complete form of protection. This is because it is an irrevocable commitment by the purchaser's bank to pay or accept the exporter's draft provided, of course, that the draft and its accompanying documents are in order; it cannot be cancelled or otherwise modified without the consent of all parties concerned.

The letter of credit provides a method of settlement between buyer and seller by providing a bank's assurance that money will be paid promptly to the exporter. At the same time, it protects the buyer by not paying the seller until the latter has performed as specified in the letter of credit. When an importer needs a letter of credit, he asks his bank to issue it "in favor of" the company from which he is buying. By its letter of credit the bank promises that it will pay the seller, on the buyer's behalf, for any and all purchases up to a certain amount if the terms of the credit are met. The letter of credit is ordinarily sent to one of the issuing bank's overseas cor-

INDUSTRIAL TRAFFIC MANAGEMENT

PNB PHILADELPHIA NATIONAL BANK
PHILADELPHIA, PA. 19101
INTERNATIONAL DIVISION

**APPLICATION AND AGREEMENT
COMMERCIAL LETTER OF CREDIT**

DATE

Gentlemen:

The undersigned hereby requests you (as its agent) to open your irrevocable credit,

☐ Via Airmail   ☐ Via Short Cable (Details Via Airmail)

through your correspondent or _____

In Favor Of: For Account Of:

up to an aggregate amount of _____
available by drafts at _____ for _____ percent of the
invoice value, drawn at your option, on you or your correspondent, and presented not later than _____
(credit expiration date)

DOCUMENTS REQUIRED AS INDICATED BY X

☐ Commercial Invoice _____

☐ U.S. Special Customs Form #5515 _____

☐ Insurance Policy/Certificate _____

☐ Other Documents _____

☐ Full set clean on board ocean Bills of Lading issued or endorsed to the order of The Philadelphia National Bank,
marked "Freight ☐ Collect ☐ Prepaid"   Notify _____
(PLEASE MENTION COMMODITY ONLY, OMITTING DETAILS ON PRICE, GRADE, QUALITY, ETC.)
Evidencing shipment of _____
From (OCEAN PORT OF DISPATCH) _____ To (OCEAN PORT OF DESTINATION) _____

Partial shipments ☐ are ☐ are not permitted. Transhipments ☐ are ☐ are not permitted.

Indicate shipping terms: (FOB, C&F, CIF) _____ . Container shipments ☐ are ☐ are not permitted.

Insurance to be effected by ourselves, if Insurance Policy/Certificate is not checked above.
The negotiating Bank is to be authorized to forward all documents in one registered airmail.
Documents must be presented for payment, acceptance or negotiation within _____ days after the date of issuance of
the Bill of Lading or other shipping documents.

INFORMATION FOR PNB: Telephone number _____

PNB account number to be debited _____

SPECIAL INSTRUCTIONS:

177 2/78   THIS APPLICATION MUST BE SIGNED ON THE REVERSE SIDE

Application and Agreement Commercial Letter of Credit.

IMPORT-EXPORT TRAFFIC

BANCO DE RESERVAS
DE LA REPUBLICA DOMINICANA
INTERNATIONAL DIVISION

DESIGNATED DEPOSITARY FOR GOVERNMENT S FUNDS

Date MAY 15, 1979
Irrevocable LC/. No. 2-79-0173

CABLE
BANRESERVA
TELEX:
3460012
3264193

Office __SANTO DOMINGO, DOM.REP.__
Advised Through

PEACOCK LABORATORIES, INC.
54 STREET AND PASCHALL AVE.
PHILADELPHIA, PA.19143, U.S.A.

THE PHILADELPHIA NATIONAL BANK
421 CHESTNUT STREET
PHILADELPHIA 1, PA. U.S.A.
P.N.B. REF. # 10-51107-2

Dear sir(s):

We hereby establish our IRREVOCABLE LETTER OF CREDIT in your favor, for account of: __FABRICA DE ESPEJOS PAJARITO,__
__SANTO DOMINGO, DOMINICAN REPUBLIC__
for the sum of sums not exceeding __TWO THOUSAND ONE HUNDRED TWENTY ONE 90/100 DOLLARS (US$2,121.90)__
Available by your draft(s) __-__ sight on __THE PHILADELPHIA NATIONAL BANK__ for __100__ %
of commercial invoice value accompanied by the following documents

(PLEASE FOLLOW THE INSTRUCTIONS MARKED "X")

[X] Original and __Six__ copies of COMMERCIAL INVOICE covering
20 FRASCOS DE 16 OZS. CONTENIENDO NITRATO DE PLATA (MATERIA PRIMA)

*Freight and other related charges appearing on the bills of lading must be in the same typeface as other details shown, either by mimeograph, duplicating machine imprinted or typewritten, but may not be handwritten. If this requirement cannot be met, detailed billing invoices or debit notes issued by the carriers indicating such charges must be presented.

[X] FULL SET CLEAN ON BOARD Ocean/XXXXX Bill of Lading-Original legalized by __DOMINICAN CONSUL__
consigned to __BANCO DE RESERVAS DE LA REP.DOM.__
Notify __BUYERS__ marked: Freight __PREPAID__ evidencing
shipmen XXXXX __C & F__ to __SANTO DOMINGO PORT__
on or before __AUGUST 13,1979__

*Please send copies of all documents by second mail.
[X] CONSULAR INVOICE and copy, written in Spanish, original legalized by the Dominican Consul.
[X] YOUR DULY SIGNED CERTIFICATION, in duplicate, stating name and amount of the commission to be received by agent/representative in the Dominican Republic or contrariwise that there is no local agent/representative. In direct purchase by agent/representative COMMERCIAL INVOICE must show commission as deduction from sales price.

[ ] Partial shipments permitted   [X] Partial shipments not permitted
[ ] Transhipment permitted   [ ] Transhipment not permitted
Insurance to be covered by: __Buyers__
SPECIAL INSTRUCTIONS:

*Amount indicated in the agent commission certificate is to be paid in the Dominican Republic, in local currency, consequently this amount must be deducted from draft amount at the time of negotiation.

Draft(s) must be drawn and negotiated on or before __AUGUST 23,1979__
The amount and date of each negotiation must be endorsed hereon and any draft(s) presented to us shall constitute a warranty by the negotiating bank that such endorsement had been effected.
We hereby agree with drawers, endorsers, and Bona Fide Holders of the drafts drawn under and negotiated in compliance with the terms of this credit that such drafts shall be duly honored on presentation.
This credit is subject to the Uniform customs and practice for documentary credits (1974 revision). International Chamber of Commerce Publication No. 290.

Very truly yours,

Authorized Signature
A-166

Authorized Signature
A-69

F-628 REF.

Irrevocable Letter of Credit.

383

respondents, with a covering letter asking that it be passed along to the beneficiary-exporter. The transaction is handled through regular banking channels until the exporter's local account is credited according to the terms of sale.

An irrevocable letter of credit is issued in two forms—"unconfirmed" and "confirmed." An unconfirmed irrevocable letter of credit is one that is issued abroad by a correspondent bank and then transmitted to an American bank for delivery to the seller. In delivering it to the seller, the correspondent bank will state that it is merely an advice and represents no guarantee of payment on its part. The seller then has an irrevocable commitment on the part of a foreign bank to honor drawings made on the letter provided, of course, that the drawings conform to the conditions of the credit.

A confirmed irrevocable letter of credit is similar to the above except that any American bank will state that it confirms the credit and thereby undertakes to honor drawings made on it in accordance with the provisions of the credit. The seller thus has the guarantee of both a foreign bank (about which he may know nothing) and an American bank with which he is familiar. Since World War II, an American exporter has usually been required by the U.S. government to have an export letter of credit confirmed by an American bank as a precaution against unfavorable exchange regulations, dollar shortages, import permits, political upheavals and similar problems in foreign lands. With the commitment of an American bank added to the credit, however, the exporter need not be concerned with any of the above contingencies—they are the bank's concern (see next page).

But while it gives the exporter protection against many eventualities, a confirmed irrevocable letter of credit is not an absolute guarantee of payment. For example, a crippling strike in an industry might prevent the exporter-beneficiary from producing the required documents within the life of the credit. Fortunately, such problems are rare and usually do not occur without warning. But the point is that although a great measure of security is provided, overseas transactions are fundamentally an exporter's risk and nothing short of prepayment can guarantee him against all hazards.

IMPORT-EXPORT TRAFFIC

PNB PHILADELPHIA NATIONAL BANK
INTERNATIONAL DIVISION
P.O. BOX 13866, PHILA., PA. 19101

CABLE ADDRESS: PHILABANK   TELEX: 84 5355

IRREVOCABLE DOCUMENTARY CREDIT

ORIGINAL

OUR CREDIT NO	CREDIT NO. - CORRESPONDENT	DATE	LETTER OF CREDIT AMOUNT
10-51107-2	2-79-0173	MAY 24, 1979	US$2,121.90

BENEFICIARY
- Peacock Laboratories, Inc.
54 Street and Paschall Avenue
Philadelphia, Penna. 19143

CORRESPONDENT MCM/dc
- Banco De Reservas De La
Republica Dominicana
International Division
Santo Domingo
Dominican Republic

Gentlemen:

Please be guided by the Clauses marked "X"

[X] At the request of our correspondent indicated above, we enclose their Credit established in your favor. Please present one additional copy of the commercial invoice with your drawing.

[ ] This refers to our previous preliminary cable advice.

[ ] We call to your attention that drafts under this Credit are NOT to be drawn on The Philadelphia National Bank or Philadelphia International Bank and therefore are not payable by P.N.B. or P.I.B. Documents for negotiation should be presented directly to the above correspondent bank.

[ ] We are in receipt of a preliminary cable advice from our correspondent indicated above. We will advise you again of the details or deliver to you the original instrument when received. Cable reads as follows:

IF DESIRED, DRAFTS AND DOCUMENTS MAY BE PRESENTED AT PHILADELPHIA INTERNATIONAL BANK; 55 BROAD STREET, NEW YORK, N.Y. 10004
Documents must conform strictly with the terms of this credit. If you are unable to comply with its terms, please communicate with your customer promptly with a view to having the conditions changed. This will eliminate difficulties and delay when your documents are presented for negotiation.

[X] THIS CREDIT IS BEING FORWARDED TO YOU AT THE REQUEST OF OUR CORRESPONDENT, AND CONVEYS NO ENGAGEMENT BY US.

[ ] WE CONFIRM THIS CREDIT AND THEREBY UNDERTAKE THAT ALL DRAFTS DRAWN IN ACCORDANCE WITH TERMS THEREOF WILL BE DULY HONORED ON PRESENTATION.

EACH DRAFT MUST INDICATE THE REFERENCE NUMBER OF THIS CREDIT.

AUTHORIZED SIGNATURE

EXCEPT-SO FAR AS OTHERWISE EXPRESSLY STATED, THIS DOCUMENTARY CREDIT IS SUBJECT TO THE "UNIFORM CUSTOMS AND PRACTICE FOR DOCUMENTARY CREDITS" (1974 REVISION), INTERNATIONAL CHAMBER OF COMMERCE, PUBLICATION 290.

American Bank Notice of Credit to American Shipper.

The exporter should examine his letter of credit thoroughly. It contains all the conditions and instructions for preparing the seller's documents. If any of the terms cannot be met, if a required document cannot be obtained, or if the description of the merchandise is not in accordance with the agreement of sale, the issuing bank or, better still, the buyer, should be notified immediately so that the letter of credit can be amended before the merchandise is shipped. This is important, because banks will not honor drafts when inconsistencies and contradictions appear in the documents.

An "authority to purchase" is a document that is issued frequently and almost exclusively by Far Eastern banks. Its purpose is to finance imports from the United States and other countries by providing a place (the advising bank) where an exporter may present a draft drawn on the buyer and obtain immediate payment. Unless otherwise specified in the authority, the exporter's draft will be negotiated by the advising bank with "recourse to the exporter" until the issuing bank is fully paid by the importer. Most Far Eastern banks, however, are willing to offer to negotiate the draft without recourse. If the authority to purchase is irrevocable and drafts may be drawn without recourse, the document becomes, in effect, a letter of credit in that it offers the seller both security and credit. The logical rejoinder is: "If everything is equal, why not cover the sale with a regular commercial credit?" The answer is simply that most Far Eastern importers are more familiar with the "authority to purchase' and as a result prefer to use it to finance their purchases. If an American exporter-beneficiary of an irrevocable "authority to purchase" entertains a doubt as to the standing of the issuing bank, he may obtain the confirmation of his American bank. In this event, the authority becomes virtually identical to a confirmed irrevocable letter of credit.

On occasion, an "authority to pay" may be issued. Generally similar to an authority to purchase, this document merely advises the beneficiary of a place of payment and carries neither the issuing nor the advising bank's guarantee of payment. Drafts under an "authority to pay" are drawn on the advising bank in the United States and, when these have been paid, the beneficiary-exporter is no longer liable. Obviously, an "authority to pay" should not be accepted by an exporter if he is seeking security over and above that afforded by the contract of sale.

Merchandise may be shipped abroad on a consignment basis. Unless the shipment is made to an overseas branch or subsidiary, however, it must be understood that the credit risk involved is considerable. As in the case of sales on open account, there is no tangible obligation on which to base legal action in the event of default. In addition, many countries impose onerous exchange restrictions that prohibit prompt conversion of local currency into the dollars that the importer needs to pay his American supplier. Common sense dictates, therefore, that sales on consignment be made to countries relatively free from foreign trade restrictions.

If sales must be made in this manner, the merchandise should be assigned, after a thorough study of all the pertinent conditions and regulations, to a responsible warehouseman in the name of the foreign correspondent bank. Arrangements may then be made to have a selling agent negotiate for the sale of the merchandise subject to certain conditions. Besides determining the responsibility of the foreign sales agent or representative, or the import house under contract, the exporter must also make certain that regulations in the destination country will permit the unsold portion of the goods to be returned to the country of origin should it be desirable or necessary.

Foreign Freight Forwarders and Customs Brokers

Foreign freight forwarders render services that are indispensable to many exporters and importers. The competent and efficient ones have a knowledge of the technicalities of international trade encompassing banking, transportation, U.S. regulations, foreign government regulations, insurance, documentation, freight rates, customs and packaging that is impossible for any one company to comprehend, short of setting up a specialized (and expensive) department of its own.

Many foreign freight forwarders are also customs brokers and hold government licenses to act as both, which allows them to serve

INDUSTRIAL TRAFFIC MANAGEMENT

as a company's agent in both importing and exporting. The normal procedure when employing a forwarder is that the shipper sends a "letter of instructions," a preprinted form that advises the forwarder that a shipment is coming and gives the information necessary for handling the paperwork (examples of these letters of instruction are given on pages 389, 390). Forwarders will book cargo space on ships and perform many other functions, such as tracing and expediting shipments from inland points to shipside, preparing export documents including customs declarations and consular invoices, and arranging for marine insurance. Charges to the shipper vary according to the services performed and the forwarder may also receive a commission from the steamship company for bringing it business.

Customs brokers, or custom house brokers as they are often referred to, are concerned with imports. They arrange for payment of the correct duty (which is often a highly technical and detailed procedure), clear shipments into and out of bonded warehouses, pay freight charges and so on. Many importers arrange for shipments to be consigned direct to their customs brokers who are frequently given power of attorney by the importer so they can endorse bills of lading or act as one of the sureties when a bond is needed. Some other functions of the custom house broker are described in the section on importing that follows.

Imports and Customs Duties

All American importers are subject to customs regulations at the port of entry (which today, of course, may be an inland airport). The United States Customs is a branch of the Treasury Department and the country is divided into customs collection districts. Legally, the U.S. Customs assumes ownership of all goods arriving from a foreign country and maintains its ownership until all duties have been paid and other requirements met. When the importer has complied with all regulations, he presents the necessary documents to the Customs office and "entry is made." "Entry" is the term used to designate the act of complying with the law and obtaining release of the goods.

IMPORT-EXPORT TRAFFIC

Callahan International, Inc.

MEMBER (IATA)

INTERNATIONAL AIR FREIGHT SPECIALISTS

P. O. Box 187, Essington, Pa. 19029
215-365-1720

DATE:_____

TO:_____ BENEFICIARY:_____

_____ _____

_____ SHIPPER'S REFERENCE:_____

YOUR ADVICE:_____ ORIGINAL CREDIT NO:_____

AMOUNT:_____

GENTLEMEN:

    IN ACCORDANCE WITH INSTRUCTIONS CONTAINED IN THE ABOVE MENTIONED LETTER OF CREDIT, WE ENCLOSE HEREWITH THE FOLLOWING DOCUMENTS:

___ORIGINAL LETTER OF CREDIT	___CERTIFICATE OF ORIGIN
___YOUR ADVICE	___INSURANCE CERTIFICATE
___AMENDMENT	___ORIGINAL AIR WAYBILL
___SIGHT DRAFT	___COPIES OF AIR WAYBILL
___SIGNED RECEIPT	___A.I.D. FORMS
___COMMERCIAL INVOICE	___SUPPLIER'S CERTIFICATE
___PACKING LIST/WEIGHT LIST	___CARRIER'S CERTIFICATE
___CONSULAR INVOICE	___INSPECTION CERTIFICATE
___CUSTOMS INVOICE	___UNUSED BALANCE STATEMENT
	___OTHER:

    WE TRUST THAT YOU WILL FIND THESE DOCUMENTS SUFFICIENTLY IN ORDER TO PERMIT PROMPT PAYMENT OF THE AMOUNT INVOLVED DIRECTLY TO THE BENEFICIARY.

VERY TRULY YOURS,
CALLAHAN INTERNATIONAL, INC.

Letter of Instruction From Shipper to Foreign Freight Forwarder.

INDUSTRIAL TRAFFIC MANAGEMENT

Letter of Instruction From Shipper to Foreign Freight Forwarder.

IMPORT-EXPORT TRAFFIC

From the laws passed by Congress has evolved the United States Customs Tariff that sets what duties must be paid on imported goods and the commodity descriptions, markings, etc., that must be used on the documents and packages.

All shipments must be cleared through customs, even if they consist of duty-free commodities. If duty is due, entry must be made for payment or an "in bond" application submitted to Customs within 48 hours of the shipment's arrival. If no bond is posted, the goods are sent to a bonded warehouse and the shipper is required to pay transportation to the warehouse and at least one month's warehouse charges. Thus as a general rule, importers make a practice of filing a "general bond," or "general term bond," which is renewed yearly, and which obligates them to conform to the import laws.

Freight in bond is, as noted earlier, still legally in the possession of the U.S. government. Hence the carriers who haul such freight are also required to give a bond to the U.S. Treasury Department which obligates them not to deliver the goods to the consignee except upon an "order" or "release" issued by the Collector of Customs. This is a convenience to importers in that it starts the goods moving but it is only possible when the shipment is destined to a city where there is a customs office. Thus the shipment is in bond to whatever destination the local customs office releases it. Carriers transporting goods in bond to "interior ports of entry" are required first to prepare a simplified form of entry referred to as an "I.T." (immediate transportation entry).

Bonded warehouses are so called because the companies running them give bonds to the U.S. Treasury Department allowing them to receive and hold imported materials not yet released by customs. This the bonded warehousing company does by filling in an "entry for warehouse" form which is a simple document designed to get the goods off the pier as quickly as possible.

Foreign trade zones are areas set aside, at or near ports or airports, for landing, storing, and ultimately shipping goods, with a minimum of customs control and without a customs bond. They are legally duty-free storage places from which imported goods may be exported without going through customs. But it is possible to mix

U.S. with foreign goods in these zones, to assemble machines and fabricate products with U.S. and foreign components, and ship them on duty free. In recent years, there has been a considerable increase in the number of applications from cities and states to open foreign trade zones.

Costs in foreign trade are influenced at times by the "free" time allowances made at the ports. These are the periods during which cargo may occupy public wharves, warehouses, transit sheds, and and other facilities without being subject to storage or demurrage charges. Free time is used for accumulating cargo for outward vessels and for removing inbound cargo from piers and forwarding it to its ultimate destination. Free time practices vary with the ports and are subject to change. Specific information on free time allowed on export and import cargo as well as wharfage, tollage, demurrage, and other charges in effect at the various U.S. ports may be found in their tariffs.

Financing Imports

The principal method of financing the importation of goods into the United States is the irrevocable import letter of credit. Certain circumstances, however, will dictate that imports be financed by means of open account, dollar drafts or on consignment. Except for the letter of credit, which will be discussed below in some detail, these methods are similar to those described in the export procedures section.

There is no fundamental difference between an import letter of credit and an export letter of credit. It simply depends on which side of the ocean or national boundary the parties to the transaction stand. Most imports into the United States are financed by means of a letter of credit, and in practically every case the irrevocable letter of credit is preferred by the foreign seller.

When an American importer has filled out an application provided by his bank for an import letter of credit, and the credit has been issued, the bank will usually forward it to its correspondent abroad, either in or near the city of the beneficiary-exporter. The foreign correspondent will be instructed to deliver the letter to the exporter who will then prepare the documents required under the letter of credit, attach a draft, and submit the documents to a bank in his city. That bank then inspects the draft and the attached documents to make certain that they conform to the specifications of the letter of credit. If all is in order, the bank will negotiate the draft and pay the beneficiary-exporter the equivalent in local currency at the rate for purchasing such drafts, or in the currency of the credit unless local exchange regulations prohibit doing so.

When the issuing bank is finally presented with the draft and its supporting documents, they are again carefully examined for conformity to the terms indicated in the letter of credit. If all is in order, one of two alternatives is open to the importer, depending on the type of draft: If the letter of credit is available by a sight draft, the importer is required to reimburse his bank immediately for payments incurred on his behalf, for which he receives the documents attached to the draft. If, however, the credit provides for drafts to be drawn at 30, 60, 90 or 180 days sight, the bank is not required to pay the drafts until they fall due, nor is the importer required to reimburse his bank until that same date of maturity.

It occasionally happens that the carrier of the goods will arrive and the documents will be required by the importer before the drafts "mature." This situation can be met either by prepayment by the importer, or the bank may agree to release the documents to the importer provided he signs a "Release of Bank's Agent and Receipt" form. Under this arrangement, the bank retains title to the goods but agrees to release them for purposes of sale or preparation for sale. The terms of the receipt also dictate that the proceeds of sale must be remitted immediately to the bank.

An import letter of credit permits the foreign beneficiary-exporter to draw on an American bank rather than the importer, which affords him a greater measure of protection. The importer, however, does not enjoy the same degree of protection, because a bank will not assume responsibility for the goods supposedly represented by the letter of credit, the genuineness of the documents, or the financial responsibility of the carrier or the insurer. Consequently, it is important that the importer trust the character and solvency of his shipper in the foreign country and that the conditions of the sales contract are thoroughly understood. Letters of credit may be opened in a foreign currency but this creates a foreign exchange risk.

The standby letter of credit is a relatively new application of letters of credit. Historically, however, the standby was probably the original letter of credit. The credit is not drawn upon unless the holder of the letter—the party who applied to the bank for it—does not perform his part of the contract; the credit is "on standby." A standby letter of credit may thus take the form of performance bonds, bid bonds, and of course, standby credits.

Standby credits may involve payments of notes, loans, mortgages, etc.; or they may prevent a bidder from altering or withdrawing his bid. They can take the place of guarantees normally covered by cash, or, they could become available when a supplier does not perform under a contract, in terms of timely shipment, quality or specifications of the merchandise, etc.

In fact, the standby letter of credit with its wide range of possible applications has probably not yet been fully exploited, although its usage has grown so enormously over the past few years, that it has become subject to rather stringent regulations by federal authorities.

Role of the Custom House Broker

As we noted earlier, all goods imported into the United States must be cleared through customs before they can be released to the consignee or owner. The Bureau of Customs, a branch of the Treasury Department, is charged with carrying out the provisions of the Tariff Acts as enacted by Congress and the customs regulations issued subsequent to these acts.

Clearing through customs entails the assembling of bills of lading, invoices, specifications, certificates, licenses, permits, and such other documents as are required by law or regulation. From this material a customs entry is prepared setting forth information so as to enable the customs authorities to verify the description of the goods and to classify them in accordance with the effective tariff schedules. Entry documents are also examined to determine whether the merchandise is subject to examination by other government agencies, such as the Food and Drug Administration, the Meat Inspection and Plant Quarantine Departments, etc., or whether it is subject to quota restrictions, import license or other controls.

The custom house broker prepares the documents necessary for declaring the arrival of merchandise within a port of entry and performs other services required to clear the goods through customs. He quite often arranges for letters of credit, and is entirely capable of handling every stage of a shipment from the foreign supplier's warehouse to that of the American buyer.

In preparing the customs entry, the broker must determine the actual value on which duty is to be paid, since the invoice total may not represent the dutiable figure. Furthermore, not all merchandise is subject to duty. Generally, all commercial importations valued at $250 and over require a formal entry, while those amounting to less than this figure are cleared on an informal entry.

When the final "port" of destination is inland, the custom house broker may arrange for warehousing the shipment at the port of entry, if necessary, and subsequent shipping to the inland destination. The broker will also often arrange warehousing for a foreign manufacturer who wishes to carry stock in this country for availability at short notice.

When commodities such as food or drugs, requiring inspection by various government agencies, are imported, the documents are usually released by the bank on receipt to the custom house broker who then files an entry and obtains the necessary inspection by the proper agency. Once the goods have been released by the authorities, the broker notifies the bank, which then presents the covering draft to the importer for payment.

Frequently the broker is called upon to act on behalf of the foreign shipper, particularly when the buyer has been quoted a price CIF at destination. In such cases, the exporter will usually request his local freight forwarder to consign the shipment to a custom house broker located at the port of entry in this country who will clear the goods through customs and deliver them to the buyer, charging back all costs to the foreign forwarding agent. If payment is by means of documentary drafts or letters of credit, the collecting bank may be instructed to honor the broker's bill, and to remit the balance to the drawer.

In addition to the functions listed above, the custom house broker may file insurance claims and arrange for insurance surveys, or become involved in the weighing, sampling, reconditioning, packing, repacking, coopering, marketing, testing, or other aspects of the physical handling of merchandise. Custom house brokers must be licensed by the Treasury Department.

Shipping by Air

The international air waybill is the basic transportation document used for international air freight shipments and is normally issued as a through waybill from any interior point by the local agent of the airline. At least ten copies are required to satisfy all needs. The non-transportation documents—customs declarations, licenses (export and import), financial documents, etc.—required for seaborne international trade are also required for international air shipments.

Most airlines provide a reservation service for air cargo space and shippers are advised to book space on the plane for their shipments—especially unusually large or out-of-the-ordinary shipments—well in advance.

The size and weight of air freight shipments today are limited only by the dimensions of airplanes' hatches and compartments, and the floorbearing ratings of the various types of aircraft. Large one-piece shipments are normally handled on all-cargo planes; standard container and unitized shipments may be transported either in all-cargo aircraft or as "belly cargo" in passenger planes. Dangerous and perishable commodities are normally restricted. Air carriers can refuse cargo if not properly marked and packed. Packaging requirements are generally less heavy than those needed for sea shipping, but temperature changes plus getting shipments to and from airports must be kept in mind.

IMPORT-EXPORT TRAFFIC

Overseas airlines, as do domestic carriers, promote use of their services with the suggestion that despite its generally higher rates, in a total-cost analysis air freight may save the international shipper money because of other factors. For example, they may ask shippers to analyze their over-all costs of packaging, insurance, domestic transport, storage enroute, brokerage, embarkation port charges, transportation overseas, discharge port costs, customs duties and onward transportation. The airlines suggest that some of these costs either don't occur when shipping by air or are so much lower that the over-all costs will favor use of air freight. This is particularly true, they say, in foreign trade on small lots, less than 1,000 pounds. The airlines go further and claim there are hidden costs in shipping by other transportation modes that do not apply to air freight—for example, overseas warehousing, inventory outlays and losses, cost of time in transit, lost sales accounts, distributor/customer dissatisfaction, and claims.

International air shipments often receive streamlined handling as regards paperwork (much use is made of computerized communications). Basic document requirements however, must be met. Export declarations (and licenses) may be processed at some inland origins when tendering cargo for shipment and the waybills used as "through" documents.

Each airline publishes and distributes its own memorandum tariff. These contain specific commodity and general cargo rates and other details on shipping overseas. Individual airlines will often offer special promotional rates. Many major overseas airlines are members of the International Air Transport Association (IATA) which publishes rates jointly for its members. Air cargo tariffs include domestic and foreign routing for joint carriage under through rates and air waybills.

Packaging

The published rules and regulations concerning packaging in foreign trade are few. Because of the varying climatic and other conditions from origin to destination, and a variety of national standards, international packaging standards have not been spelled out coherently in any one place. Proper packaging, nevertheless, is important in successful exporting and importing.

The marine package must of course be designed to withstand all of the hazards of salt water, condensation, dock handling, changes in climate, and warehousing in foreign countries. At the same time, it must be as light as possible to keep both freight costs and customs duties to a minimum. (Some countries levy duties on gross, rather than on net or legal, weight of goods. Even when duties are levied on value, the levy may be on the value of the package as well as on the goods, or even on the total transportation charges paid from the originating port to destination.)

Any steamship official will tell you, if pressed, of international shippers that do not either package or mark their freight adequately for the conditions and handling it must endure. They will say that too much attention is paid to the "cost-of-packaging" factor and not enough to the "safe-arrival-at-destination" factor. The fact is that over-all losses due to faulty packaging of goods in foreign trade are in the millions each year.

The average dry cargo freighter, depending of course on size, will transport anywhere from 4,000 to 10,000 tons and up of cargo. Packages loaded into a ship may be subject to pressure of as much as 1,000 pounds per square foot. The rolling and pitching of the vessel is bound to cause the strain to shift. Weak packaging gives way under these conditions.

The use of containers, cribs, platforms and pallets (even though the pallets are removed once the cargo is in the hold of the ship) and other forms of unitizing, greatly improve ship-to-pier and pier-to-ship handling.

Even if freight is handled properly, loaded properly and stowed into the ship properly, it must then face the often much greater hazard of unloading, handling, and reshipment at foreign ports. Many foreign ports have facilities requiring unloading on open piers or from the ship anchored off shore into lighters or barges.

Pilferage and theft are age-old problems in international shipping. Packaging should be designed to prevent them as much as is reasonably possible. The many handlings foreign shipments receive

IMPORT-EXPORT TRAFFIC

gives opportunities for freight handlers to break a carton or open a package, remove part of the contents, and hide any evidence of tampering. Steamship companies can refuse to pay claims for concealed loss or damage. Various types of "pilfer-proof" packaging have been designed so that if broken into, the packages cannot be resealed or repaired with the result that the theft is discovered at the next inspection, and the approximate time and place of the pilferage can be determined.

Weather at destination or enroute is another packaging factor that should be given attention. High humidity and rain on goods stored on open piers cause rust, mildew, or fungus growth, and destroy the binding strength of cartons fastened with glue, as well as causing other damage. A ship moving through the tropics enroute to destination builds up considerable heat in its holds, and this can cause damage from melting, sweating and condensation. Packaging in waterproof solid and corrugated cartons, or lining cases with heavy tar paper or other waterproof material will help. So will plastic wrappings, oiling tools, machinery and similar items, vacuum packaging and enclosing silica gel or some other moisture-absorbing agent, of which there are several on the market today. Inhibitors that will prevent corrosion can be placed in packages.

Good business demands getting the product to destination in good condition. Competition for foreign markets quite often means that exporters doing the best packaging job will get the business. Foreign importers often know the types of packaging required to get goods to their countries safely and may specify packing requirements in the sales agreement or as part of the specifications laid down in the letter of credit. Experience plus the advice of the receiver will provide the right answer sooner or later.

Some countries assess a duty on the finished product but not on parts. Therefore, if the item can be shipped as parts, or "knocked-down" (KD), with instructions to the receiver as to how to assemble, there may be a saving in the duty levied.

Steamship conferences have not issued any standardized packaging regulations for several reasons. One reason is the obvious fact that some foreign steamship lines would be reluctant to accept any standards, feeling that it might jeopardize their competitive situation. The Federal Maritime Commission has not mandated packaging standards for the shipment of overseas cargo for similar reasons. The Bureau of Foreign and Domestic Commerce of the Department of Commerce has issued some approved packaging methods which are optional for the shipper. Some military packaging standards are excellent, but are often considered to be too rigid and expensive for commercial purpose.

Firms shipping measurement cargo (i.e., cargo that is charged for by its "cube" rather than by its weight; see earlier mention in this chapter) in foreign commerce would do well to study their package designs, as it is sometimes possible to change the dimensions only slightly and reduce the exterior cube, resulting in a decrease in the ocean freight costs.

Hazardous and dangerous cargo is subject to special packaging, marking and stowage rules and regulations. Agent R.M. Graziano's Water Carrier Tariff republishes United States Coast Guard regulations governing the transportation or storage of explosives, combustible liquids, and other dangerous articles or substances on board vessels together with restrictions covering the acceptance and transportation of explosives and other dangerous articles by carrier parties to the tariff. A copy of this publication may be obtained from the Bureau of Explosives, Room 620, American Railroads Building, Washington, D.C.

The National Committee on International Trade Documentation

The National Committee on International Trade Documentation (NCITD) is a non-profit, privately financed, membership organization dedicated to simplifying and improving international trade documentation and procedures.

Working through individuals and companies, members and non-members, United States and overseas governmental departments and agencies, and national and international committees and organizations, it serves as a coordinator and as a central source of information, reference and recommendations on problems of international trade information exchange and procedures.

IMPORT-EXPORT TRAFFIC

One of the most promising programs that NCITD is promoting is CARDIS (Cargo Data Interchange System). The aim of the CARDIS program is to create a "paperless" system for exchanging the information that shippers and carriers need for handling international shipments efficiently. CARDIS has focused efforts on finding standardized systems, languages and formats so that computerized communications networks can replace the myriads of documents that are such a problem and disincentive for exporters and importers today. At this writing, the program is still undergoing tests which have so far proved its great potential. The service will be made available to anyone involved in international trade, shippers, forwarders, ocean and air carriers, land carriers, banks, insurance companies, port authorities and consignees. In 1979, the CARDIS program was given a successful test run at twelve ports throughout the world—Norfolk, New York, Elizabeth, Houston, Long Beach, Rochester, Battle Creek, Wilmington, Rotterdam, Bremen, Yokohama and Tokyo. The companies in this pilot test made about 1,000 shipments over a four month period.

Anyone involved in international trade will find it worthwhile to become acquainted with NCITD, at 30 East 42nd Street, New York, N.Y. 10017.

New Ships and Shipping Systems

The last twenty-five years have seen big changes in ocean shipping. The spread of containerization over the world's trade routes has been the most dramatic change, but other methods, some similar in principle, some altogether different, but all based on the concepts of consolidating, unitizing and standardizing in ever larger measures, have found their advocates and users. International efforts towards standardizing containers used in world trade continue. The vast majority of containers in use today are 20 or 40 feet long, by eight feet wide, by eight or eight-and-one-half feet high but there are still variants of 24, 27 or 35 feet to be found.

Concurrent with the large scale introduction of containers and container ships was the "systems concept" of using them, which is often called in the intermodalism transportation context. Inter-

INDUSTRIAL TRAFFIC MANAGEMENT

modalism demands the loading of a container at the originating shipper's dock and the transporting of that container undisturbed until it is "unstuffed" at final destination. The idea is to use large standardized "boxes" as the basic handling units for all connecting modes of transportation. Ideally, for carriers the box *is* the cargo and its contents should be a matter of indifference to them.

The degree of available container service still varies from trade to trade. General cargo liner service between the United States and Europe, the main ports of the Far East and the Mediterranean is almost completely containerized whereas on other trade routes growth has been slower. Traffic rules vary significantly from trade to trade and often times between operators in the same trade. Generally speaking, ocean carriers offer tariff provisions which cover containers owned or rented by the carrier and with the tariff rules based on classifications of pier-to-pier, house-to-house, pier-to-house or house-to-pier rates and terms. Each classification often includes minimum revenue requirements per container and frequently includes a demurrage scale of charges after a designated free time period for containers that move to an originating plant or similar location for loading by the shipper. Similar provisions for demurrage cover delivery and unstuffing locations.

In the late 1960's and early 1970's came ships designed so that

Vessel "Anders Maersk" About to Dock to Unload Containers From the Far East.

IMPORT-EXPORT TRAFFIC

they can take aboard and carry loaded barges. Known as LASH (lighter aboard ship), the sytem calls for the barges to be loaded at various points upriver or at a port's trading area and then assembled offshore where they are met by the ocean going "mother ship." The barges are taken on board by the ship's stern lift and unloaded at destination ports by the same method. This operation permits ships to call at ports where they would not normally be able to because of inadequate or crowded facilities. The barges can await their turn for unloading and reloading while the mother ship continues on to its next port of call without delay. LASH barges measure 61 feet 6 inches long by 31 feet 2 inches wide with 19,900 cubic feet of space and a capacity for 370 long tons. Some LASH ships are constructed for both barge and container operations.

In addition to the LASH, there is the proposed SEABEE system on somewhat the same principle. SEABEE barges are 97 feet 6 inches long by 35 feet wide, with 39,100 cubic feet and 833 long tons capacity. The SEABEE system is said to show quicker turn around and the mother ship would call at fewer ports because the barges will be moved to the final port by tug.

Ro-Ro (Roll on-Roll off) Vessel Loading Heavy Equipment and Tall Vehicle.

INDUSTRIAL TRAFFIC MANAGEMENT

Other ship loading and unloading methods are perhaps not as technically glamorous but are none the less important in certain trades. The roll-on/roll-off (RO-RO) system has shown growth in recent years especially in relatively short-voyage trades such as to and from the Caribbean area and mainland United States. Limited RO-RO service between the U.S. and Europe is also available, however, and RO-RO service from the U.S. West Coast to the Orient is planned. RO-RO ships range from relatively small vessels to very large ships with up to 150,000 square feet on which to park wheeled vehicles. RO-RO ships are in fact floating parking lots, with multiple flat decks and connecting ramps so that loaded trailers and trucks can be driven directly on board and off at destination ports. Special facilities such as refrigeration connections are generally provided and these services run on a regular schedule between specific ports.

Another method of shipping general cargo that has shown recent growth is the ocean-going tug-barge system. Tug-barge systems use large multiple-deck enclosed barges towed by very powerful sea-going tugs. Some of these huge barges are designed for palletized cargo and provide regularly scheduled service on relatively short hauls.

17

SHIPPING HAZARDOUS MATERIAL

The safe shipping of hazardous materials over the transportation systems of this country for many years now has been deemed a requirement in the public interest and hence a concern of governmental agencies as well as of shippers and transportation people. A wide variety of new chemicals, insecticides, pesticides, gasses, and radioactive products in commercial quantities are being transported long distances, often in railroad cars, motor vehicles, and ships of greatly expanded size and capacity. The Department of Transportation has been made responsible for the federal regulations covering the packing, labeling and shipping of such materials. State, municipal, and other requirements have become more stringent over the past decade. And the U.S. Coast Guard now exercises responsibility for those hazardous materials regulations affecting the use of ports and the loading of ships.

Potentially dangerous products demand particular care both in transportation and in their packaging/handling. The regulations covering them are binding on both shippers and carriers; penalties may be as high as $25,000 per violation or five years in prison, or both, if death or bodily injury result from violations. The regulations have a wide and complicated range and they affect a firm's internal and external handling and shipping procedures. Fundamentally, hazardous materials should be dealt with so that the risk of accidents is eliminated as far as humanly possible. Employees should know, understand, and comply with the regulations, the reasons for each of which should be made clear.

The "Transportation Safety Act of 1974" was passed by Congress and became the law of the land in 1975 as Public Law 93-633 of the 93rd Congress (H.R. 15223). Its purpose is "to regulate commerce by improving the protections afforded the public against risks connected with the transportation of hazardous materials, and for other purposes".

The Law

This law, P.L. 93-633. expanded DOT's responsibility and authority and established its right to regulate container manufacturers and repairers for the first time. It clearly places the regulatory authority for the transportation of hazardous materials in DOT, removing it from the various agencies where it formerly rested. (Currently, there is conflict between DOT and environmental agencies on the transportation of certain materials that may have to be resolved by Congress.)

The law applies to domestic shippers and to domestic private, common and contract carriers by all modes of transportation, and the regulations based on it govern the loading, unloading and storage of hazardous materials immediately adjacent to the point of actual transportation as well as their packaging and movement in vehicles. The regulations also spell out reporting requirements for carriers and shippers when accidents or "incidents" occur.

The Secretary of DOT is authorized to "issue . . . regulations for the safe transportation in commerce of hazardous materials . . . which may govern any safety aspect of the transportation of hazardous materials which the Secretary deems necessary or appropriate . . . .". "Hazardous materials" is generally interpreted as including the following classifications of items: explosives, combustible liquids, flammable liquids, corrosive material, flammable gases, flammable solids, organic peroxides, oxidizers, poisons, irritating materials, etiological agents and radioactive material.

The full definitions of hazardous materials are to be found in Title 49—Transportation, of the Code of Federal Regulations. Since these definitions are subject to modifications, additions and eliminations, it is best to contact the Department of Trans-

SHIPPING HAZARDOUS MATERIAL

portation, Materials Transportation Bureau, Office of Hazardous Materials Operations, Operations Division MTH-30, 2100 2nd Street, S.W., Washington, D.C. 20590, for the latest regulatory definition and listing. The act is given in its entirety as appendix 8 and is followed by a DOT abstract of the basic definitions of hazardous materials in appendix 9.

The traffic manager should pay particular attention to the following parts of Title 49—Transportation:
- Part 171—General information, regulations and definitions
- Part 172—Hazardous materials table and hazardous materials communications regulations
- Part 173—Shippers—General requirements for shipments and packaging
- Part 174—Carriage by rail
- Part 175—Carriage by aircraft
- Part 176—Carriage by vessel
- Part 177—Carriage by public highway
- Part 178—Shipping container specifications
- Part 179—Specifications for tank cars.

Part 172.101, the table of hazardous materials, lists descriptions and proper shipping names, required labels, packaging specification references and other specifications as they are affected by the mode of transportation used. For an excerpt from Part 172.101 see overleaf.

Establishing and running training programs on the requirements of hazardous materials shipping has become an important traffic management responsibility in many firms. The Department of Transportation says that it is the duty of any company that offers hazardous materials for transportation to instruct each of its officers, agents and employees having any responsibility for preparing hazardous materials for shipment as to applicable regulations (Sec. 173.1 (6), HM-103/HM-112).

But while such training is mandatory for everyone involved, including corporate officers, the selection of materials used is left up to the instructor. There is a considerable choice of "packaged" courses in the requirements of hazardous materials handling and shipping now commercially available (including some that supply

INDUSTRIAL TRAFFIC MANAGEMENT

Hazardous Materials

(1) •/W/A	(2) Hazardous materials descriptions and proper shipping names	(3) Hazard class	(4) Label(s) required (if not excepted)	(5) Packaging (a) Exceptions	(5) Packaging (b) Specific requirements
	Accumulator, pressurized (pneumatic or hydraulic), containing nonflammable gas	Nonflammable gas	Nonflammable gas	173.306	
	Acetal	Flammable liquid	Flammable liquid	173.118	173.119
	Acetaldehyde (ethyl aldehyde)	Flammable liquid	Flammable liquid	None	173.119
A	Acetaldehyde ammonia	ORM-A	None	173.505	173.510
•	Acetic acid (aqueous solution)	Corrosive material	Corrosive	173.244	173.245
	Acetic acid, glacial	Corrosive material	Corrosive	173.244	173.245
	Acetic anhydride	Corrosive material	Corrosive	173.244	173.245
	Acetone	Flammable liquid	Flammable liquid	173.112	173.119
	Acetone cyanohydrin	Poison B	Poison	None	173.346
	Acetone oil	Flammable liquid	Flammable liquid	173.118	173.119
	Acetonitrile	Flammable liquid	Flammable liquid	173.118	173.119
	Acetyl benzoyl peroxide, solid	Forbidden			
	Acetyl benzoyl peroxide solution, *not over 40% peroxide*	Organic peroxide	Organic peroxide	None	173.222
	Acetyl bromide	Corrosive material	Corrosive	173.244	173.247
	Acetyl chloride	Flammable	Flammable	173.244	173.247

SHIPPING HAZARDOUS MATERIAL

Table, Excerpt

(6) Maximum net quantity in one package		(7) Water shipments		
(a) Passenger carrying aircraft or railcar	(b) Cargo only aircraft	(a) Cargo vessel	(b) Passenger vessel	(c) Other requirements
No limit		1,2	1,2	
1 quart	10 gallons	1,3	4	
Forbidden	10 gallons	1,3	5	
1 quart	10 gallons	1,2	1,2	Stow separate from nitric acid or oxidizing materials
1 quart	10 gallons	1,2	1,2	Stow separate from nitric acid or oxidizing materials. Segregation same as for flammable liquids
1 quart	1 gallon	1,2	1,2	
1 quart	10 gallons	1,3	4	
Forbidden	55 gallons	1	5	Shade from radiant heat. Stow away from corrosive materials
1 quart	10 gallons	1,2	1	
1 quart	10 gallons	1	4	Shade from radiant heat
Forbidden	1 quart	1,2	1	
quart	1 gallon	1	1	Keep dry. Glass carboys not permitted on passenger vessels
quart	1 gallon	1	1	Stow away from alcohols. Keep cool and

409

INDUSTRIAL TRAFFIC MANAGEMENT

instructors as well as course materials). Many companies have opted to run their own courses or seminars.

Carriers, too, are required under the regulations to give their employees training in the handling and transportation of hazardous materials:

—Rail—each carrier, including a connecting carrier, shall perform the duties specified and comply with each applicable requirement . . . and shall instruct his employees. (Sec. 174.7, HM-103/HM-112)

—Air—each operator shall comply with all the regulations in Part 102, 171, 172 and 175 and shall thoroughly instruct his employees in relation thereto. (Sec. 14CFR, Sec. 121.135, 121.401, 121.433a, 135.27 and 135.140. Sec. 175.20, HM-103/HM112)

—Water—each carrier, including a connecting carrier, shall comply with all regulations in this part, and shall thoroughly instruct his employees in relation thereto. (Sec. 176.13, HM-103/HM112)

—Highway—it is the duty of each such highway carrier to make the prescribed regulations effective and to thoroughly instruct employees in relation thereto (Title 49CFR, Sec. 177.800).

The traffic manager should make himself familiar with these regulations, and as the handling of hazardous materials shipments in an industrial company almost always involves a number of departments in addition to traffic, the TM is the logical candidate to coordinate the necessary company-wide training programs with warehousing, purchasing, packaging, sales, marketing and production.

Setting Up a Compliance Program

The company's compliance program must be the core of any training in the handling and transportation of hazardous materials. While most firms will want to tailor their compliance programs to their own operations, the outline of a practical program should include a manual or handbook describing the policy and giving an overview of the regulations that cover the company's handling and shipping procedures. The manual should show who has specific responsibilities for hazardous materials handling within the various company departments and where that person's office is.

An effective compliance program must ensure that company departments know what their responsibilities are. Production should know the hazardous materials packaging and labeling requirements and the specifications and regulations covering the different containers used for different materials. (A point to emphasize is that when repacking, the same materials and labels must be used as for an original package.)

Some further examples of the "do's" and "don'ts" of hazardous materials handling from the viewpoints of the different parties involved, are given below.

For the distribution department:

—Keep copies of the regulations on hand and make sure department members are familiar with those covering labeling, placarding, packing and transportation.

—Be sure DOT-approved shipping names and hazard class are on the documents and that proper documents, proper counts and correct weights are used.

—Be sure the driver is made aware your shipment contains hazardous material and that the loading, blocking and bracing is approved by the carrier's representative when the loading is done by the shipper.

—Be sure the carrier placards the vehicle properly before it is moved.

—Be sure the dispatcher is aware of the hazardous material shipment when calling for transportation.

—Check documents against the shipment itself and inspect all materials for possible leaks or damage.

For the forwarder or warehouseman:

—Do not assume the shipper has met all hazardous materials shipping requirements. You are as responsible as the shipper, so have copies of the regulations ready for reference.

—Require a written statement from the shipper that he has met all current regulations for packaging, labeling and shipping if he is shipping hazardous materials. (But you should also have sufficient

INDUSTRIAL TRAFFIC MANAGEMENT

knowledge to be able to check that statement.)

—Know the shipper, his products, his packaging and his documentation.

For the carrier:

—Have all regulations available to all operating personnel.

—Review procedures for handling hazardous material shipments and update those procedures as necessary.

—Maintain lists of hazardous materials shippers, including types of materials they ship, and make that list available to dispatchers, dock, traffic and sales personnel.

—Keep a current list of sources of help and information to be used when contamination and other emergencies occur and when unrecognizable chemicals are encountered.

—Have an isolated area for leaking containers, and refuse to move any until the shipper has put the container in proper condition or has repacked the material.

—List what products can and cannot be loaded in the same vehicle with hazardous materials.

The value of training in hazardous materials regulations is best exemplified by the following recent cases before DOT:

—A carrier transported a package bearing a poison label in the same vehicle with foodstuffs and other edible products. Fine—$3,000.

—A shipper offered an air freight shipment of electrolyte packaged in polyethylene bags. Fine—$4,000.

—A container manufacturer failed to retain the prescribed hydrostatic and drop test samples on a DOT specification container. Fine—$1,500.

—A carrier transported a shipment of hazardous materials in non-specification containers and without the proper labels. Fine—$5,000.

—A carrier employee failed to mark or placard a vehicle transporting hazardous materials. Fine—$250.

A logical routine to follow when preparing the shipping documents for a hazardous materials shipment is to first identify the product and make-up in the hazardous materials tables and then classify it by DOT regulations. This should also give the shipping name authorized by DOT. Select the proper package, label and markings. Prepare the bill of lading and certify the propriety of the shipment while preparing the other paperwork required by the carrier. Finally, determine what placards are needed and make sure they are put on the vehicles.

The DOT regulations for labeling and placarding require that hazardous materials shipments must be labeled in accordance with the list in Sec. 172.101 (Title 49 CFR) and when shipped in rail cars or trucks, these must have placards on four sides while the material is in the vehicles. Labels must be diamond shaped and be at least four inches by four inches. Mixed packages must be labeled in accordance with Sec. 172.404 of Title 49 CFR.

There are exemptions to some of the hazardous materials handling regulations and the procedures for seeking exemptions is described in Part 107 of Title 49 CFR. But be sure to consult the DOT Hazardous Materials Regulation (HM)-127 before making application, or petition for exemptions.

Exporting and Importing Hazardous Materials

Hazardous materials in transit across United States territory come under the DOT regulations, so that exports and imports moving to and from ports and airports must be in compliance with 49 CFR. Other countries have their own codes and regulations for potentially dangerous shipments, however, and the shipper who needs to know them will have to contact the organizations and associations that operate under them. Some of these organizations are:

—OCTI (Office Centrale) (Central Office for International Railway Transport) includes most of the railroads of Western Europe, some railroads in North Africa and a few in the Near East.

—OSJD (Organization for the Collaboration of Railways) includes the Cuban railroads and East European nations not members of OCTI. OSJD is in fact principally an organization of railroads of the "Iron Curtain" countries, with some Asian countries included.

—ECOSOC (Economic and Social Council of the United Nations General Assembly). This council has several committees working on codes of regulations covering the transportation of dangerous goods overseas.

—IATA (Internation Air Transport Association) publishes its own list of restricted articles and regulations which can be purchased from its office at P.O. Box 160, 1216 Cointrin, Geneva, Switzerland.

—IMCO (International Maritime Dangerous Goods Code) requirements and regulations are enforced in the national waters of the United States by the U.S. Coast Guard. This code, which regulates the shipping of dangerous goods in international maritime commerce, is also subscribed to by 23 other nations. It is available from the New York Nautical Inst. & Service Corp., 140 W. Broadway, New York, N.Y. 10013.

Transporting Hazardous Materials by Air

The transportation of hazardous materials via domestic airlines is also governed by Title 49 CFR. A copy of this law and the CAB82 and IATA regulations belong in every traffic library.

If such a shipment originates and terminates in the United States, the governing law is the 49 CFR. But airlines, airfreight forwarders and brokers may not:
- Identify or classify products
- Interpret regulations for the shipper
- Repack restricted articles
- Sign shippers certificates (except for forwarders, who may only sign for shippers in the U.S. In so doing they assume *all* responsibilities of the shipper).

All other conditions of 49 CFR apply to air shipments. Air shippers will find a copy of "Airfreight Guide for Hazardous Materials" (published by Flying Tigers, P.O. Box 3979, North Hollywood, CA 91609) a useful aid.

Some Other Aids

In addition to the publications already mentioned in this

SHIPPING HAZARDOUS MATERIAL

chapter, the "Red Book on Transportation of Hazardous Material" by L.W. Bierlein is a comprehensive reference. And membership in COSTHA—Council for Safe Transportation of Hazardous Articles (685 Third Ave., New York, N.Y. 10017) is worth considering. COSTHA is an organization that is constantly examining the hazardous materials regulations and appears before the agencies involved to clarify (and if need be, to propose modifying) the rules for a better understanding by its members.

The Hazardous Material Advisory Council (HMAC) at 1100 17th St., N.W., Washington, D.C. 20036, is a valuable source of information, and various trade associations stand ready to give expert and knowledgeable help to industrial companies engaged in shipping hazardous materials. The American Petroleum Institute (API), The Fertilizer Institute (TFI), of Washington, D.C., and the Compressed Gas Association (CGA), of New York, are among these.

Two particularly helpful programs are "Chem-Card" and "Chemtrec," originated by the Manufacturing Chemists Association (MCA), of Washington, D.C. The "Chem Card" program is a transportation emergency guide consisting of a manual and a collection of cards, each of which covers a chemical in common use. In short, non-technical words the cards describe the hazards (fire, exposure) of particular chemicals and what to do—and not to do—if accidents (leaks, spills, fires and so on) occur. If a tank truck overturns loaded with anhydrous ammonia for example, the driver has a quick reference on what to do from the card or the manual. Though designed for tank truck carriers, the manual and cards are now quite widely used by other modes.

"Chemtrec" gets its name from the "Chemical Transportation Emergency Center," an office operated around the clock every day of the year to provide immediate information about the hazards of chemicals involved in transportation accidents anywhere in the continental United States. Firemen, police, carriers, shippers, or anyone dealing with an emergency involving chemicals can get immediate help by dialing a toll-free, nationwide number carried on the shipping papers issued by most MCA members. When an emergency call comes into "Chemtrec," the person on duty tells the caller if the chemical involved is hazardous and what to do in case

INDUSTRIAL TRAFFIC MANAGEMENT

of spills, fires, leaks, or human exposures—enough information for the first steps needed to control the emergency. The "Chemtrec" office also relays details of the accident to the shipper who can often provide additional expert help. "Chemtrec"'s files, covering thousands of chemicals and cross-referenced by manufacturer, generic and trade names, are constantly updated with new information and products from member companies.

Finally, tariffs covering explosives and other dangerous articles may be bought from the carriers' associations. The Bureau of Explosives of the Association of American Railroads (1920 L Street N.W., Washington, D.C. 20036) sells such a tariff for rail shipments, as does the American Trucking Associations (1616 P Street, N.W., Washington, D.C.) for motor carrier shipments, while the air transport "Restricted Articles" tariff may be bought from the Airline Tariff Publishers, Inc. (1825 K Street, N.W., Washington, D.C. 20006).

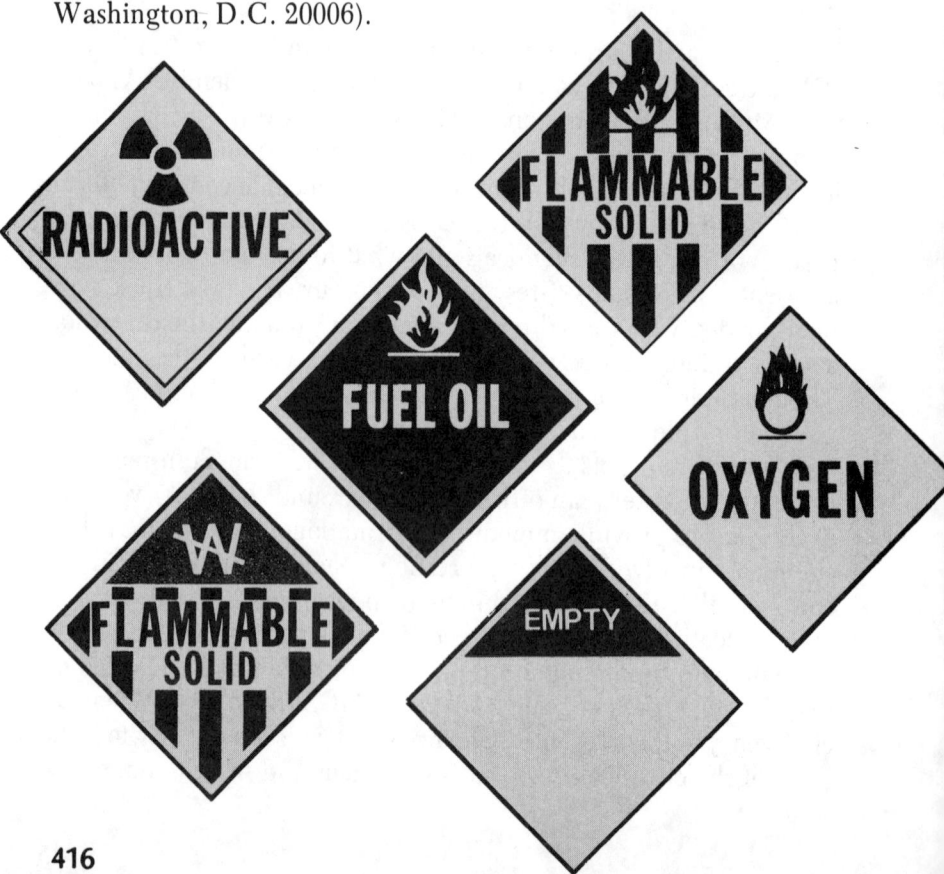

18

PHYSICAL DISTRIBUTION MANAGEMENT

In previous chapters we have said, in a number of different ways, that the traffic manager must develop and exercise an expertise in transportation. This is primarily a matter of *operational* knowledge, of developing the "know-how" to do the job and to teach others to perform it.

But there is a step beyond this level of competence. In traffic management, there is, and must always be, a very close working understanding with those involved in warehousing, order processing, inventory control, customer relations, site selection, sales forecasting, packaging and material handling. In each of these functions, traffic management is a vital factor that if ignored (so that each department tries to work as an entity unto itself) creates transportation havoc.

Coordinating these functions into a cohesive operation is called physical distribution management (PDM), and whatever title the individual responsible for it holds, it is both a "line" and "staff" job, with emphasis more on the managerial than the operational side.

In the early 1930s, the U.S. Department of Commerce recognized the value of coupling "traffic management" and "physical distribution" when it used the terms interchangeably in a published analysis of industrial traffic management. But the concept expressed in the term lay dormant for many years until after World War II, when the value of physical distribution management began to be recognized by both the academic and industrial worlds. It was the establishment of the "total cost" concept in the late 1950's that triggered the removal from marketing of certain functions and that opened the "distribution era" as against the "marketing era" of the '30s and '40s.

About 1963, the National Council of Physical Distribution Management (NCPDM) came into being and defined what it understood by physical distribution management. It was, said NCPDM:

> "A term employed in manufacturing and commerce to describe the broad range of activities concerned with efficient movement of finished products from the end of the production line to the consumer, and in some cases includes the movement of raw materials from the source of supply to the beginning of the production line. These activities include freight transportation, warehousing, material handling, protective packaging, inventory control, plant and warehouse site selection, order processing, marketing forecasting and customer service."

Since that time the terms "materials management" and "business logistics" have appeared and in 1976 NCPDM made some changes in its definition. Now it reads as follows:

> "Physical distribution management is the term describing the integration of two or more activities for the purpose of planning, implementing and controlling the efficient flow of raw materials, in-process inventory, and finished goods from point of origin to point of consumption. These activities may include, but are not limited to, customer service, demand forecasting, distribution communications, inventory control, material handling, order processing, parts and service support, plant and warehouse site selection, procurement, packaging, return goods handling, salvage and scrap disposal, traffic and transportation and warehousing and storage."

The potential of PDM has appealed to many other parts of the industrialized world and PDM organizations have made their appearance in Canada, Great Britain, Continental Europe, Japan and Australia. Many papers, articles and books have been published in a number of languages and a brisk exchange of ideas goes on today.

The Centre for Physical Distribution Management (CPDM), organized in 1970 in Great Britain, under the wing of the British Institute of Management, defines physical distribution as follows:

> "Physical distribution is the broad range of activities within a company concerned with the efficient movement of goods and

material both inwards to the point of manufacture and outwards from the end of the production line to the customer.

"These constituent activities include, for example, transport, warehousing, inventory management, materials handling, packaging and order processing, all of which are closely interrelated."

The British CPDM distinguished between PDM as a "field" and as an "area of management":

"PDM's aim is to achieve the highest possible measure of efficiency in the physical distribution activity. That efficiency can be measured in two ways—cost and quality of service. These are, in turn, interdependent, since the higher the level of service, the greater the cost. PDM is essentially a matter of coordination—both internally as between its constituent activities, and externally with marketing, production and purchasing so as to ensure the best overall balance in the interests of the company as a whole."

But there is general international agreement as to the functions that comprise PDM no matter what semantics are used. These functions are: Transportation, traffic management, order processing, warehousing, material handling, packaging, forecasting, inventory control, customer service, site location—and of course the data processing that makes it possible to analyze the interrelationships between them.

PDM coordinates those activities that provide time and place utility to the commodities needed by the manufacturing corporation and then also to the goods the corporation itself manufactures. The flow is first to manufacturing and then from the end of the production line to the consumer.

At various points and times in this flow, opportunities occur to trade the costs of one function for the value of another, with the aim of reducing total costs. Thus "normal" transportation may be traded for "premium" transportation despite its higher cost because the cash flow generated by faster delivery billing adds up to a lower total cost. A change in packaging or the elimination of distribution centers may show the same total-cost arithmetic.

PDM also attempts to coordinate its own function and those of production and marketing. Yet trading-off for the total cost is not as new an operation as many seem to think. It has always been going on but in many variations. Now PDM tries to bring the varieties under one roof.

Organization

The place of PDM in the corporate structure depends on the philosophy of the company. No standard organization has been adopted by the industrial world. However, we offer several ideas that may serve as a basis for organizing a department. In Figure 1, we show the functions of the three major units of a corporation before adopting PDM. Common practice is to scatter the PDM functions among the production, marketing and finance departments.

Figure 2 shows PDM as a staff organization with no line responsibilities. Five functions are managed by one person. Figure 3 shows what the *operations* responsibilities of this manager would be if they were given to him. Of course, the company might employ two managers of equal status, but this would open the potential for conflict between the producer of procedures and the doer of the job.

Thus in Figure 4 we show the combination of line and staff functions under one manager. This is a healthier situation, allowing the manager to interchange personnel so that everyone understands

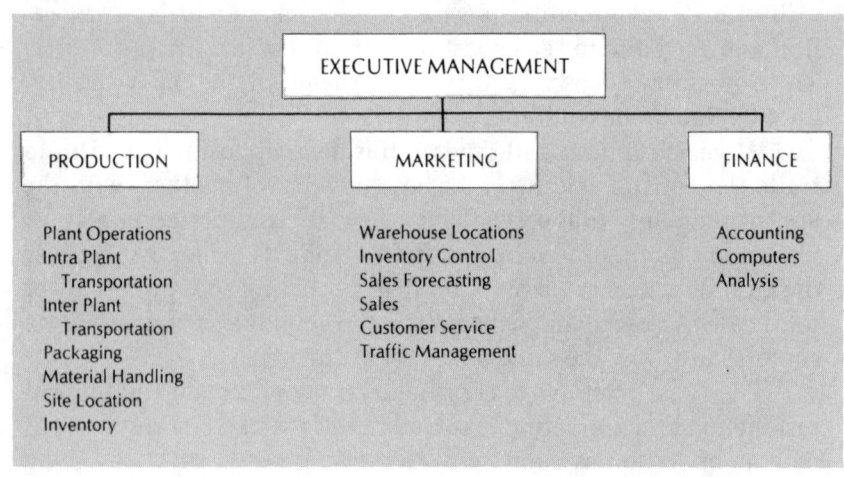

```
                    EXECUTIVE MANAGEMENT
                              |
        ┌─────────────────────┼─────────────────────┐
   PRODUCTION              MARKETING               FINANCE

Plant Operations        Warehouse Locations      Accounting
Intra Plant             Inventory Control        Computers
   Transportation       Sales Forecasting        Analysis
Inter Plant             Sales
   Transportation       Customer Service
Packaging               Traffic Management
Material Handling
Site Location
Inventory
```

Figure 1. Physical distribution functions scattered among other divisions prior to establishing a physical distribution department.

PHYSICAL DISTRIBUTION MANAGEMENT

and appreciates others responsibilities. Here we only have need for four sub-managers to cover all functions shown in Figure 1.

Once established, where should PDM be placed in the cor-

```
                      PHYSICAL DISTRIBUTION MANAGER
        ┌──────────────┬──────────────┬──────────────┬──────────────┐
   CUSTOMER       DISTRIBUTION      COST         ENGINEERING    FORECASTING
   SERVICE         PLANNING       ANALYSIS

   Standards       Long Term       Budgets        Packaging        Sales
                   Short Term      Expenses       Material       Inventory
                                                  Handling        Control
```

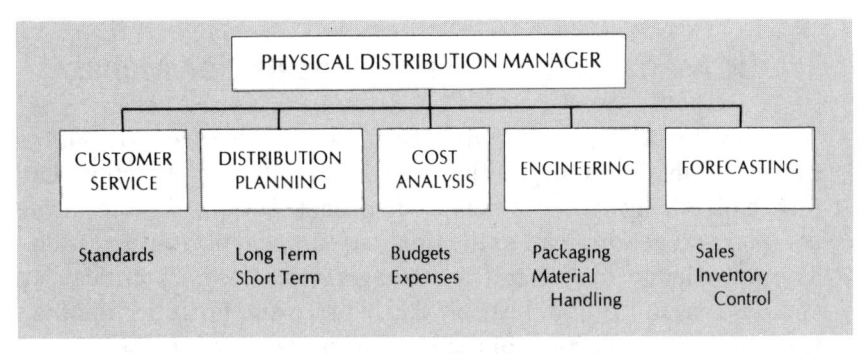

Figure 2. Physical distribution as a staff position.

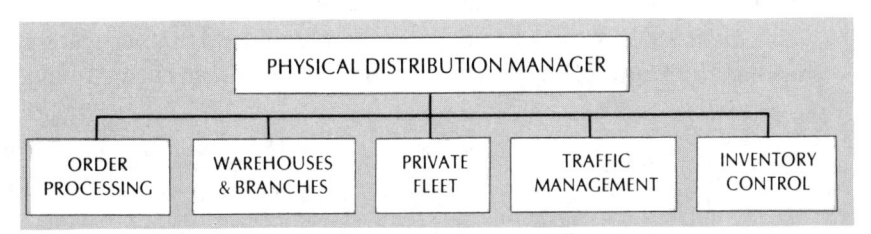

```
                      PHYSICAL DISTRIBUTION MANAGER
        ┌──────────────┬──────────────┬──────────────┬──────────────┐
    ORDER        WAREHOUSES       PRIVATE         TRAFFIC        INVENTORY
  PROCESSING     & BRANCHES        FLEET        MANAGEMENT        CONTROL
```

Figure 3. Physical distribution as a line position.

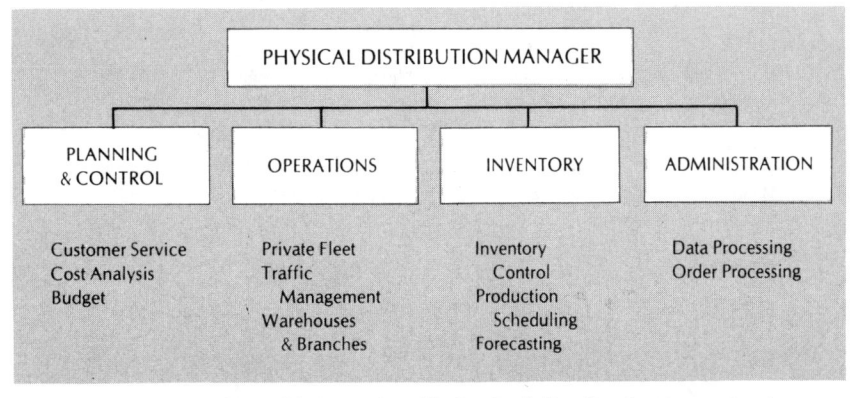

```
                      PHYSICAL DISTRIBUTION MANAGER
        ┌──────────────┬──────────────┬──────────────┐
     PLANNING        OPERATIONS      INVENTORY     ADMINISTRATION
    & CONTROL

  Customer Service   Private Fleet    Inventory     Data Processing
  Cost Analysis      Traffic          Control       Order Processing
  Budget             Management       Production
                     Warehouses       Scheduling
                     & Branches       Forecasting
```

Figure 4. Combined line and staff physical distribution organization.

421

porate structure? The responsibilities are such that it should have equal status with the other major corporate functions of production, finance, marketing and administration at the vice-president level. This is shown in Figure 5.

Traffic Manager into Physical Distribution Manager

Traffic management is the one activity of physical distribution that interfaces with all its other components. It is the function that physically brings the customers his orders. Within the industrial complex, all actions lead to the final requirement of transportation. The knowledge of the traffic manager thus becomes critical. It enables him to bring to bear his skills in rate and tariff negotiating, transport law, carrier operations, site location, mode usage, consolidation and routing while taking account of the interests of marketing, production and purchasing.

This is PDM's most important advantage; his is the operation that has the trade-off opportunities to meet customer needs, to im-

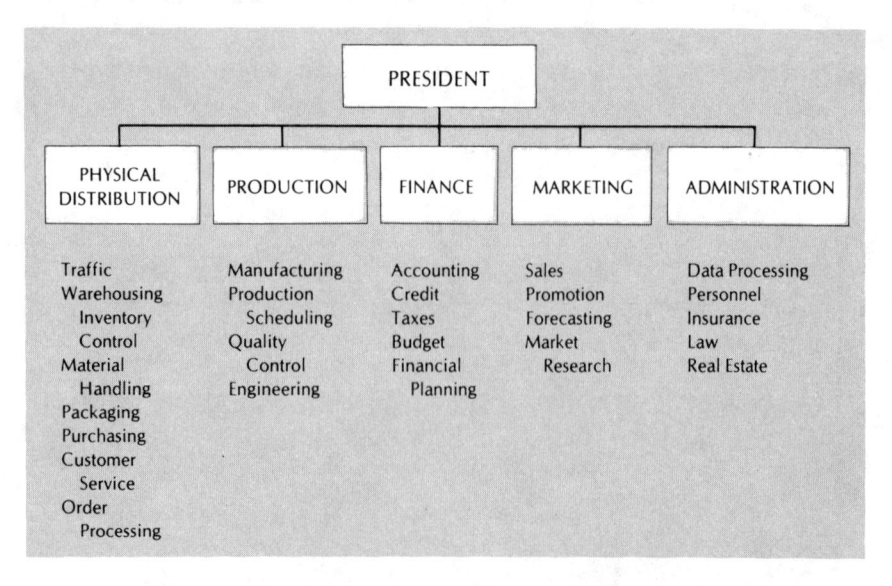

```
                              PRESIDENT
        ┌──────────┬──────────┬──────────┬──────────┐
    PHYSICAL    PRODUCTION  FINANCE   MARKETING  ADMINISTRATION
  DISTRIBUTION

  Traffic        Manufacturing  Accounting   Sales        Data Processing
  Warehousing    Production     Credit       Promotion    Personnel
  Inventory      Scheduling     Taxes        Forecasting  Insurance
    Control      Quality        Budget       Market       Law
  Material         Control      Financial      Research   Real Estate
    Handling     Engineering      Planning
  Packaging
  Purchasing
  Customer
    Service
  Order
    Processing
```

Figure 5. Physical distribution as a major corporate activity.

PHYSICAL DISTRIBUTION MANAGEMENT

prove inventory control to its most efficient level for the financial investment in that inventory. The traffic manager *knows* the size and weight factors and transit times; he *knows* the competitive services for small packages. He *knows* how to evaluate LTL or volume shipments to regional small package operators for break-bulk and delivery service. He *knows* how to negotiate volume commodity rates, column commodity LTL rates, aggregate rates, intransit fabrication and/or storage rates.

The PDM can select transport means—common, contract or private carriage—make trade-offs among the transport modes—motor, rail, air or water—all the while taking account of customer service standards, financial necessities and security considerations.

Inventory Control

Finance wants a low inventory so that the investment is small. Marketing wants a large inventory to avoid stock-outs. Production, too, wants a large inventory so it can lower its per-unit cost by manufacturing larger quantities. Traffic management wants an inventory that allows for volume shipments to lower the transportation cost of each unit shipped. The inventory control people want an inventory large enough to satisfy every order received with enough of an overage to cover sudden sales surges or emergencies

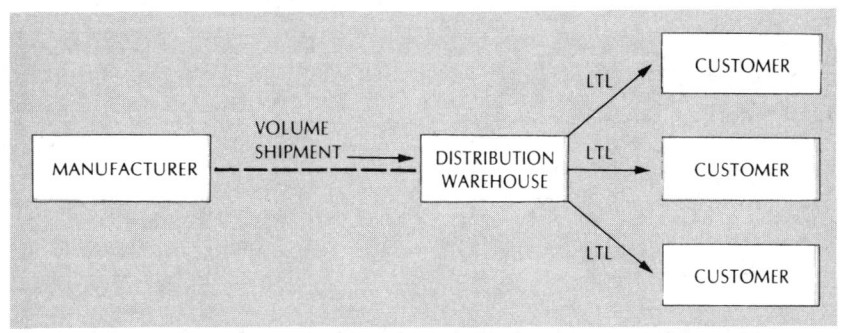

Figure 6. Manufacturer shipping volume load to a warehouse for break-bulk operations.

423

such as transportation breakdowns or production down-time. The warehouse manager wants just enough inventory to reduce crowding and handling costs.

The PDM is confronted with all these conflicting demands. His final judgments will of course be determined by corporate philosophy and sales forecasting but the firm's economic order quantity (EOQ), coupled with the cost of maintaining the inventory and the cost of being out-of-stock will usually override the other considerations.

Warehousing

The cost of maintaining an inventory (whether it is in finished goods or raw materials) includes the price of warehousing that inventory. So warehousing, too, becomes an integral part of traffic management, especially when the warehouse lies between manufacturing plant and market.

The traffic manager's training and experience in rates and routes is all-important in locating a distribution warehouse. Warehousing today covers a variety of operations in addition to simple storage. For example, the warehouse may be used as a break-bulk

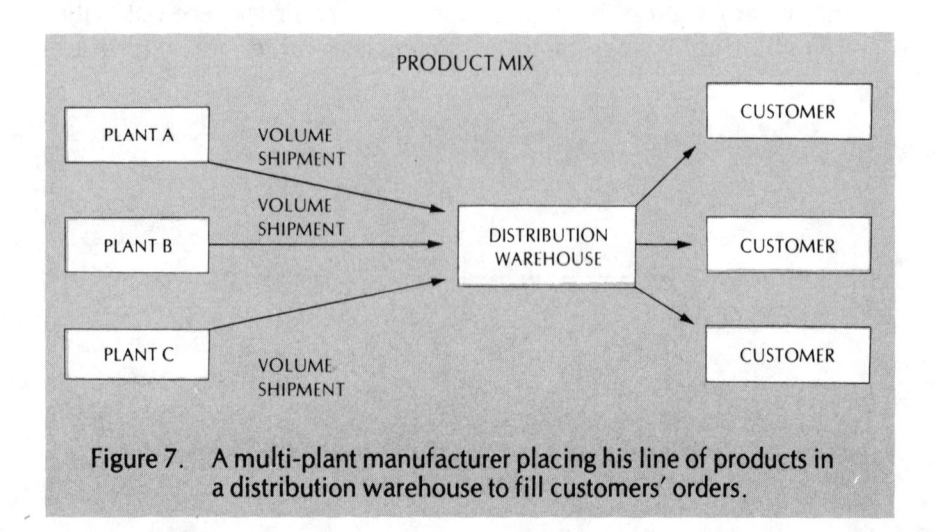

Figure 7. A multi-plant manufacturer placing his line of products in a distribution warehouse to fill customers' orders.

and distribution point where trailer loads may be broken down into small shipments through the simple expedient of sending the manifest and bills of lading along with the volume shipment (Figure 6). Or the warehouse may be used for product mixing by a company having multiple manufacturing plants. Each plant forwards trailer loads of its products to the warehouse which fills orders for the full range of company products (Figure 7). Or it may be used for consolidation of raw and semi-manufactured items for volume movement to the manufacturer.

The PDM has available to him cold-storage warehouses, and warehouses that specialize in everything from liquid bulk and dry bulk items, steel and agricultural products, to protecting things of extraordinary value, such as paintings and jewelry. He may use the warehouse for consolidating, distributing, storing, packing, labeling, coopering, product display, in-transit storage and filling orders.

Material Handling

Material handling costs are an important consideration in freight rate negotiations. The right material handling equipment can reduce loading and unloading time—thus aiding the carrier in making better use of his equipment and manpower.

The PDM today, besides using the usual fork lifts, conveyors, pallets, tow trucks and so on, may also make use of unit load systems and container systems for water and rail shipments. All of these, coupled with the conveyors and fork lift trucks, increase the effective capacity of a warehouse, improve operating efficiency by reducing the amount of handling that shipments must undergo and improve efficiency and cost-saving in distribution.

Packaging

Marketing's interest in packaging is in eye-appealing containers often with little regard to the effects on transportation claims and shipping weight and dimensions. Production's interest is in the size and shape of the package as it affects the efficiency of filling the shipping carton.

The PDM has to arbitrate these conflicting interests and stand up for his own interest in damage-free transportation—cartons that fit pallets or containers neatly, packages that meet the size requirements of small package transportation companies.

Customer Service

In the modern corporation, the customer service department handles the customers' orders and through a series of steps (often using computers), determines if the items ordered are available, prices them, adjusts inventory, determines the customer's credit status and verifies such items as terms of sale, shipping addresses, codes and descriptions.

Establishing a level of service gives a company's customers what they need most—predictability. Customers "build-in" to *their* procedures the time it takes to process an order and make delivery. Thus the customer can establish his re-order conditions. Established customer service levels also enable the PDM to recommend his own ordering priorities—for example, case lots instead of single units, truckload quantities instead of LTL—which affect order processing and transit time.

Site Location

In searching for a site for a production plant the production department must have the decisive say as to lay-out and other factors affecting its ability to perform. But if the site is for a warehouse, then production would not be a part of the committee evaluating the site, which would, however, need tax information from finance, structural information from the engineers, capacity and layout information from the warehouse manager and transportation availability and costs from the traffic manager. In addition to information on the availability of labor, the committee would want the facts and figures on schools and medical facilities, utility services, security (police and fire) services, sanitation services, and other items the company's own maintenance crews could not handle.

Summary

PDM focuses on the total cost. In this, the managerial knowledge of moving or transferring products to their point of use is the most important ingredient. PDM does not make profits for a company but it does make sure the company keeps on making them.

Appendix 1

Glossary of Abbreviations and Terms Commonly Used in Foreign Trade

Acceptance. A time draft, or bill of exchange, which the drawee has accepted but which has not yet matured. The drawee, now known as the "acceptor," writes the word "accepted" above his signature on the face of the draft as well as the date and place of payment.

Ad valorem. A term meaning "according to value," which means a fixed percentage of the value of dutiable goods, computed for duty assessment.

Advance against documents. This form of lending is often used in lieu of an acceptance when, for example, the underlying documents do not satisfy Federal Reserve requirements for the creation of acceptances. Instead of creating an acceptance, therefore, an advance is made on the security of the documents covering the shipment.

Arrival draft. A sight draft which is modified to the extent that it does not require payment until after the merchandise has arrived at the destination port.

At sight. A term indicating that a negotiable instrument is to be paid upon presentation or demand.

Authority to pay. An advice from a buyer, sent by his bank to the seller's bank, authorizing the seller's bank to pay the seller-exporter's drafts up to a fixed amount. The seller has no protection against cancellation or modification of the instrument until the issuing bank pays the drafts drawn on it, in which case the seller is no longer liable. These instruments are usually not confirmed by the seller's American bank.

Authority to purchase. Similar to an authority to pay except that under this arrangement drafts are drawn directly on the buyer. The correspondent bank purchases them with or without recourse against the drawer and, as in the case of the authority to pay, they are usually not confirmed by an American bank. This type of transaction is peculiar to Far Eastern trade.

Back-to-back letter of credit. Popularized soon after World War II when merchandise was scarce, this is a domestic instrument which complements or supports a commercial letter of credit. For example, ABC Company in this country is designated the beneficiary of an irrevocable letter of credit, confirmed by the American bank. But the merchandise which XYZ Company in Germany (whose bank issued the letter of credit) wishes to buy is not in stock at ABC; therefore, they will have to purchase the goods from a third company. This third company, however, will not fill ABC's order unless it is "sweetened" by some form of prepayment, either in cash or through some type of financing. If ABC is unable to prepay in cash, it will come to the American bank and ask it to issue a domestic letter of credit in favor of the third company. If the American bank agrees, a situation is created whereby the domestic credit is "backed" by the foreign credit and a "back-to-back" commercial credit transaction exists.

Balance of payments. A statement indicating a country's foreign economic transactions over a specified period of time.

Balance of trade. The difference between the total exports and the total imports of a country during the course of a fixed period of time. A "favorable" balance of trade is said to exist when exports exceed imports.

Bankers' acceptance. A time bill, or acceptance, which has been drawn on, and accepted, by a bank or bankers.

Beneficiary. The individual or company in whose favor a letter of credit is opened or a draft is drawn.

Bilateral trade. Commerce between two countries, usually based on a formal reciprocal trade agreement, under which the quantity of goods to be traded and the duration of the pact are clearly defined, and which specifies that balances due are to be remitted directly between the two nations.

Bill of exchange. Generally used interchangeably with the word "draft," this is an unconditional order written from one person (the drawer) to another person (the drawee) directing the drawee to pay a certain sum at a fixed or future determinable date, to the order of a third party.

Carnet, ATA. Issued under an international treaty and administered by the International Chamber of Commerce, this is a document which permits the duty-free temporary importation of commercial samples and professional equipment into over 30 countries. Carnets are issued by authorized chambers of commerce who, in addition, guarantee payment of full duties if the goods in question are not re-exported within a specified period of time. The U.S. Council of the International Chamber of Commerce, 1212 Avenue of the Americas, New York, N.Y.

FOREIGN TRADE GLOSSARY

is the issuing and guaranteeing association for ATA Carnets in the United States.

Cash letter of credit. A letter addressed from one bank to one of its correspondents making available to the party named in the letter a fixed sum of money up to some future specific date. The sum indicated in the letter is equal to the amount deposited in the issuing bank by the party before the letter is issued—hence, a "cash letter of credit."

Certificate of manufacture A document accompanying a letter of credit stating that the goods named thereon have been made ready for shipment and are being held in safekeeping.

Certificate of origin. A document issued by the exporter certifying as to the country of origin of the merchandise in question.

Clean draft. A draft to which nothing whatsoever is attached.

Confirmed letter of credit. A letter of credit issued by a foreign bank which bears the confirmation of the seller's American bank. The purpose of a confirmed letter of credit is to lend a degree of prestige and responsibility to the credit because the status of the foreign or issuing bank may not be known to the purchaser.

Consignment. Is the physical transfer of goods from a seller (consignor) with whom the title remains, to another legal entity (consignee) who acts as a selling agent, selling the goods and remitting the net proceeds to the consignor.

Consular invoice. Covering a shipment of goods certified by a Consul of the destination country, its principal function is to record accurately the types of merchandise and their quantity, grade and value, for import duty and general statistical purposes.

Convertibility. Refers to the ability of holders of a currency to exchange it for foreign currency or for gold in the open market.

Correspondent bank A bank which is a depository for another bank and which performs various banking services for its depositor throughout the world.

Customs union. An agreement between two or more countries in which they arrange to abolish tariffs and other import restrictions on each other's goods and establish a common tariff for the imports of all other countries.

Del credere agent. A sales agent who, for a certain percentage over and above his sales commission, will guarantee payment to the person for whom he is selling, on shipments made to the seller's customers.

Devaluation. Is effected by a government decision to lower the trade value of a nation's currency vis-a-vis other currencies by reducing the gold content or by revising the ratio to a new standard.

Documentary draft. A draft to which documents are attached, which are to be delivered to the drawee when he accepts or pays the accompanying draft, and which ordinarily controls title to the merchandise indicated thereon.

Dollar area. A loosely-knit, general area comprising the Western Hemisphere, United States territories and possessions, and the Republic of the Philippines.

Draft. See Bill of exchange.

Duty. A government tax levied on imports and exports, or on the use and consumption of goods.

Escape clause. An arrangement whereby the Federal law granting tariff concessions on certain commodities may be revised upward if the importation of those commodities becomes so high as to threaten or actually cause serious damage to American producers.

First of exchange. The original copy of a draft when two or more are drawn.

Foreign exchange. Refers to the buying and selling of foreign currencies, in relation to United States funds, for immediate future delivery; these funds are used to pay for imports or exports in the currency required by the seller of the merchandise.

Free port. A port which is a foreign trade zone, open to all traders on equal terms; more specifically, a port where merchandise may be stored duty-free, pending re-export or sale within that country.

Free trade area. An agreement whereby two or more countries conduct free trade among themselves.

Hard currency. This phrase describes a currency which is sufficiently sound so that it is generally accepted at face value internationally. In another context this term means metallic money as opposed to paper money.

Indent merchant. One who gathers scattered orders from other local merchants and places a single order on his account with a foreign manufacturer. The indent merchant assumes the credit risk and is compensated by charging a commission to those on whose bahelf he placed the orders.

Par value. The official value assigned to a nation's currency, determined by the quantity of gold backing it up or by its relation to another currency.

Peril point. A term from the U.S. Trade Agreements Act which refers to that point below which a tariff cannot be lowered without threatening or actually causing serious damage to American producers.

Public Law 480. The most common reference to The Agricultural Trade Development and Assistance Act of 1954. Generally, P.L. 480

FOREIGN TRADE GLOSSARY

authorizes the President to make available various types of assistance to American agricultural exporters, such as making sales in the currency of the destination country.

Recourse. A term which defines the rights of a holder in due course of a negotiable instrument to force prior endorsers to meet their legal obligations by making payment on the instrument should it be dishonored by the maker or the acceptor.

Second of exchange. A duplicate copy of a draft.

Soft currency. Is not freely convertible into other currencies.

Spot exchange. Foreign exchange for immediate delivery by the seller and immediate payment by the buyer.

Sterling area. Includes the United Kingdom, all Commonwealth nations except Canada, British colonies and some non-British areas whose currencies are related to sterling and whose monetary reserves are largely in sterling.

Tenor. Refers to the date agreed upon for the payment of a draft.

Trust receipt. Used extensively in letter of credit financing, this is a document or receipt in which the buyer promises to hold the property received in the name of the bank releasing it to him, while the bank retains title to the goods. The merchant is called the trustee, the bank the entruster. Trust receipts are used primarily to permit an importer of good standing to take possession of the merchandise for resale, before he pays the issuing bank.

Warehouse receipt. An instrument which lists, and is a receipt for, goods or commodities deposited in the warehouse which issues the receipt. These receipts may be negotiable or non-negotiable. A negotiable warehouse receipt is made to the "bearer," while a non-negotiable warehouse receipt specifies precisely to whom the goods shall be delivered.

Appendix 2
Uniform Customs and Practice for Documentary Credits (1974 Revision) International Chamber of Commerce Publication No. 290

"The Rules have been drawn up by the Commission on Banking Practices and Technique of the International Chamber of Commerce in consultation with the banking associations of many countries. The International Chamber of Commerce acts to promote business interests at international levels, to foster the greater freedom of international trade, and to harmonize and facilitate business and trade practices. Paris based, the Chamber has National Committees in fifty-one countries and is represented in over thirty others. In the United States, the International Chamber of Commerce is represented by its United States Council at 1212 Avenue of the Americas, New York, N.Y., 10036 (telephone (212) 582-4850), which has printed and distributed the American version of the 1974 version of the 1974 revision of Uniform Customs and Practice for Documentary Credits, International Chamber of Commerce Publication No. 290, and which holds the copyright thereto."

Foreword

Considerable changes have since taken place in international trading and transport techniques. Terms of purchase and sale have swung from the traditional FOB and CIF towards "Delivered to Buyer's Premises," and the through, multi-modal movement of unitized cargo is increasingly competing with the traditional single-mode carriage of break-bulk cargo. Consequential changes have become necessary in documentary credit practice.

Therefore we have taken a careful and critical look at the 1962 rules, amending them as appropriate to fit the 1980's. The changes made particularly concern the documentary aspects of multi-modal transport and unitized cargoes, the easier production and processing of documents in "short form," and the problem of "stale" documents.

Revision has been greatly assisted by the cooperation of the United

Nations Commission on International Trade Law (UNCITRAL). Banks in the Socialist countries have also contributed through an ad hoc Working Party.

General Provisions and Definitions

a. These provisions and definitions and the following articles apply to all documentary credits and are binding upon all parties thereto unless otherwise expressly agreed.
b. For the purposes of such provisions, definitions and articles the expressions "documentary credit(s)" and "credit(s)" used therein mean any arrangement, however named or described, whereby a bank (the issuing bank), acting at the request and in accordance with the instructions of a customer (the applicant for the credit),
   i. is to make payment to or to the order of a third party (the beneficiary), or is to pay, accept or negotiate bills of exchange (drafts) drawn by the beneficiary, or
   ii. authorizes such payments to be made or such drafts to be paid, accepted or negotiated by another bank,
against stipulated documents, provided that the terms and conditions of the credit are complied with.
c. Credits, by their nature, are separate transactions from the sales or other contracts on which they may be based and banks are in no way concerned with or bound by such contracts.
d. Credit instructions and the credits themselves must be complete and precise.
   In order to guard against confusion and misunderstanding, issuing banks should discourage any attempt by the applicant for the credit to include excessive detail.
e. The bank first entitled to exercise the option available under Article 32 b. shall be the bank authorized to pay, accept or negotiate under a credit. The decision of such bank shall bind all parties concerned.
   A bank is authorized to pay or accept under a credit by being specifically nominated in the credit.
   A bank is authorized to negotiate under a credit either
   i. by being specifically nominated in the credit, or
   ii. by the credit being freely negotiable by any bank.
f. A beneficiary can in no case avail himself of the contractual relationships existing between banks or between the applicant for the credit and the issuing bank.

INTERNATIONAL CHAMBER OF COMMERCE DOCUMENTARY CREDITS

A. Form and Notification of Credits

Article 1—
a. Credits may be either
   i. revocable, or
   ii. irrevocable.
b. All credits, therefore, should clearly indicate whether they are revocable or irrevocable.
c. In the absence of such indication the credit shall be deemed to be revocable.

Article 2—A revocable credit may be amended or cancelled at any moment without prior notice to the beneficiary. However, the issuing bank is bound to reimburse a branch or other bank to which such a credit has been transmitted and made available for payment, acceptance or negotiation, for any payment, acceptance or negotiation complying with the terms and conditions of the credit and any amendments received up to the time of payment, acceptance or negotiation made by such branch or other bank prior to receipt by it of notice of amendment or of cancellation.

Article 3—
a. An irrevocable credit constitutes a definite undertaking of the issuing bank, provided that the terms and conditions of the credit are complied with:
   i. to pay, or that payment will be made, if the credit provides for payment, whether against a draft or not;
   ii. to accept drafts if the credit provides for acceptance by the issuing bank or to be responsible for their acceptance and payment at maturity if the credit provides for the acceptance of drafts drawn on the applicant for the credit or any other drawee specified in the credit;
   iii. to purchase/negotiate, without recourse to drawers and/or bona fide holders, drafts drawn by the beneficiary, at sight or at a tenor, on the applicant for the credit or on any other drawee specified in the credit, or to provide for purchase/negotiation by another bank, if the credit provides for purchase/negotiation.
b. An irrevocable credit may be advised to a beneficiary through another bank (the advising bank) without engagement on the part of that bank, but when an issuing bank authorizes or requests another bank to confirm its irrevocable credit and the latter does so, such confirmation constitutes a definite undertaking of the confirming bank in addition to the undertaking of the issuing bank, provided that the terms and conditions of the credit are complied with:

i. to pay, if the credit is payable at its own counters, whether against a draft or not, or that payment will be made if the credit provides for payment elsewhere;

ii. to accept drafts if the credit provides for acceptance by the confirming bank, at its own counters, or to be responsible for their acceptance and payment at maturity if the credit provides for the acceptance of drafts drawn on the applicant for the credit or any other drawee specified in the credit;

iii. to purchase/negotiate, without recourse to drawers and/or bona fide holders, drafts drawn by the beneficiary, at sight or at a tenor, on the issuing bank, or on the applicant for the credit or on any other drawee specified in the credit, if the credit provides for purchase/negotiation.

c. Such undertakings can neither be amended nor cancelled without the agreement of all parties thereto. Partial acceptance of amendments is not effective without the agreement of all parties thereto.

Article 4—
a. When an issuing bank instructs a bank by cable, telegram or telex to advise a credit, and intends the mail confirmation to be the operative credit instrument, the cable, telegram or telex must state that the credit will only be effective on receipt of such mail confirmation. In this event, the issuing bank must send the operative credit instrument (mail confirmation) and any subsequent amendments to the credit to the beneficiary through the advising bank.
b. The issuing bank will be responsible for any consequences arising from its failure to follow the procedure set out in the preceding paragraph.
c. Unless a cable, telegram or telex states "details to follow" (or words of similar effect), or states that the mail confirmation is to be the operative credit instrument, the cable, telegram or telex will be deemed to be the operative credit instrument and the issuing bank need not send the mail confirmation to the advising bank.

*Article 5—*When a bank is instructed by cable, telegram or telex to issue, confirm or advise a credit similar in terms to one previously established and which has been the subject of amendments, it shall be understood that the details of the credit being issued, confirmed or advised will be transmitted to the beneficiary excluding the amendments, unless the instructions specify clearly any amendments which are to apply.

*Article 6—*If incomplete or unclear instructions are received to issue, confirm or advise a credit, the bank requested to act on such instruc-

INTERNATIONAL CHAMBER OF COMMERCE DOCUMENTARY CREDITS

tions may give preliminary notification of the credit to the beneficiary for information only and without responsibility; in this event the credit will be issued, confirmed or advised only when the necessary information has been received.

B. Liabilities and Responsibilities

Article 7—Banks must examine all documents with reasonable care to ascertain that they appear on their face to be in accordance with the terms and conditions of the credit. Documents which appear on their face to be inconsistent with one another will be considered as not appearing on their face to be in accordance with the terms and conditions of the credit.

Article 8—
a. In documentary credit operations all parties concerned deal in documents and not in goods.
b. Payment, acceptance or negotiation against documents which appear on their face to be in accordance with the terms and conditions of a credit by a bank authorized to do so, binds the party giving the authorization to take up the documents and reimburse the bank which has effected the payment, acceptance or negotiation.
c. If, upon receipt of the documents, the issuing bank considers that they appear on their face not to be in accordance with the terms and conditions of the credit, that bank must determine, on the basis of the documents alone, whether to claim that payment, acceptance or negotiation was not effected in accordance with the terms and conditions of the credit.
d. The issuing bank shall have a reasonable time to examine the documents and to determine as above whether to make such a claim.
e. If such claim is to be made, notice to that effect, stating the reasons therefor, must, without delay, be given by cable or other expeditious means to the bank from which the documents have been received (the remitting bank) and such notice must state that the documents are being held at the disposal of such bank or are being returned thereto.
f. If the issuing bank fails to hold the documents at the disposal of the remitting bank, or fails to return the documents to such bank, the issuing bank shall be precluded from claiming that the relative payment, acceptance or negotiation was not effected in accordance with the terms and conditions of the credit.
g. If the remitting bank draws the attention of the issuing bank to any irregularities in the documents or advises such bank that it has paid,

accepted or negotiated under reserve or against a guarantee in respect of such irregularities, the issuing bank shall not thereby be relieved from any of its obligations under this article. Such guarantee or reserve concerns only the relations between the remitting bank and the beneficiary.

Article 9—Banks assume no liability or responsibility for the form, sufficiency, accuracy, genuineness, falsification or legal effect of any documents, or for the general and/or particular conditions stipulated in the documents or superimposed thereon; nor do they assume any liability or responsibility for the description, quantity, weight, quality, condition, packing, delivery, value or existence of the goods represented thereby, or for the good faith or acts and/or omissions, solvency, performance or standing of the consignor, the carriers or the insurers of the goods or any other person whomsoever.

Article 10—Banks assume no liability or responsibility for the consequences arising out of delay and/or loss in transit of any messages, letters or documents, or for delay, mutilation or other errors arising in the transmission of cables, telegrams or telex. Banks assume no liability or responsibility for errors in translation or interpretation of technical terms, and reserve the right to transmit credit terms without translating them.

Article 11—Banks assume no liability or responsibility for consequences arising out of the interruption of their business by Acts of God, riots, civil commotions, insurrections, wars or any other causes beyond their control or by any strikes or lockouts. Unless specifically authorized, banks will not effect payment, acceptance or negotiation after expiration under credits expiring during such interruption of business.

Article 12—
a. Banks utilizing the services of another bank for the purpose of giving effect to the instructions of the applicant for the credit do so for the account and at the risk of the latter.
b. Banks assume no liability or responsibility should the instructions they transmit not be carried out, even if they have themselves taken the initiative in the choice of such other bank.
c. The applicant for the credit shall be bound by and liable to indemnify the banks against all obligations and responsibilities imposed by foreign laws and usages.

Article 13—A paying or negotiating bank which has been authorized to claim reimbursement from a third bank nominated by the issuing bank and which has effected such payment or negotiation shall not be required to confirm to the third bank that it has done so in accordance with the terms and conditions of the credit.

C. Documents

Article 14—
a. All instructions to issue, confirm or advise a credit must state precisely the documents against which payment, acceptance or negotiation is to be made.
b. Terms such as "first class," "well known," "qualified" and the like shall not be used to describe the issuers of any documents called for under credits and if they are incorporated in the credit terms banks will accept documents as tendered.

C.1—Documents evidencing shipment or dispatch or taking in charge (shipping documents).

Article 15—Except as stated in Article 20, the date of the Bill of Lading, or the date of any other document evidencing shipment or dispatch or taking in charge, or the date indicated in the reception stamp or by notation on any such document, will be taken in each case to be the date of shipment or dispatch or taking in charge of the goods.

Article 16—
a. If words clearly indicating payment or prepayment of freight, however named or described, appear by stamp or otherwise on documents evidencing shipment or dispatch or taking in charge they will be accepted as constituting evidence of payment of freight.
b. If the words "freight pre-payable" or "freight to be prepaid" or words of similar effect appear by stamp or otherwise on such documents they will not be accepted as constituting evidence of the payment of freight.
c. Unless otherwise specified in the credit or inconsistent with any of the documents presented under the credit, banks will accept documents stating that freight or transportation charges are payable on delivery.
d. Banks will accept shipping documents bearing reference by stamp or otherwise to costs additional to the freight charges, such as costs of, or disbursements incurred in connection with loading, unloading or similar operations, unless the conditions of the credit specifically prohibit such reference.

INDUSTRIAL TRAFFIC MANAGEMENT

Article 17—Shipping documents which bear a clause on the face thereof such as "shipper's load and count" or "said by shipper to contain" or words of similar effect, will be accepted unless otherwise specified in the credit.

Article 18—
a. A clean shipping document is one which bears no superimposed clause or notation which expressly declares a defective condition of the goods and/or the packaging.
b. Banks will refuse shipping documents bearing such clauses or notations unless the credit expressly states the clauses or notations which may be accepted.

C.1.1—Marine Bills of Lading.

Article 19—
a. Unless specifically authorized in the credit, Bills of Lading of the following nature will be rejected:
   i. Bills of Lading issued by forwarding agents.
   ii. Bills of Lading which are issued under and are subject to the conditions of a Charter-Party.
   iii. Bills of Lading covering shipment by sailing vessels.
However, subject to the above and unless otherwise specified in the credit, Bills of Lading of the following nature will be accepted:
   i. "Through" Bills of Lading issued by shipping companies or their agents even though they cover several modes of transport.
   ii. Short Form Bills of Lading (i.e., Bills of Lading issued by shipping companies or their agents which indicate some or all of the conditions of carriage by reference to a source or document other than the Bill of Lading).
   iii. Bills of Lading issued by shipping companies or their agents covering unitized cargoes, such as those on pallets or in containers.

Article 20—
a. Unless otherwise specified in the credit, Bills of Lading must show that the goods are loaded on board a named vessel or shipped on a named vessel.
b. Loading on board a named vessel or shipment on a named vessel may be evidenced either by a Bill of Lading bearing wording indicating loading on board a named vessel or shipment on a named vessel, or by means of a notation to that effect on the Bill of Lading signed or initialled and dated by the carrier or his agent, and the date of this notation shall be regarded as the date of loading on board the named vessel or shipment on the named vessel.

INTERNATIONAL CHAMBER OF COMMERCE DOCUMENTARY CREDITS

Article 21—
a. Unless transshipment is prohibited by the terms of the credit, Bills of Lading will be accepted which indicate that the goods will be transshipped en route, provided the entire voyage is covered by one and the same Bill of Lading.
b. Bills of Lading incorporating printed clauses stating that the carriers have the right to transship will be accepted notwithstanding the fact that the credit prohibits transshipment.

Article 22—
a. Banks will refuse a Bill of Lading stating that the goods are loaded on deck, unless specifically authorized in the credit.
b. Banks will not refuse a Bill of Lading which contains a provision that the goods may be carried on deck, provided it does not specifically state that they are loaded on deck.

C.1.2—Combined transport documents

Article 23—
a. If the credit calls for a combined transport document, i.e., one which provides for a combined transport by at least two different modes of transport, from a place at which the goods are taken in charge to a place designated for delivery, or if the credit provides for a combined transport, but in either case does not specify the form of document required and/or the issuer of such document, banks will accept such documents as tendered.
b. If the combined transport includes transport by sea the document will be accepted although it does not indicate that the goods are on board a named vessel, and although it contains a provision that the goods, if packed in a container, may be carried on deck, provided it does not specifically state that they are loaded on deck.

C.1.3—Other shipping documents, etc.

Article 24—Banks will consider a Railway or Inland Waterway Bill of Lading or Consignment Note, Counterfoil Waybill, Postal Receipt, Certificate of Mailing, Air Mail Receipt, Air Waybill, Air Consignment Note or Air Receipt, Trucking Company Bill of Lading or any other similar document as regular when such document bears the reception stamp of the carrier or his agent, or when it bears a signature purporting to be that of the carrier or his agent.

Article 25—Where a credit calls for an attestation or certification of weight in the case of transport other than by sea, banks will accept a

INDUSTRIAL TRAFFIC MANAGEMENT

weight stamp or declaration of weight superimposed by the carrier on the shipping document unless the credit calls for a separate or independent certificate of weight.

C.2—Insurance documents.

Article 26—
a. Insurance documents must be as specified in the credit, and must be issued and/or signed by insurance companies or their agents or by underwriters.
b. Cover notes issued by brokers will not be accepted, unless specifically authorized in the credit.

Article 27—Unless otherwise specified in the credit, or unless the insurance documents presented establish that the cover is effective at the latest from the date of shipment or dispatch or, in the case of combined transport, the date of taking the goods in charge, banks will refuse insurance documents presented which bear a date later than the date of shipment or dispatch or, in the case of combined transport the date of taking the goods in charge, as evidenced by the shipping documents.

Article 28—
a. Unless otherwise specified in the credit, the insurance document must be expressed in the same currency as the credit.
b. The minimum amount for which insurance must be effected is the CIF value of the goods concerned. However, when the CIF value of the goods cannot be determined from the documents on their face, banks will accept as such minimum amount the amount of the drawing under the credit or the amount of the relative commercial invoice, whichever is the greater.

Article 29—
a. Credits should expressly state the type of insurance required and, if any, the additional risks which are to be covered. Imprecise terms such as "usual risks" or "customary risks" should not be used; however, if such imprecise terms are used, banks will accept insurance documents as tendered.
b. Failing specific instructions, banks will accept insurance cover as tendered.

Article 30—Where a credit stipulates "insurance against all risks," banks will accept an insurance document which contains any "all risks" notation or clause, and will assume no responsibility if any particular risk is not covered.

Article 31—Banks will accept an insurance document which indicates that the cover is subject to a franchise or an excess (deductible), unless it is specifically stated in the credit that the insurance must be issued irrespective of percentage.

C.3—Commercial invoices.

Article 32—
a. Unless otherwise specified in the credit, commercial invoices must be made out in the name of the applicant for the credit.
b. Unless otherwise specified in the credit, banks may refuse commercial invoices issued for amounts in excess of the amount permitted by the credit.
c. The description of the goods in the commercial invoice must correspond with the description in the credit. In all other documents the goods may be described in general terms not inconsistent with the description of the goods in the credit.

C.4—Other documents.

Article 33—When other documents are required, such as Warehouse Receipts, Delivery Orders, Consular Invoices, Certificates of Origin, of Weight, of Quality or of Analysis, etc. and when no further definition is given, banks will accept such documents as tendered.

D. Miscellaneous Provisions

Quantity and amount.

Article 34—
a. The words "about," "circa" or similar expressions used in connection with the amount of the credit or the quantity or the unit price of the goods are to be construed as allowing a difference not to exceed ten percent more or ten percent less.
b. Unless a credit stipulates that the quantity of the goods specified must not be exceeded or reduced a tolerance of three percent more or three percent less will be permissible, always provided that the total amount of the drawings does not exceed the amount of the credit. This tolerance does not apply when the credit specifies quantity in terms of a stated number of packing units or individual items.

INDUSTRIAL TRAFFIC MANAGEMENT

Partial shipments.

Article 35—
a. Partial shipments are allowed, unless the credit specifically states otherwise.
b. Shipments made on the same ship and for the same voyage, even if the Bills of Lading evidencing shipment "on board" bear different dates and/or indicate different ports of shipment, will not be regarded as partial shipments.

Article 36—If shipment by installments within given periods is stipulated and any installment is not shipped within the period allowed for that installment, the credit ceases to be available for that or any subsequent installments, unless otherwise specified in the credit.

Expiry date.

Article 37—All credits, whether revocable or irrevocable, must stipulate an expiry date for presentation of documents for payment, acceptance or negotiation, notwithstanding the stipulation of a latest date for shipment.

Article 38—The words "to," "until," "till" and words of similar import applying to the stipulated expiry date for presentation of documents for payment, acceptance or negotiation, or to the stipulated latest date for shipment, will be understood to include the date mentioned.

Article 39—
a. When the stipulated expiry date falls on a day on which banks are closed for reasons other than those mentioned in Article 11, the expiry date will be extended until the first following business day.
b. The latest date for shipment shall not be extended by reason of the extension of the expiry date in accordance with this Article. Where the credit stipulates a latest date for shipment, shipping documents dated later than such stipulated date will not be accepted. If no latest date for shipment is stipulated in the credit, shipping documents dated later than the expiry date stipulated in the credit or amendments thereto will not be accepted. Documents other than the shipping documents may, however, be dated up to and including the extended expiry date.
c. Banks paying, accepting or negotiating on such extended expiry date must add to the documents their certification in the following wording: "Presented for payment (or acceptance or negotiation as the

case may be) within the expiry date extended in accordance with Article 39 of the *Uniform Customs.*"

Shipment, loading or dispatch.

Article 40—
a. Unless the terms of the credit indicate otherwise, the words "departure," "dispatch," "loading" or "sailing" used in stipulating the latest date for shipment of the goods will be understood to be synonymous with "shipment."
b. Expressions such as "prompt," "immediately," "as soon as possible" and the like should not be used. If they are used, banks will interpret them as a request for shipment within thirty days from the date on the advice of the credit to the beneficiary by the issuing bank or by an advising bank, as the case may be.
c. The expression "on or about" and similar expressions will be interpreted as a request for shipment during the period from five days before to five days after the specified date, both end days included.

Presentation

Article 41—Notwithstanding the requirement of Article 37 that every credit must stipulate an expiry date for presentation of documents, credits must also stipulate a specified period of time after the date of issuance of the Bills of Lading or other shipping documents during which presentation of documents for payment, acceptance or negotiation must be made. If no such period of time is stipulated in the credit, banks will refuse documents presented to them later than 21 days after the date of issuance of the Bills of Lading or other shipping documents.

Article 42—Banks are under no obligation to accept presentation of documents outside their banking hours.

Date terms.

Article 43—The terms "first half," "second half" of a month shall be construed respectively, as from the 1st to the 15th, and the 16th to the last day of each month, inclusive.

Article 44—The terms "beginning," "middle," or "end" of a month shall be construed respectively as from the 1st to the 10th, the 11th to the 20th, and the 21st to the last day of each month, inclusive.

Article 45—When a bank issuing a credit instructs that the credit be confirmed or advised as available "for one month," "for six months" or the like, but does not specify the date from which the time is to run, the confirming or advising bank will confirm or advise the credit as expiring at the end of such indicated period from the date of its confirmation or advice.

E. Transfer

Article 46—
a. A transferable credit is a credit under which the beneficiary has the right to give instructions to the bank called upon to effect payment or acceptance or to any bank entitled to effect negotiation to make the credit available in whole or in part to one or more third parties (second beneficiaries).
b. The bank requested to effect the transfer, whether it has confirmed the credit or not, shall be under no obligation to effect such transfer except to the extent and in the manner expressly consented to by such bank, and until such bank's charges in respect of transfer are paid.
c. Bank charges in respect of transfers are payable by the first beneficiary unless otherwise specified.
d. A credit can be transferred only if it is expressly designated as "transferable" by the issuing bank. Terms such as "divisible," "fractionable," "assignable" and "transmissible" add nothing to the meaning of the term "transferable" and shall not be used.
e. A transferable credit can be transferred once only. Fractions of a transferable credit (not exceeding in the aggregate the amount of the credit) can be transferred separately, provided partial shipments are not prohibited, and the aggregate of such transfers will be considered as constituting only one transfer of the credit. The credit can be transferred only on the terms and conditions specified in the original credit, with the exception of the amount of the credit, of any unit prices stated therein, and of the period of validity or period for shipment, any or all of which may be reduced or curtailed.

Additionally, the name of the first beneficiary can be substituted for that of the applicant for the credit, but if the name of the applicant for the credit is specifically required by the original credit to appear in any document other than the invoice, such requirement must be fulfilled.
f. The first beneficiary has the right to substitute his own invoices for those of the second beneficiary, for amounts not in excess of the original amount stipulated in the credit and for the original unit prices if stipulated in the credit, and upon such substitution of in-

voices the first beneficiary can draw under the credit for the difference, if any, between his invoices and the second beneficiary's invoices. When a credit has been transferred and the first beneficiary is to supply his own invoices in exchange for the second beneficiary's invoices but fails to do so on first demand, the paying, accepting or negotiating bank has the right to deliver to the issuing bank the documents received under the credit, including the second beneficiary's invoices, without further responsibility to the first beneficiary.

g. The first beneficiary of a transferable credit can transfer the credit to a second beneficiary in the same country or in another country unless the credit specifically states otherwise. The first beneficiary shall have the right to request that payment or negotiation be effected to the second beneficiary at the place to which the credit has been transferred, up to and including the expiry date of the original credit, and without prejudice to the first beneficiary's right subsequently to substitute his own invoices for those of the second beneficiary and to claim any difference due to him.

Article 47—The fact that a credit is not stated to be transferable shall not affect the beneficiary's rights to assign the proceeds of such credit in accordance with the provisions of the applicable law.

Uniform Rules for the Collection of Commercial Paper (1967 Revision) International Chamber of Commerce Publication No. 254

General Provisions and Definitions

a. These provisions and definitions and the following articles apply to all collections of commercial paper and are binding upon all parties thereto unless otherwise expressly agreed or unless contrary to the provisions of a national, state or local law and/or regulation which cannot be departed from.

b. For the purposes of such provisions, definitions and articles:

    i. "commercial paper" consists of clean remittances and documentary remittances.

    "Clean remittances" means items consisting of one or more bills of exchange, whether already accepted or not, promissory notes, cheques, receipts, or other similar documents for obtaining the payment of money (there being neither invoices, shipping documents, documents of title, or other similar documents, nor any other documents whatsoever attached to the said items).

    "Documentary remittances" means all other commercial paper, with documents attached to be delivered against payment, acceptance, trust receipt or other letter of commitment, free or on other terms and conditions.

    ii. The "parties thereto" are the principal who entrusts the operation of collection to his bank (the customer), the said bank (the remitting bank), and the correspondent commissioned by the remitting bank to see to the acceptance or collection of the commercial paper (the collecting bank).

    iii. The "drawee" is the party specified in the remittance letter as the one to whom the commercial paper is to be presented.

c. All commercial paper sent for collection must be accompanied by a remittance letter giving complete and precise instructions. Banks are only permitted to act upon the instructions given in such remittance letter.

INTERNATIONAL CHAMBER OF COMMERCE DOCUMENTARY CREDITS

If the collecting bank cannot, for any reason, comply with the instructions given in the remittance letter received by it, it must advise the remitting bank immediately.

Presentation

Article 1—Commercial paper is to be presented to the drawee in the form in which it is received from the customer, except that the collecting bank is to affix any necessary stamps, at the expense of the customer unless otherwise instructed.

Remitting and collecting banks have no obligation to examine the commercial paper or the accompanying documents if any, and assume no responsibility for the form and/or regularity thereof.

Article 2—Commercial paper should bear the complete address of the drawee or of the domicile at which the collecting bank is to make the presentation. If the address is incomplete or incorrect, the collecting bank may, without obligation and responsibility on its part, endeavor to ascertain the proper address.

Article 3—In the case of commercial paper payable at sight the collecting bank must make presentation for payment without delay.

In the case of commercial paper payable at a usance other than sight the collecting bank must, where acceptance is called for, make presentation for acceptance without delay, and must, in every instance, make presentation for payment not later than the appropriate maturity date.

Article 4—In respect of a documentary remittance accompanied by a bill of exchange payable at a future date, the remittance letter should state whether the documents are to be released to the drawee against acceptance (D/A) or against payment (D/P).

In the absence of instruction, the documents will be released only against payment.

Payment

Article 5—In the case of commercial paper expressed to be payable in the currency of the country of payment (local currency) the collecting bank will only release the commercial paper to the drawee against payment in local currency which can immediately be disposed of in accordance with the instructions given in the remittance letter.

Article 6—In the case of commercial paper expressed to be payable in a currency other than that of the country of payment (foreign currency) the collecting bank will only release the commercial paper to the drawee against payment in the relative foreign currency which can immediately be remitted in accordance with the instructions given in the remittance letter.

Article 7—In respect of clean remittances partial payments may be accepted if and to the extent to which and on the conditions on which partial payments are authorized by the law in force in the place of payment. The clean remittance will only be released to the drawee when full payment thereof has been received.

In respect of documentary remittances partial payments will only be accepted if specifically authorized in the remittance letter, but unless otherwise instructed the collecting bank will only release the documents to the drawee after full payment has been received.

In all cases where partial payments are acceptable, either by reason of a specific authorization or in accordance with the provisions of this Article, such partial payments will be received and dealt with in accordance with the provisions of Article 5 or 6.

Acceptance

Article 8—The collecting bank is responsible for seeing that the form of the acceptance appears to be complete and correct, but is not responsible for the genuineness of any signature or for the authority of any signatory to sign the acceptance.

Protest

Article 9—The remittance letter should give specific instructions regarding legal process in the event of non-acceptance or non-payment.

In the absence of such specific instructions the banks concerned with the collection are not responsible for any failure to have the commercial paper protested (or subjected to legal process in lieu thereof) for non-payment or non-acceptance.

The collecting bank is not responsible for the regularity of the form of the protest (or other legal process).

Case-of-need (Customer's representative) and protection of goods

Article 10—If the customer nominates a representative to act as case-of-need in the event of non-acceptance and/or non-payment the remittance letter should clearly and fully indicate the powers of such case-of-need.

Whether a case-of-need is nominated or not, in the absence of

INTERNATIONAL CHAMBER OF COMMERCE DOCUMENTARY CREDITS

specific instructions the collecting bank has no obligation to take any action in respect of the goods represented by a documentary remittance.

Advice of Fate, etc.

Article 11—The collecting bank is to send advice of payment or advice of acceptance, with appropriate detail, to the remitting bank without delay.

Article 12—The collecting bank is to send advice of non-payment or advice of non-acceptance, with appropriate detail, to the remitting bank without delay.

Article 13—In the absence of specific instructions, the collecting bank is to send all advices or information to the remitting bank by quickest mail.

If, however, the collecting bank considers the matter to be urgent, it may advise by other quicker methods at the expense of the customer.

Charges and expenses

Article 14—If the remittance letter includes an instruction that collection charges and/or expenses are to be for account of the drawee and the drawee refuses to pay them, the collecting bank, unless expressly instructed to the contrary, may deliver the commercial paper against payment or acceptance as the case may be without collecting charges and/or expenses. In such a case collection charges and/or expenses will be for account of the customer.

Article 15—In all cases where in the express terms of a collection, or under these Rules, disbursements and/or expenses and/or collection charges are to be borne by the customer, the collecting bank is entitled to recover its outlay in respect of disbursements and expenses and its charges from the remitting bank and the remitting bank has the right to recover from the customer any amount so paid out by it, together with its own disbursements, expenses and charges.

Liabilities and responsibilities

Article 16—Banks utilizing the services of another bank for the purpose of giving effect to the instructions of the customer do so for the account of and at the risk of the latter.

Banks are free to utilize as the collecting bank any of their correspondent banks in the country of payment or acceptance as the case may be.

If the customer nominates the collecting bank, the remitting bank nevertheless has the right to direct the commercial paper to such nominated collecting bank through a correspondent bank of its own choice.

Article 17—Banks concerned with a collection of commercial paper assume no liability or responsibility for the consequences arising out of delay and/or loss in transit of any messages, letters or documents, or for delay, mutilation or other errors arising in the transmission of cables, telegrams or telex, or for errors in translation or interpretation of technical terms.

Article 18—Banks concerned with a collection of commercial paper assume no liability or responsibility for consequences arising out of the interruption of their business by strikes, lock-outs, riots, civil commotions, insurrections, wars, Acts of God or any other causes beyond their control.

Article 19—In the event of goods being despatched direct to the address of a bank for delivery to a drawee against payment or acceptance or upon other terms without prior agreement on the part of that bank, the bank has no obligation to take delivery of the goods, which remain at the risk and responsibility of the party despatching the goods.

Appendix 3

Revised American Foreign Trade Definitions—1941

Adopted July 30, 1941, by a Joint Committee representing the Chamber of Commerce of the United States of America, the National Council of American Importers, Inc., and the National Foreign Trade Council, Inc.

Foreward

The following *Revised American Foreign Trade Definitions—1941* are recommended for general use by both exporters and importers. These revised definitions have no status at law unless there is specific legislation providing for them, or unless they are confirmed by court decisions. Hence, it is suggested that sellers and buyers agree to their acceptance as part of the contract of sale. These revised definitions will then become legally binding upon all parties.

Adoption by exporters and importers of these revised terms will impress on all parties concerned their respective responsibilities and rights.

General Notes of Caution

1. As foreign trade definitions have been issued by organizations in various parts of the world, and as the courts of countries have interpreted these definitions in different ways, it is important that sellers and buyers agree that their contracts are subject to the *Revised American Foreign Trade Definitions—1941* and that the various points listed are accepted by both parties.

2. In addition to the foreign trade terms listed herein, there are terms that are at times used, such as Free Harbor, C.I.F.&C. (Cost, Insurance, Freight, and Commission), C.I.F.C.&I. (Cost, Insurance, Freight, Commission, and Interest), C.I.F. Landed (Cost, Insurance, Freight, Landed),

INDUSTRIAL TRAFFIC MANAGEMENT

and others. None of these should be used unless there has first been a definite understanding as to the exact meaning thereof. It is unwise to attempt to interpret other terms in the light of the terms given herein. Hence, whenever possible, one of the terms defined herein should be used.

3. It is unwise to use abbreviations in quotations or in contracts which might be subject to misunderstanding.

4. When making quotations, the familiar terms "hundredweight" or "ton" should be avoided. A hundredweight can be 100 pounds of the short ton, or 112 pounds of the long ton. A ton can be a short ton of 2,000 pounds, or a metric ton of 2,204.6 pounds, or a long ton of 2,240 pounds. Hence, the type of hundredweight or ton should be clearly stated in quotations and in sales confirmations. Also, all terms referring to quantity weight, volume, length, or surface should be clearly defined and agreed upon.

5. If inspection, or certificate of inspection, is required, it should be agreed, in advance, whether the cost thereof is for account of seller or buyer.

6. Unless otherwise agreed upon, all expenses are for the account of seller up to the point at which the buyer must handle the subsequent movement of goods.

7. There are a number of elements in a contract that do not fall within the scope of these foreign trade definitions. Hence, no mention of these is made herein. Seller and buyer should agree to these separately when negotiating contracts. This particularly applies to so-called "customary" practices.

Definitions of Quotations

(I) Ex (Point of Origin)

"Ex Factory," "Ex Mill," "Ex Mine," "Ex Plantation," "Ex Warehouse," etc. (named point of origin)

Under this term, the price quoted applies only at the point of origin, and the seller agrees to place the goods at the disposal of the buyer at the agreed place on the date or within the period fixed.

Under this quotation:
Seller must
(1) bear all costs and risks of the goods until such time as the buyer is obliged to take delivery thereof;
(2) render the buyer, at the buyer's request and expense, assistance in obtaining the documents issued in the country of origin, or of shipment, or of both, which the buyer may require either for pur-

AMERICAN FOREIGN TRADE DEFINITIONS

poses of exportation, or of importation at destination.
Buyer must
(1) take delivery of the goods as soon as they have been placed at his disposal at the agreed place on the date or within the period fixed;
(2) pay export taxes, or other fees or charges, if any, levied because of exportation;
(3) bear all costs and risks of the goods from the time when he is obligated to take delivery thereof;
(4) pay all costs and charges incurred in obtaining the documents issued in the country of origin, or of shipment, or of both which may be required either for purposes of exportation, or of importation at destination.

(II) F.O.B. (Free on Board)

Note: *Seller and buyer should consider not only the definitions but also the "Comments on All F.O.B. Terms" given at end of this section in order to understand fully their respective responsibilities and rights under the several classes of "F.O.B." terms.*

(II-A) "F.O.B. (named inland carrier at named inland point of departure)"*

Under this term, the price quoted applies only at inland shipping point, and the seller arranges for loading of the goods on, or in, railway cars, trucks, lighters, barges, aircraft, or other conveyance furnished for transportation.

Under this quotation:
Seller must
(1) place goods on, or in, conveyance, or deliver to inland carrier for loading;
(2) provide clean bill of lading or other transportation receipt, freight collect;
(3) be responsible for any loss or damage, or both, until goods have been placed in, or on, conveyance at loading point, and clean bill of lading or other transportation receipt has been furnished by the carrier;
(4) render the buyer at the buyer's request and expense, assistance in obtaining the documents issued in the country of origin, or of shipment, or of both, which the buyer may require either for purposes of exportation, or of importation at destination.
Buyer must
(1) be responsible for all movement of the goods from inland point

---
**See Note and Comments on all F.O.B. Terms*

INDUSTRIAL TRAFFIC MANAGEMENT

of loading, and pay all transportation costs;

(2) pay export taxes, or other fees or charges, if any, levied because of exportation;

(3) be responsible for any loss or damage, or both, incurred after loading at named inland point of departure;

(4) pay all costs and charges incurred in obtaining the documents issued in the country of origin, or of shipment, or of both, which may be required either for purposes of exportation, or of importation at destination.

(II-B) "F.O.B. (named inland carrier at named inland point of departure) Freight Prepaid To (named point of exportation)"*

Under this term, the seller quotes a price including transportation charges to the named point of exportation, and prepays freight to named point of exportation, without assuming responsibility for the goods after obtaining a clean bill of lading or other transportation receipt at named inland point of departure.

Under this quotation:

Seller must

(1) assume the seller's obligations as under II-A, except that under

(2) he must provide clean bill of lading or other transportation receipt, freight prepaid to named point of exportation.

Buyer must

(1) assume the same buyer's obligations as under II-A, except that he does not pay freight from loading point to named point of exportation.

(II-C) "F.O.B. (named carrier at named inland point of departure) Freight Allowed To (named point)"*

Under this term, the seller quotes a price including the transportation to the named point, shipping freight collect and deducting the cost of transportation, without assuming responsibility for the goods after obtaining a clean bill of lading or other transportation receipt at named inland point of departure.

Under this quotation:

Seller must

(1) assume the same seller's obligations as under II-A, but deducts from his invoice the transportation cost to named point.

Buyer must

(1) assume the same buyer's obligations as under II-A, including payment of freight from inland loading point to named point, for which seller has made deduction.

---

*See Note and Comments on all F.O.B. Terms

AMERICAN FOREIGN TRADE DEFINITIONS

(II-D) "F.O.B. (named inland carrier at named point of exportation)"*

Under this term, the seller quotes a price including the costs of transportation of the goods to named point of exportation, bearing any loss or damage, or both, incurred up to that point.

Under this quotation:

Seller must

(1) place goods on, or in, conveyance, or deliver to inland carrier for loading;

(2) provide clean bill of lading or other transportation receipt, paying all transportation costs from loading point to named point of exportation;

(3) be responsible for any loss or damage, or both, until goods have arrived in, or on, inland conveyance at the named point of exportation;

(4) render the buyer, at the buyer's request and expense, assistance in obtaining the documents issued in the country of origin, or of shipment, or of both, which the buyer may require either for purposes of exportation, or of importation at destination.

Buyer must

(1) be responsible for all movement of the goods from inland conveyance at named point of exportation;

(2) pay export taxes, or other fees or charges, if any, levied because of exportation;

(3) be responsible for any loss or damage, or both, incurred after goods have arrived in, or on, inland conveyance at the named point of exportation;

(4) pay all costs and charges incurred in obtaining the documents issued in the country of origin, or of shipment, or of both, which may be required either for purposes of exportation, or of importation at destination.

(II-E) "F.O.B. Vessel (named port of shipment)"*

Under this term, the seller quotes a price covering all expenses up to, and including, delivery of the goods upon the overseas vessel provided by, or for, the buyer at the named port of shipment.

Under this quotation:

Seller must

(1) pay all charge incurred in placing goods actually on board the vessel designated and provided by, or for, the buyer on the date or within the period fixed;

(2) provide clean ship's receipt or on-board bill of lading;

---

*See Note and Comments on all F.O.B. Terms.

INDUSTRIAL TRAFFIC MANAGEMENT

(3) be responsible for any loss or damage, or both, until goods have been placed on board the vessel on the date or within the period fixed;
(4) render the buyer, at the buyer's request and expense, assistance in obtaining the documents issued in the country of origin, or of shipment, or of both, which the buyer may require either for purposes of exportation, or of importation at destination.
Buyer must
(1) give seller adequate notice of name, sailing date, loading berth of, and delivery time to, the vessel;
(2) bear the additional costs incurred and all risks of the goods from the time when the seller has placed them at his disposal if the vessel named by him fails to arrive or to load within the designated time;
(3) handle all subsequent movement of the goods to destination:
  (a) provide and pay for insurance;
  (b) provide and pay for ocean and other transportation;
(4) pay export taxes, or other fees or charges, if any, levied because of exportation;
(5) be responsible for any loss or damage, or both, after goods have been loaded on board the vessel;
(6) pay all costs and charges incurred in obtaining the documents, other than clean ship's receipt or bill of lading, issued in the country of origin, or of shipment, or of both, which may be required either for purposes of exportation, or of importation at destination.

(II-F) "F.O.B. (named inland point in country of importation)"*

Under this term, the seller quotes a price including the cost of the merchandise and all costs of transportation to the named inland point in the country of importation.
Under this quotation:
Seller must
(1) provide and pay for all transportation to the named inland piont in the country of importation;
(2) pay export taxes, or other fees or charges, if any, levied because of exportation;
(3) provide and pay for marine insurance;
(4) provide and pay for war risk insurance, unless otherwise agreed upon between the seller and buyer;
(5) be responsible for any loss or damage, or both, until arrival of goods on conveyance at the named inland point in the country of importation;
(6) pay the costs of certificates of origin, consular invoices, or any other documents issued in the country of origin, or of shipment, or of both, which the buyer may require for the importation of goods into

---

*See Note and Comments on all F.O.B. Terms.

AMERICAN FOREIGN TRADE DEFINITIONS

the country of destination and, where necessary, for their passage in transit through another country;
(7) pay all costs of landing, including wharfage, landing charges, and taxes, if any;
(8) pay all costs of customs entry in the country of importation;
(9) pay customs duties and all taxes applicable to imports, if any, in the country of importation.

Note: *The seller under this quotation must realize that he is accepting important responsibilities, costs, and risks, and should therefore be certain to obtain adequate insurance. On the other hand, the importer or buyer may desire such quotations to relieve him of the risks of the voyage and to assure him of his landed costs at inland point in country of importation. When competition is keen, or the buyer is accustomed to such quotations from other sellers, seller may quote such terms, being careful to protect himself in an appropriate manner.*

Buyer must
(1) take prompt delivery of goods from conveyance upon arrival at destination;
(2) bear any costs and be responsible for all loss or damage, or both, after arrival at destination.

Comments on All F.O.B. Terms

In connection with F.O.B. terms, the following points of caution are recommended:

1. The method of inland transportation, such as trucks, railroad cars, lighters, barges, or aircraft should be specified.

2. If any switching charges are involved during the inland transportation, it should be agreed, in advance, whether these charges are for account of the seller or the buyer.

3. The term "F.O.B. (named port)," without designating the exact point at which the liability of the seller terminates and the liability of the buyer begins, should be avoided. The use of this term gives rise to disputes as to the liability of the seller or the buyer in the event of loss or damage arising while the goods are in port, and before delivery to or on board the ocean carrier. Misunderstandings may be avoided by naming the specific point of delivery.

4. If lighterage or trucking is required in the transfer of goods from the inland conveyance to ship's side, and there is a cost therefor, it should be understood, in advance, whether this cost is for account of the seller or the buyer.

5. The seller should be certain to notify the buyer of the minimum

---

See Note and Comments on all F.O.B. Terms.

quantity required to obtain a carload, a truckload, or a barge-load freight rate.

6. Under F.O.B. terms, excepting "F.O.B. (named inland point in country of importation)," the obligation to obtain ocean freight space, and marine and war risk insurance, rests with the buyer. Despite this obligation on the part of the buyer, in many trades the seller obtains the ocean freight space, and marine and war risk insurance, and provides for shipment on behalf of the buyer. Hence, seller and buyer must have an understanding as to whether the buyer will obtain the ocean freight space, and marine, and war risk insurance, as is his obligation, or whether the seller agrees to do this for the buyer.

7. For the seller's protection, he should provide in his contract of sale that marine insurance obtained by the buyer include standard warehouse to warehouse coverage.

(III) F.A.S. (Free Along Side)

Note: *Seller and buyer should consider not only the definitions but also the "Comments" given at the end of this section in order to understand fully their respective responsibilities and rights under "F.A.S." terms.*

"F.A.S. Vessel (named port of shipment)"

Under this term, the seller quotes a price including delivery of the goods along side overseas vessel and within reach of its loading tackle.

Under this quotation:

Seller must

(1) place goods along side vessel or on dock designated and provided by, or for, buyer on the date or within the period fixed; pay any heavy lift charges, where necessary, up to this point;

(2) provide clean dock or ship's receipt;

(3) be responsible for any loss or damage, or both, until goods have been delivered along side the vessel or on the dock;

(4) render the buyer, at the buyer's request and expense, assistance in obtaining the documents issued in the country of origin, or of shipment, or of both, which the buyer may require either for purposes of exportation, or of importation at destination.

Buyer must

(1) give seller adequate notice of name, sailing date, loading berth of, and delivery time to, the vessel;

(2) handle all subsequent movement of the goods from along side the vessel:

    (a) arrange and pay for demurrage or storage charges, or both, in warehouse or on wharf, where necessary;

    (b) provide and pay for insurance;

    (c) provide and pay for ocean and other transportation;

(3) pay export taxes, or other fees or charges, if any, levied because

of exportation;

(4) be responsible for any loss or damage, or both, while the goods are on a lighter or other conveyance along side vessel within reach of its loading tackle, or on the dock awaiting loading, or until actually loaded on board the vessel, and subsequent thereto;

(5) pay all costs and charges incurred in obtaining the documents, other than clean dock or ship's receipt, issued in the country of origin, or of shipment, or both, which may be required either for purposes of exportation, or of importation at destination.

F.A.S. Comments

1. Under F.A.S. terms, the obligation to obtain ocean freight space, and marine and war risk insurance, rests with the buyer. Despite this obligation on the part of the buyer, in many trades the seller obtains ocean freight space, and marine and war risk insurance, and provides for shipment on behalf of the buyer. In others, the buyer notifies the seller to make delivery along side a vessel designated by the buyer and the buyer provides his own marine and war risk insurance. Hence, seller and buyer must have an understanding as to whether the buyer will obtain the ocean freight space, and marine and war risk insurance, as is his obligation, or whether the seller agrees to do this for the buyer.

2. For the seller's protection, he should provide in his contract of sale that marine insurance obtained by the buyer include standard warehouse to warehouse coverage.

(IV) C. & F. (Cost and Freight)

Note: *Seller and buyer should consider not only the definitions but also the "C. & F. Comments" and the "C. & F. and C.I.F. Comments," in order to understand fully their respective responsibilities and rights under "C. & F." terms.*

"C. F. (named point of destination)"

Under this term, the seller quotes a price including the cost of transportation to the named point of destination.

Under this quotation:

Seller must

(1) provide and pay for transportation to named point of destination;

(2) pay export taxes, or other fees or charges, if any, levied because of exportation;

(3) obtain and dispatch promptly to buyer, or his agent, clean bill of lading to named point of destination;

(4) where received-for-shipment ocean bill of lading may be tendered, be responsible for any loss or damage, or both, until the goods have been delivered into the custody of the ocean carrier;

(5) where on-board ocean bill of lading is required, be responsible

for any loss or damage, or both, until the goods have been delivered on board the vessel;

(6) provide, at the buyer's request and expense, certificates of origin, consular invoices, or any other documents issued in the country of origin, or of shipment, or of both, which the buyer may require for importation of goods into country of destination and, where necessary, for their passage in transit through another country.

Buyer must

(1) accept the documents when presented;

(2) receive goods upon arrival, handle and pay for all subsequent movement of the goods, including taking delivery from vessel in accordance with bill of lading clauses and terms; pay all costs of landing, including any duties, taxes, and other expenses at named point of destination;

(3) provide and pay for insurance;

(4) be responsible for loss of or damage to goods, or both, from time and place at which seller's obligations under (4) or (5) above have ceased;

(5) pay the costs of certificates of origin, consular invoices, or any other documents issued in the country of origin, or of shipment, or of both, which may be required for the importation of goods into the country of destination and, where necessary, for their passage in transit through another country.

C. & F. Comments

1. For the seller's protection, he should provide in his contract of sale that marine insurance obtained by the buyer include standard warehouse to warehouse coverage.

2. The comments listed under the following C.I.F. terms in many cases apply to C. & F. Terms as well, and should be read and understood by the C. & F. seller and buyer.

(V) C.I.F. (Cost, Insurance, Freight)

Note: *Seller and buyer should consider not only the definitions but also the "Comments" at the end of this section, in order to understand fully their respective responsibilities and rights under "C.I.F." terms.*

"C.I.F. (named point of destination)"

Under this term, the seller quotes a price including the cost of the goods, the marine insurance, and all transportation charges to the named point of destination.

Under this quotation:

Seller must

(1) provide and pay for transportation to named point of destination;

AMERICAN FOREIGN TRADE DEFINITIONS

(2) pay export taxes, or other fees or charges, if any, levied because of exportation;
(3) provide and pay for marine insurance;
(4) provide war risk insurance as obtainable in seller's market at time of shipment at buyer's expense, unless seller has agreed that buyer provide for war risk coverage;
(5) obtain and dispatch promptly to buyer, or his agent, clean bill of lading to named point of destination, and also insurance policy or negotiable insurance certificate;
(6) where received-for-shipment ocean bill of lading may be tendered, be responsible for any loss or damage, or both, until the goods have been delivered into the custody of the ocean carrier;
(7) where on-board ocean bill of lading is required, be responsible for any loss or damage, or both, until the goods have been delivered on board the vessel;
(8) provide, at the buyer's request and expense, certificates of origin, consular invoices, or any other documents issued in the country of origin, or of shipment, or both, which the buyer may require for importation of goods into country of destination and, where necessary, for their passage in transit through another country.

Buyer must
(1) accept the documents when presented;
(2) receive the goods upon arrival; handle and pay for all subsequent movement of the goods, including taking delivery from vessel in accordance with bill of lading clauses and terms; pay all costs of landing, including any duties, taxes, and other expenses at named point of destination;
(3) pay for war risk insurance provided by seller;
(4) be responsible for loss of or damage to goods, or both, from time and place at which seller's obligations under (6) or (7) above have ceased;
(5) pay the cost of certificates of origin, consular invoices, or any other documents issued in the country of origin, or of shipment, or both, which may be required for importation of the goods into the country of destination and, where necessary, for their passage in transit through another country.

C. & F. and C.I.F. Comments

Under C. & F. and C.I.F. contracts there are the following points on which the seller and the buyer should be in complete agreement at the time that the contract is concluded:
1. It should be agreed upon, in advance, who is to pay for miscellaneous expenses, such as weighing or inspection charges.
2. The quantity to be shipped on any one vessel should be agreed upon, in advance, with a view to the buyer's capacity to take delivery

INDUSTRIAL TRAFFIC MANAGEMENT

upon arrival and discharge of the vessel, within the free time allowed at the port of importation.

3. Although the terms C. & F. and C.I.F. are generally interpreted to provide that charges for consular invoices and certificates of origin are for the account of the buyer, and are charged separately, in many trades these charges are included by the seller in his price. Hence, seller and buyer should agree, in advance, whether these charges are part of the selling price, or will be invoiced separately.

4. The point of final destination should be definitely known in the event the vessel discharges at a port other than the actual destination of the goods.

5. When ocean freight space is difficult to obtain, or forward freight contracts cannot be made at firm rates, it is advisable that sales contracts, as an exception to regular C. & F. or C.I.F. terms, should provide that shipment within the contract period be subject to ocean freight space being available to the seller, and should also provide that changes in the cost of ocean transportation between the time of sale and the time of shipment be for account of the buyer.

6. Normally, the seller is obligated to prepay the ocean freight. In some instances, shipments are made freight collect and the amount of the freight is deducted from the invoice rendered by the seller. It is necessary to be in agreement on this, in advance, in order to avoid misunderstanding which arises from foreign exchange fluctuations which might affect the actual cost of transportation, and from interest charges which might accrue under letter of credit financing. Hence, the seller should always prepay the ocean freight unless he has a specific agreement with the buyer, in advance, that goods can be shipped freight collect.

7. The buyer should recognize that he does not have the right to insist on inspection of goods prior to accepting the documents. The buyer should not refuse to take delivery of goods on account of delay in the receipt of documents, provided the seller has used due diligence in their dispatch through the regular channels.

8. Sellers and buyers are advised against including in a C.I.F. contract any indefinite clause at variance with the obligations of a C.I.F. contract as specified in these Definitions. There have been numerous court decisions in the United States and other countries invalidating C.I.F. contracts because of the inclusion of indefinite clauses.

9. Interest charges should be included in cost computations and should not be charged as a separate item in C.I.F. contracts, unless otherwise agreed upon, in advance, between the seller and buyer; in which case, however, the term C.I.F. and I. (Cost, Insurance, Freight, and Interest) should be used.

10. In connection with insurance under C.I.F. sales, it is necessary

AMERICAN FOREIGN TRADE DEFINITIONS

that seller and buyer be definitely in accord upon the following points:

(a) The character of the marine insurance should be agreed upon in so far as being W.A. (With Average) or F.P.A. (Free of Particular Average), as well as any other special risks that are covered in specific trades, or against which the buyer may wish individual protection. Among the special risks that should be considered and agreed upon between seller and buyer are theft, pilferage, leakage, breakage, sweat, contact with cargoes, and other peculiar to any particular trade. It is important that contingent or collect freight and customs duty should be insured to cover Particular Average losses, as well as total loss after arrival and entry but before delivery.

(b) The seller is obligated to exercise ordinary care and diligence in selecting an underwriter that is in good financial standing. However, the risk of obtaining settlement of insurance claims rests with the buyer.

(c) War risk insurance under this term is to be obtained by the seller at the expense and risk of the buyer. It is important that the seller be in definite accord with the buyer on this point, particularly as to the cost. It is desirable that the goods be insured against both marine and war risk with the same underwriter, so that there can be no difficulty arising from the determination of the cause of the loss.

(d) Seller should make certain that in his marine or war risk insurance, there be included the standard protection against strikes, riots and civil commotions.

(e) Seller and buyer should be in accord as to the insured valuation, bearing in mind that merchandise contribute, in General Average on certain bases of valuation which differ in various trades. It is desirable that a competent insurance broker be consulted, in order that full value be covered and trouble avoided.

(VI) "Ex Dock (named port of importation)"

Note: *Seller and buyer should consider not only the definitions but also the "Ex Dock Comments" at the end of this section, in order to understand fully their respective responsibilities and rights under "Ex Dock" terms.*

Under this term, seller quotes a price including the cost of the goods and all additional costs necessary to place the goods on the dock at the named port of importation, duty paid, if any.

Under this quotation:
Seller must
(1) provide and pay for transportation to named port of importation;
(2) pay export taxes, or other fees or charges, if any, levied because of exportation;
(3) provide and pay for marine insurance;

(4) provide and pay for war risk insurance, unless otherwise agreed upon between the buyer and seller;
(5) be responsible for any loss or damage, or both, until the expiration of the free time allowed on the dock at the named port of importation;
(6) pay the costs of certificates of origin, consular invoices, legalization of bill of lading, or any other documents issued in the country of origin, or of shipment, or of both, which the buyer may require for the importation of goods into the country of destination and, where necessary, for their passage in transit through another country;
(7) pay all costs of landing, including wharfage, landing charges, and taxes, if any;
(8) pay all costs of customs entry in the country of importation;
(9) pay customs duties and all taxes applicable to imports, if any, in the country of importation, unless otherwise agreed upon.

Buyer must
(1) take delivery of the goods on the dock at the named port of importation within the free time allowed;
(2) bear the cost and risk of the goods if delivery is not taken within the free time allowed.

Ex Dock Comments

This term is used principally in United States import trade. It has various modifications, such as "Ex Quay," "Ex Pier," etc., but it is seldom, if ever, used in American export practice. Its use in quotations for export is not recommended.

Appendix 4

Uniform Straight Bill of Lading

Contract Terms and Conditions

Sec. 1. (a) The carrier or party in possession of any of the property herein described shall be liable as at common law for any loss thereof or damage thereto, except as hereinafter provided.

(b) No carrier or party in possession of all or any of the property herein described shall be liable for any loss thereof or damage thereto or delay caused by the act of God; the public enemy; the authority of law; or the act or default of the shipper or owner, or for natural shrinkage. The carrier's liability shall be that of warehouseman, only, for loss, damage, or delay caused by fire occurring after the expiration of the free time allowed by tariffs lawfully on file (such free time to be computed as therein provided) after notice of arrival of the property at destination or at the port of export (if intended for export) has been duly sent or given, and after placement of the property for delivery at destination, or tender of delivery of the property to the party entitled to receive it, has been made. Except in case of negligence of the carrier or party in possession (and the burden to prove freedom from such negligence shall be on the carrier or party in possession), the carrier or party in possession shall not be liable for loss, damage, or delay occurring while the property is stopped and held in transit upon the request of the shipper, owner, or party entitled to make such request, or resulting from a defect or vice in the property, or for country damage to cotton, or from riots or strikes. Except in case of carrier's negligence, no carrier or party in possession of all or any of the property herein described shall be liable for delay caused by highway obstruction, faulty or impassable highway or lack of capacity of any highway, bridge or ferry, and the burden to prove freedom from such negligence shall be on the carrier or party in possession.

(c) In case of quarantine the property may be discharged at risk and expense of owners into quarantine depot or elsewhere, as required by quarantine regulations or authorities, or for the carrier's dispatch at nearest available point in carrier's judgment, and in any such case carrier's responsibility shall cease when property is so discharged, or property may be returned by carrier at owner's expense to shipping point, earning freight both ways. Quarantine expenses of whatever nature or kind upon or in respect to property shall be borne by the owners of the property to be in lien thereon. The carrier shall not be liable for loss or damage occasioned by fumigation or disinfection or other acts required or done by quarantine regulations or authorities even though the same may have been done by carrier's officers, agents, or employees, nor for detention, loss, or damage of any kind occasioned by quarantine or the enforcement thereof. No carrier shall be liable, except in case of negligence, for any mistake or inaccuracy in any information furnished by the carrier, its agents, or officers, as to quarantine laws or regulations. The shipper shall hold the carriers harmless from any expense they may incur, or damages they may incur, or damages they may be required to pay, by reason of the introduction of the property covered by this contract into any place against the quarantine laws or regulations in effect at such place.

Sec. 2. (a) No carrier is bound to transport said property by any particular train or vessel, or in time for any particular market or otherwise than with reasonable dispatch. Every carrier shall have the right in case of physical necessity to forward said property by any carrier or route between the point of shipment and the point of destination. In all cases not prohibited by law, where a lower value than actual value has been represented in writing by the shipper or has been agreed upon in writing as the released value of the property as determined by the classification or tariffs upon which the rate is based, such lower value plus freight charges if paid shall be the maximum amount to be recovered, whether or not such loss or damage occurs from negligence.

(b) As a condition precedent to recovery, claims must be filed in writing with the receiving or delivering carrier, or carrier issuing this bill of lading, or carrier on whose line the loss, damage, injury or delay occurred, within nine months after delivery of the property (or in case of export traffic, within nine months after delivery at port of export) or, in case of failure to make delivery, then within nine months after a reasonable time for delivery has elapsed; and suits shall be instituted against any carrier only within two years and one day from the day when notice in writing is given by the carrier to the claimant that the carrier has disallowed the claim or any part or parts thereof specified in the notice. Where claims are not filed or suits are not instituted thereon in accordance with the foregoing provisions, no carrier hereunder shall be liable, and such claims will not be paid.

(c) Any carrier or party liable on account of loss of or damage to any of said property shall have the full benefit of any insurance that may have been effected upon

UNIFORM STRAIGHT BILL OF LADING

or on account of said property, so far as this shall not avoid the policies or contracts of insurance: Provided, That the carrier reimburse the claimant for the premium paid thereon.

Sec. 3. Except where such service is required as the result of carrier's negligence, all property shall be subject to necessary cooperage and baling at owner's cost. Each carrier over whose route cotton or cotton linters is to be transported hereunder shall have the privilege, at its own cost and risk, of compressing the same for greater convenience in handling or forwarding, and shall not be held responsible for deviation or unavoidable delays in procuring such compression. Grain in bulk consigned to a point where there is a railroad, public or licensed elevator, may (unless otherwise expressly noted herein, and then if it is not promptly unloaded) be there delivered and placed with other grain of the same kind and grade without respect to ownership (and prompt notice thereof shall be given to the consignor), and if so delivered shall be subject to a lien for elevator charges in addition to all other charges hereunder.

Sec. 4. (a) Property not removed by the party entitled to receive it within the free time allowed by tariffs, lawfully on file (such free time to be computed as therein provided), after notice of the arrival of the property at destination or at the port of export (if intended for export) has been duly sent or given, and after placement of the property for delivery at destination has been made, may be kept in vessel, car, depot, warehouse or place of delivery of the carrier, subject to the tariff charge for storage and to carrier's responsibility as warehouseman, only, or at the option of the carrier, may be removed to and stored in a public or licensed warehouse at the place of delivery or other available place, at the cost of the owner, and there held without liability on the part of the carrier, and subject to a lien for all freight and other lawful charges, including a reasonable charge for storage. In the event consignee cannot be found at address given for delivery, then in that event, notice of the placing of such goods in warehouse shall be mailed to the address given for delivery and mailed to any other address given on the bill of lading for notification, showing the warehouse in which such property has been placed, subject to the provisions of this paragraph.

(b) Where nonperishable property which has been transported to destination hereunder is refused by consignee or the party entitled to receive it, or said consignee or party entitled to receive it fails to receive it within 15 days after notice of arrival shall have been duly sent or given, the carrier may sell the same at public auction to the highest bidder, at such place as may be designated by the carrier: Provided, That the carrier shall have first mailed, sent, or given to the consignor notice that the property has been refused or remains unclaimed, as the case may be, and that it will be subject to sale under the terms of the bill of lading if disposition be not arranged for, and shall have published notice containing a description of the property, the name of the party to whom consigned, or, if shipped order notify, the name of the party to be notified, and the time and place of sale, once a week for two successive weeks, in a newspaper of general circulation at the place of sale or nearest place where such newspaper is published: Provided, That 30 days shall have elapsed before publication of notice of sale after said notice that the property was refused or remains unclaimed was mailed, sent, or given.

(c) Where perishable property which has been transported hereunder to destination is refused by consignee or party entitled to receive it, or said consignee or party entitled to receive it shall fail to receive it promptly, the carrier may, in its discretion, to prevent deterioration or further deterioration, sell the same to the best advantage at private or public sale: Provided, That if time serves for notification to the consignor or owner of the refusal of the property or the failure to receive it and request for disposition of the property, such notification shall be given, in such manner as the exercise of due diligence requires, before the property is sold.

(d) Where the procedure provided for in the two paragraphs last preceding is not possible, it is agreed that nothing contained in said paragraphs shall be construed to abridge the right of the carrier at its option to sell the property under such circumstances and in such manner as may be authorized by law.

(e) The proceeds of any sale made under this section shall be applied by the carrier to the payment of freight, demurrage, storage, and any other lawful charges and the expense of notice, advertisement, sale, and other necessary expense and of caring for and maintaining the property, if proper care of the same requires special expense, and should there be a balance it shall be paid to the owner of the property sold hereunder.

(f) Property destined to or taken from a station, wharf, or landing at which there is no regularly appointed freight agent shall be entirely at risk of owner after unloaded from cars or vessels or until loaded into cars or vessels, and, except in case of carrier's negligence, when received from or delivered to such stations, wharves, landings or other places, shall be at owner's risk until the cars are attached to and after they are detached from locomotive or train or until loaded into and after unloaded from vessels, or if property is transported in motor vehicle trailers or semi-trailers until such trailers or semi-trailers are attached to and after they are detached from power units. Where a carrier is directed to unload or deliver property transported by motor vehicle at a particular location where consignee or consignee's agent is not regularly located, the risk after unloading or delivery shall be that of the owner.

Sec. 5. No carrier hereunder will carry or be liable in any way for any documents, specie, or for any articles of extraordinary value not specifically rated in the

UNIFORM STRAIGHT BILL OF LADING

published classifications or tariffs unless a special agreement to do so and a stipulated value of the articles are indorsed hereon.

Sec. 6. Every party, whether principal or agent, shipping explosives or dangerous goods, without previous full written disclosure to the carrier of their nature, shall be liable for and indemnify the carrier against all loss or damage caused by such goods, and such goods may be warehoused at owner's risk and expense or destroyed without compensation.

Sec. 7. The owner or consignee shall pay the freight and average, if any, and all other lawful charges accruing on said property; but, except in those instances where it may lawfully be authorized to do so, no carrier by railroad shall deliver or relinquish possession at destination of the property covered by this bill of lading until all tariff rates and charges thereon have been paid. The consignor shall be liable for the freight and all other lawful charges, except that if the consignor stipulates, by signature, in the space provided for that purpose on the face of this bill of lading that the carrier shall not make delivery without requiring payment of such charges and the carrier, contrary to such stipulation, shall make delivery without requiring such payment, the consignor (except as hereinafter provided) shall not be liable for such charges. Provided, that, where the carrier has been instructed by the shipper or consignor to deliver said property to a consignee other than the shipper or consignor, such consignee shall not be legally liable for transportation charges in respect to the transportation of said property (beyond those billed against him at the time of delivery for which he is otherwise liable) which may be found to be due after the property has been delivered to him, if the consignee (a) is an agent only and has no beneficial title in said property and (b) prior to delivery of said property has notified the delivering carrier in writing of the fact of such agency and absence of beneficial title, and, in the case of a shipment reconsigned or diverted to a point other than that specified in the original bill of lading, has also notified the delivering carrier in writing of the fact of such agency and absence of beneficial title, and, in the case of a shipment reconsigned or diverted to a point other than that specified in the original bill of lading, has also notified the delivering carrier in writing of the name and address of the beneficial said property; and in such cases the shipper or consignor, or, in the case of a shipment so reconsigned or diverted, the beneficial owner, shall be liable for such additional charges. If the consignee has given to the carrier erroneous information as to who the beneficial owner is, such consignee shall be liable for such additional charges. Nothing herein shall limit the right of the carrier to require at time of shipment the prepayment or guarantee of the charges. If upon inspection it is ascertained that the articles shipped are not those described in this bill of lading, the freight charges must be paid upon the articles actually shipped.

Sec. 8. If this bill of lading is issued on the order of the shipper, or his agent, in exchange or in substitution for another bill of lading, the shipper's signature to the prior bill of lading as to the statement of value or otherwise, or election of common law or bill of lading liability, in or in connection with such prior bill of lading shall be considered a part of this bill of lading as fully as if the same were written or made in or in connection with this bill of lading.

Sec. 9 (a) If all or any part of said property is carried by water over any part of said route, such water carriage shall be performed subject to all the terms and provisions of, and all the exemptions from liability contained in the Act of the Congress of the United States, approved on February 13, 1893, and entitled "An act relating to the navigation of vessels, etc.," and of other statutes of the United States according carriers by water the protection of limited liability, and to the conditions contained in this bill of lading not inconsistent therewith or with this section.

(b) No such carrier by water shall be liable for any loss or damage resulting from any fire happening to or on board the vessel, or from explosion, bursting of boilers or breakage of shafts, unless caused by the design or neglect of such carrier.

(c) If the owner shall have exercised due diligence in making the vessel in all respects seaworthy and properly manned, equipped, and supplied, no such carrier shall be liable for any loss or damage resulting from the perils of the lakes, seas, or other waters, or from latent defects in hull, machinery, or appurtenances whether existing prior to, at the time of, or after sailing, or from collision, stranding, or other accidents of navigation, or from prolongation of the voyage. And, when for any reason it is necessary, any vessel carrying any or all of the property herein described shall be at liberty to call at any port or ports, in or out of the customary route, to tow and be towed, to transfer, trans-ship, or lighter, to load and discharge goods at any time, to assist vessels in distress, to deviate for the purpose of saving life or property, and for docking and repairs. Except in case of negligence such carrier shall not be responsible for any loss or damage to property if it be necessary or is usual to carry the same upon deck.

(d) General Average shall be payable according to the York-Antwerp Rules of 1924, Sections 1 to 15 inclusive, and Sections 17 to 22, inclusive, and as to matters not covered thereby according to the laws and usages of the Port of New York. If the owners shall have exercised due diligence to make the vessel in all respects seaworthy and properly manned, equipped and supplied, it is hereby agreed that in case of danger, damage or disaster resulting from faults or errors in navigation, or in the management of the vessel, or from any latent or other defects in the vessel, her machinery or appurtenances, or from unseaworthiness, whether existing at the time of shipment or at the beginning of the voyage (provided the latent or other defects or the unseaworthiness was not discoverable by the exercise of due diligence), the shippers, consignees and/or owners of the cargo shall nevertheless

UNIFORM STRAIGHT BILL OF LADING

pay salvage and any special charges incurred in respect of the cargo, and shall contribute with the shipowner in general average to the payment of any sacrifices, losses or expenses of a general average nature that may be made or incurred for the common benefit or to relieve the adventure from any common peril.

(e) If the property is being carried under a tariff which provides that any carrier or carriers party thereto shall be liable for loss from perils of the sea, then as to such carrier or carriers the provisions of this section shall be modified in accordance with the tariff provisions, which shall be regarded as incorporated into the conditions of this bill of lading.

(f) The term "water carriage" in this section shall not be construed as including lighterages in or across rivers, harbors, or lakes, when performed by or on behalf of carriers other than water.

Sec. 10. Any alteration, addition, or erasure in this bill of lading which shall be made without the special notation hereon of the agent of the carrier issuing this bill of lading, shall be without effect, and this bill of lading shall be enforceable according to its original tenor.

Advancing Charges

Except as provided in tariffs of carrier at point of origin or destination or transit station (as the case may be), no charges of any description will be advanced to shippers, owners, consignees, their warehousemen or agents.

Item 310
Advertising or Premiums (see Note 1)

Advertising Matter or Store Display Racks or Stands (see Note 2)
Sec. 1. Advertising matter (other than figures or images or electric signs) which advertise, or store display racks or stands or show or display cases which display commodities contained in a shipment may constitute part of a shipment and shall be charged for at the rate or class applicable to the commodities advertised or to be displayed, providing the following conditions are met:

(a) In LTL shipments, the advertising matter or store display racks or stands or show or display cases are in the same package with and do not exceed ten percent of the gross weight of the commodity advertised or to be displayed, nor more than

INDUSTRIAL TRAFFIC MANAGEMENT

25 percent of the shipping package cubage (cubage to be determined from outside measurements of shipping package).

(b) In TL shipments, the advertising matter or store display racks or stands or show or display cases are in the same package with or in separate packages from the commodity advertised or to be displayed and do not exceed ten percent of the minimum weight applicable to such commodity, or do not exceed ten percent of the actual weight of such commodity when such weight is greater than the applicable minimum weight, as the case may be.

(c) Any weight of advertising matter or store display racks or stands or show or display cases in excess of the ten percent allowed under (a) or (b) will be charged for at the rate or class applicable to such advertising matter or store display racks or stands or show or display cases.

(d) The description and weight of the advertising matter or store display racks or stands or show or display cases must be shown by shipper on shipping orders and bills of lading.

Premiums (see Note 3)

Sec. 2. Premiums may constitute part of a shipment, chargeable at the rate or class applicable to the commodity for which they are premiums, providing they are in the same shipping package LTL, or the same or separate packages TL, as such commodity, and providing one of the conditions in (a), (b) or (c) is met, and that the condition in (d) is Met:

(a) Premiums do not exceed one for each shipping package, or one in or for each inner package of commodity.

(b) Premiums do not exceed the total exchange value of coupons included in package with commodity.

(c) Premiums consist of a container for the commodity shipped.

(d) The density (weight per cubic foot) of premiums and the articles with which they are given or sold (together), or of commodities in premium containers, as tendered for shipment, must be not less than 50 percent of the lightest density as tendered for shipment of any sales unit subject to the same rates or classes used by the same shipper for the same quantity of the same commodity without a premium or not in a premium container.

The shipper must state on the bill of lading and shipping order that premiums are included.

UNIFORM STRAIGHT BILL OF LADING

Note 1—If a lower charge results from the application of items 640 or 645, than under the provisions of this rule, apply items 640 or 645, as the case may be.

Note 2—Advertising matter shipped under the provisions of this rule need not be prepaid.

Note 3—Monetary coins will be accepted as premiums with other articles only when the total value of such coin(s) does not exceed 12 cents in United States funds per retail sales unit.

Item 360
Bills of Lading, Freight Bills and Statements of Charges

Sec. 1. Issuance and Requirements.

Sec. 1 (a). Rates subject to the provisions of this classification are conditioned upon the use of the appropriate bill of lading required by this rule.

Sec. 1 (b). When property is transported subject to the provisions of this classification, the acceptance and use are required, respectively, of the bills of lading, domestic or export, Uniform Straight, Straight or Order, as set forth on pages 291 through 304, or as amended, of the classification.

The format of such bills of lading must contain all the information as outlined in the examples set forth on pages 291 through 304, or as amended, of the classification.

Sec. 1 (c). The Uniform Straight or Straight Bills of Lading are to be used for any shipment not consigned to the order of any corporation, firm, institution or person.

Sec. 1 (d). The "Order" bill of lading is to be used for any shipment consigned to order of any corporation, firm, institution or person.

Sec. 1 (e). Except as otherwise provided, carriers shall not furnish:

(1) bill of lading sets that consist of more than an "Original," a "Shipping Order" and a "Memorandum" per shipment.

(2) more than one original freight bill on its own standard form and one duplicate thereof, exclusive of the consignee's memo copy, per shipment.

(3) more than one original and one copy of its statement of transportation charges on its own standard form.

Sec. 1 (f). When payor of freight or other lawful charges requires or requests, as a prerequisite to payment (see Notes 2 and 3):

(1) the return of any part of bill of lading sets or copies thereof, other than one

INDUSTRIAL TRAFFIC MANAGEMENT

shipper-furnished copy, a charge of 50 cents for each such document or copy will be made; Or.

(2) copies of freight bills or statements of transportation charges in excess of the number specified in Sec. 1 (c), a charge of 50 cents for each such document or copy will be made; or.

(3) the preparation by the carrier of any forms requiring itemization, listing or description of single or multiple freight bills for submittal with freight bills or statements of charges, a charge of ten cents per line of itemization, listing or description (or portion thereof) subject to a minimum charge of 50 cents per page, per copy, will be made; or.

(4) any forms or copies of forms, other than those described in Sec. 1 (f)(1) or 1 (f)(2), to be submitted with freight bills or statements of charges, a charge of 50 cents for each such form or copy will be made; or.

(5) that information not shown on the shipping order at time of shipment be shown on freight bills or statements of charges, a charge of 50 cents per shipment will be made.

(6) that proof of delivery be furnished in any form, a charge of 50 cents for each such document or copy will be made.

Sec. 1 (g). Carriers are not obligated to furnish bills of lading containing information beyond that shown in the examples set forth, on pages 291 through 304, or as amended, of the classification.

Sec. 1 (h). Consignors may elect to have printed their own bills of lading, in which case, all requirements of Sec. 1(a) through 1(d) and Sec. 2 of this item must be observed. These forms may also contain such information as (1) identification or location of consignor or consignee, (2) commodity descriptions, (3) rates or classes, or (4) other information pertinent to the shipment.

Note 1—Not currently used.

Note 2—The charges set forth in Sec. 1(f) will not apply to:

(a) Bank Payment Plans when documentation is limited to (1) deposit ticket(s) supplied by the bank, (2) supporting freight bills not in excess of the number set forth in Sec. 1(e), or (3) the return of a copy of the bill of lading furnished by shipper.

(b) Sight Draft Plans when documentation is limited to (1) sight drafts, (2) supporting freight bill(s) and statement(s) of charges not in excess of number set forth in Sec. 1(e), or (3) the return of a copy of the bill of lading furnished by shipper.

Note 3—The provisions set forth in Sec. 1(e) and 1(f) will not apply to shipments moving on United States government bills of lading.

Sec. 2. Information to be shown on a Bill of Lading.

Sec. 2 (a). The name and address of only one consignor and one consignee and only

UNIFORM STRAIGHT BILL OF LADING

one destination shall appear on a bill of lading. When a shipment is consigned to a point of which there are two or more of the same name in the same state, the name of the county must be shown.

Sec. 2 (b). An "Order" bill of lading will not be issued unless the name of the corporation, firm, institution or person to whose order the shipment is consigned is plainly shown thereon after the words "Consigned to Order of."

Sec. 2 (c). To insure the assessment of correct freight charges and avoid infractions of federal and state laws, shippers should acquaint themselves with the descriptions of articles in the tariff under which they ship. Commodity word descriptions must be used in bills of lading and shipping orders and must conform to those in the applicable tariff. Appropriate word abbreviations may be used. Appropriate abbreviated descriptions are permitted provided the NMFC item number and appropriate Sub number thereof are shown. The kind of package used must be shown. Bills of lading and shipping orders must specify number of articles, packages or pieces.

Sec. 2 (d). Articles indicated as explosives or as dangerous articles in Dangerous Articles Tariff (DAT) must be described on bills of lading and shipping orders as shown in that tariff. Abbreviations must not be used. When DAT descriptions differ from the tariff description in connection with which the applicable class or rate is published, the tariff description must also be shown on bills of lading and shipping orders.

Sec. 3. Inspection of Property. When carrier's agent believes it necessary that the contents of packages be inspected, he shall make or cause such inspection to be made, or require other sufficient evidence to determine the actual character of the property. When found to be incorrectly described, freight charges must be collected according to proper description.

Sec. 4. Delivery of Shipments on Order Bills of Lading:

Sec. 4 (a). The surrender to the carrier of the original Order Bill of Lading, properly endorsed, is required before the delivery of the property; but, if such bill of lading be lost or delayed, Sec. 4 (b) will govern.

Sec. 4 (b). The property may be delivered in advance of the surrender of the bill of lading upon receipt of a certified check, money order or bank cashier's check (or cash at carrier's option) by the carrier's agent for an amount equal to one hundred and twenty-five percent of the invoice or value of the property, or at carrier's option, upon receipt of a bond, acceptable to the carrier, in an amount for twice the invoice or value of the property, or a blanket bond may be accepted when satisfactory to the carrier as to surety, amount and form. Amounts of money deposited by

INDUSTRIAL TRAFFIC MANAGEMENT

certified check, money order, bank cashier's check or in cash shall be refunded in full:
    Immediately upon surrender of bill of lading properly endorsed; or when
    The carrier has received a bond, acceptable to the carrier, in an amount twice the invoice or value of the property.

Sec. 5. Insurance Against Marine Risk: The cost of insurance against marine risk will not be assumed by the carrier unless so provided, specifically, in its tariffs.

Item 365
Straight Bill of Lading—Contract Terms and Conditions

The contract terms and conditions as set forth on pages 292 and 293 of the classification are applicable for all shipments moving on the:
(1) Straight Bill of Lading—Short Form as illustrated on page 303 of the classification,
(2) Straight Bill of Lading as illustrated on page 304 of the classification.

Appendix 5
Railroad Bill of Lading

(The same terms, slightly modified, also apply for motor carriers and freight forwarders.)

Contract Terms and Conditions

Sec. 1. (a) The carrier or party in possession of any of the property herein described shall be liable as at common law for any loss thereof or damage thereto, except as hereinafter provided.

(b) No carrier or party in possession of all or any of the property herein described shall be liable for any loss thereof or damage thereto or delay caused by the act of God; the public enemy; the authority of law; or the act or default of the shipper or owner, or for natural shrinkage. The carrier's liability shall be that of warehouseman only for loss, damage, or delay caused by fire occurring after the expiration of the free time allowed by tariffs lawfully on file (such free time to be computed as therein provided) after notice of arrival of the property at destination or at the port of export (if intended for export) has been duly sent or given, and after placement of the property for delivery at destination, or tender of delivery of the property to the party entitled to receive it, has been made. Except in case of negligence of the carrier or party in possession (and the burden to prove freedom from such negligence shall be on the carrier or party in possession), the carrier or party in possession shall not be liable for loss, damage, or delay occurring while the property is stopped and held in transit upon the request of the shipper, owner, or party entitled to make such request, or resulting from a defect or vice in the property, or for country damage to cotton, or from riots or strikes.

(c) In case of quarantine the property may be discharged at risk and expense of owners into quarantine depot or elsewhere, as required by quarantine regulations or authorities, or for the carrier's dispatch at nearest available point in carrier's judgment, and in any case carrier's responsibility shall cease when property is so discharged, or property may be returned by carrier at owner's expense to shipping point, earning freight both ways. Quarantine expenses of whatever nature or kind upon or in respect to property shall be borne by the owners of the property or be in lien thereon. The carrier shall not be liable for loss or damage occasioned by fumigation or disinfection or other acts required or done by quarantine regulations or authorities even though the same may have been done by carrier's officers, agents, or employees, nor for detention, loss, or damage of any kind occasioned by quarantine or the enforcement thereof. No carrier shall be liable, except in case of negligence, for any mistake or inaccuracy in any information furnished by the carrier, its agents, or officers, as to quarantine laws or regulations. The shipper shall hold the carriers harmless from any expense they may incur, or damages they may incur, or damages they may be required to pay, by reason of the introduction of the property covered by this contract into any place against the quarantine laws or regulations in effect at such place.

Sec. 2 (a) No carrier is bound to transport said property by any particular train or vessel, or in time for any particular market or otherwise than with reasonable dispatch. Every carrier shall have the right in case of physical necessity to forward said property by any carrier or route between the point of shipment and the point of destination. In all cases not prohibited by law, where a lower value than actual value has been represented in writing by the shipper or has been agreed upon in writing as the released value of the property as determined by the classification or tariffs upon which the rate is based, such lower value plus freight charges if paid shall be the maximum amount to be recovered, whether or not such loss or damage occurs from negligence.

(b) As a condition precedent to recovery, claims must be filed in writing with the receiving or delivering carrier, or carrier issuing this bill of lading, or carrier on whose line the loss, damage, injury or delay occurred, within nine months after delivery of the property (or, in case of export traffic, within nine months after delivery at port of export) or, in case of failure to make delivery, then within nine months after a reasonable time for delivery has elapsed; and suits shall be instituted against any carrier only within two years and one day from the day when notice in writing is given by the carrier to the claimant that the carrier has disallowed the claim or any part or parts thereof specified in the notice. Where claims are not filed or suits are not instituted thereon in accordance with the foregoing provisions, no carrier hereunder shall be liable, and such claims will not be paid.

(c) Any carrier or party liable on account of loss of or damage to any of said property shall have the full benefit of any insurance that may have

RAILROAD BILL OF LADING

been effected upon or on account of said property, so far as this shall not avoid the policies or contracts of insurance: Provided, That the carrier reimburse the claimant for the premium paid thereon.

Sec. 3. Except where such service is required as the result of carrier's negligence, all property shall be subject to necessary cooperage and baling at owner's cost. Each carrier over whose route cotton or cotton linters is to be transported hereunder shall have the privilege, at its own cost and risk, of compressing the same for greater convenience in handling or forwarding, and shall not be held responsible for deviation or unavoidable delays in procuring such compression. Grain in bulk consigned to a point where there is a railroad, public or licensed elevator, may (unless otherwise expressly noted herein, and then if it is not promptly unloaded) be there delivered and placed with other grain of the same kind and grade without respect to ownership (and prompt notice thereof shall be given to the consignor), and if so delivered shall be subject to a lien for elevator charges in addition to all other charges hereunder.

Sec. 4. (a) Property not removed by the party entitled to receive it within the free time allowed by tariffs, lawfully on file (such free time to be computed as therein provided), after notice of the arrival of the property at destination or at the port of export (if intended for export) has been duly sent or given, and after placement of the property for delivery at destination has been made, may be kept in vessel, car, depot, warehouse or place of delivery of the carrier, subject to the tariff charge for storage and to carrier's responsibility as warehouseman only, or at the option of the carrier, may be removed to and stored in a public or licensed warehouse at the place of delivery or other available place, at the cost of the owner, and there held without liability on the part of the carrier, and subject to a lien for all freight and other lawful charges, including a reasonable charge for storage.

(b) Where non-perishable property which has been transported to destination hereunder is refused by consignee or the party entitled to receive it, or said consignee or party entitled to receive it fails to receive it within 15 days after notice of arrival shall have been duly sent or given, the carrier may sell the same at public auction to the highest bidder, at such place as may be designated by the carrier: Provided, That the carrier shall have first mailed, sent, or given to the consignor notice that the property has been refused or remains unclaimed, as the case may be, and that it will be subject to sale under the terms of the bill of lading if disposition be not arranged for, and shall have published notice containing a description of the property, the name of the party to whom consigned, or, if shipped order notify, the name of the party to be notified, and the time and place of sale, once a week for two successive weeks, in a newspaper of general circulation at the place of sale or nearest place where such newspaper is published: Provided, That 30 days shall have elapsed before publication of notice of sale after said notice that the property was refused or remains unclaimed was mailed, sent, or given.

(c) Where perishable property which has been transported hereunder to destination is refused by consignee or party entitled to receive it, or said consignee or party entitled to receive it shall fail to receive it promptly, the carrier may, in its discretion, to prevent deterioration or further deterioration, sell the same to the best advantage at private or public sale: Provided, That if time serves for notification to the consignor or owner, of the refusal of the property or the failure to receive it and request for disposition of the property, such notification shall be given, in such manner as the exercise of due diligence requires, before the property is sold.

(d) Where the procedure provided for in the two paragraphs last preceding is not possible, it is agreed that nothing contained in said paragraphs shall be construed to abridge the right of the carrier at its option to sell the property under such circumstances and in such manner as may be authorized by law.

(e) The proceeds of any sale made under this section shall be applied by the carrier to the payment of freight, demurrage, storage, and any other lawful charges and the expense of notice, advertisement, sale, and other necessary expense and of caring for and maintaining the property, if proper care of the same requires special expense, and should there be a balance it shall be paid to the owner of the property sold hereunder.

(f) Property destined to or taken from a station, wharf, or landing at which there is no regularly appointed freight agent shall be entirely at risk of owner after unloaded from cars or vessels or until loaded into cars or vessels, and, except in case of carrier's negligence, when received from or delivered to such stations, wharves, or landings, shall be at owner's risk until the cars are attached to and after they are detached from locomotive or train or until loaded into and after unloaded from vessels.

Sec. 5. No carrier hereunder will carry or be liable in any way for any documents, specie, or for any articles of extraordinary value not specifically rated in the published classifications or tariffs unless a special agreement to do so and a stipulated value of the articles are indorsed hereon.

Sec. 6. Every party, whether principal or agent, shipping explosives or dangerous goods, without previous full written disclosure to the carrier of their nature, shall be liable for and indemnify the carrier against all loss or damage caused by such goods, and such goods may be warehoused at owner's risk and expense or destroyed without compensation.

Sec. 7. The owner or consignee shall pay the freight and average, if any, and all other lawful charges accruing on said property; but, except in those instances where it may lawfully be authorized to do so, no carrier by railroad shall deliver or relinquish possession at destination of the property

RAILROAD BILL OF LADING

covered by this bill of lading until all tariff rates and charges thereon have been paid. The consignor shall be liable for the freight and all other lawful charges, except that if the consignor stipulates, by signature, in the space provided for that purpose on the face of this bill of lading that the carrier shall not make delivery without requiring payment of such charges and the carrier, contrary to such stipulation, shall make delivery without requiring such payment, the consignor (except as hereinafter provided) shall not be liable for such charges. Provided, that, where the carrier has been instructed by the shipper or consignor to deliver said property to a consignee other than the shipper or consignor, such consignee shall not be legally liable for transportation charges in respect to the transportation of said property (beyond those billed against him at the time of delivery for which he is otherwise liable) which may be found to be due after the property has been delivered to him, if the consignee (a) is an agent only and has no beneficial title in said property and (b) prior to delivery of said property has notified the delivering carrier in writing of the fact of such agency and absence of beneficial title, and, in the case of a shipment reconsigned or diverted to a point other than that specified in the original bill of lading, has also notified the delivering carrier in writing of the name and address of the beneficial owner of said property; and in such cases the shipper or consignor, or, in the case of a shipment so reconsigned or diverted, the beneficial owner, shall be liable for such additional charges. If the consignee has given to the carrier erroneous information as to who the beneficial owner is, such consignee shall be liable for such additional charges. On shipments reconsigned or diverted by an agent who has furnished the carrier in the reconsignment or diversion order with a notice of agency and the proper name and address of the beneficial owner, and where such shipments are refused or abandoned at ultimate destination, the said beneficial owner shall be liable for all legally applicable charges in connection therewith, If the reconsignor or diverter has given to the carrier erroneous information as to who the beneficial owner is, such reconsignor or diverter shall himself be liable for all such charges.

If a shipper or consignor of a shipment of property (other than a prepaid shipment) is also the consignee named in the bill of lading and, prior to the time of delivery, notifies, in writing, a delivering carrier by railroad (a) to delivery such property at destination to another party, (b) that such party is the beneficial owner of such property, and (c) that delivery is to be made to such party only upon payment of all transportation charges in respect of the transportation of such property, and delivery is made by the carrier to such party without such payment, such shipper or consignor shall not be liable (as shipper, consignor, consignee, or otherwise) for such transportation charges but the party to whom delivery is so made shall in any event be liable for transportation charges billed against the property at the time of such delivery, and also for any additional charges which may be found to be due after delivery of the property, except that if such party prior to such delivery has notified in writing the delivering carrier that he is not the beneficial owner of the property, and has given in writing to such delivering carrier the name and address of such beneficial owner, such party shall not be liable for any additional charges which may be found to be due after delivery of the property; but if the party to whom delivery is made has given to the carrier erroneous information as to the beneficial owner, such party shall nevertheless be liable for such additional charges. If the shipper or consignor has given to the delivering carrier erroneous information as to who the beneficial owner is, such shipper or consignor shall himself be liable for such transportation charges, notwithstanding the foregoing provisions of this paragraph and irrespective of any provisions to the contrary in the bill of lading or in the contract of transportation under which the shipment was made. The term "delivering carrier" means the line-haul carrier making ultimate delivery.

Nothing herein shall limit the right of the carrier to require at time of shipment the prepayment or guarantee of the charges. If upon inspection it is ascertained that the articles shipped are not those described in this bill of lading, the freight charges must be paid upon the aritcles actually shipped.

Where delivery is made by a common carrier by water the foregoing provisions of this action shall apply, except as may be inconsistent with Part III of the Interstate Commerce Act.

Sec. 8. If this bill of lading is issued on the order of the shipper, or his agent, in exchange or in substitution for another bill of lading, the shipper's signature to the prior bill of lading as to the statement of value or otherwise, or election of common law or bill of lading liability, in or in connection with such prior bill of lading shall be considered a part of this bill of lading as fully as if the same were written or made in or in connection with this bill of lading.

Sec. 9. (a) If all or any part of said property is carried by water over any part of said route, and loss, damage or injury to said property occurs while the same is in the custody of a carrier by water, the liability of such carrier shall be determined by the bill of lading of the carrier by water (this bill of lading being such bill of lading if the property is transported by such water carrier thereunder) and by and under the laws and regulations applicable to transportation by water. Such water carriage shall be performed subject to all the terms and provisions of, and all the exemptions from liability contained in the Act of Congress of the United States, approved on February 13, 1893, and entitled "An act relating to the navigation of vessels, etc.," and of other statutes of the United States according carriers by water the protection of limited liability, as well as the following subdivisions of this section; and to the conditions contained in this bill of

RAILROAD BILL OF LADING

lading not inconsistent with this section, when this bill of lading becomes the bill of lading of the carrier by water.

(b) No such carrier by water shall be liable for any loss or damage resulting from any fire happening to or on board the vessel, or from explosion, bursting of boilers or breakage of shafts, unless caused by the design or neglect of such carrier.

(c) If the owner shall have exercised due diligence in making the vessel in all respects seaworthy and properly manned, equipped, and supplied, no such carrier shall be liable for any loss or damage resulting from the perils of the lakes, seas, or other waters, or from latent defects in hull, machinery, or appurtenances whether existing prior to, at the time of, or after sailing, or from collision, stranding, or other accidents of navigation, or from prolongation of the voyage. And, when for any reason it is necessary, any vessel carrying any or all of the property herein described shall be at liberty to call at any port or ports, in or out of the customary route, to tow and be towed, to transfer, trans-ship, or lighter, to load and discharge goods at any time, to assist vessels in distress, to deviate for the purpose of saving life or property, and for docking and repairs. Except in case of negligence such carrier shall not be responsible for any loss or damage to property if it be necessary or is usual to carry the same upon deck.

(d) General Average shall be payable according to the York-Antwerp Rules of 1924, Sections 1 to 15 inclusive, and Sections 17 to 22, inclusive, and as to matters not covered thereby according to the laws and usages of the Port of New York. If the owners shall have exercised due diligence to make the vessel in all respects seaworthy and properly manned, equipped and supplied, it is hereby agreed that in case of danger, damage or disaster resulting from faults or errors in navigation, or in the management of the vessel, or from any latent or other defects in the vessel, her machinery or appurtenances, or from unseaworthiness, whether existing at the time of shipment or at the beginning of the boyage (provided the latent or other defects or the unseaworthiness was not discoverable by the exercise of due diligence), the shippers, consignees and/or owners of the cargo shall nevertheless pay salvage and any special charges incurred in respect of the cargo, and shall contribute with the shipowner in general average to the payment of any sacrifices, losses or expenses of a general average nature that may be made or incurred for the common benefit or to relieve the adventure from any common peril.

(e) If the property is being carried under a tariff which provides that any carrier or carriers party thereto shall be liable for loss from perils of the sea, then as to such carrier or carriers the provisions of this section shall be modified in accordance with the tariff provisions, which shall be regarded as incorporated into the conditions of this bill of lading.

INDUSTRIAL TRAFFIC MANAGEMENT

(f) The term "water carriage" in this section shall not be construed as including lighterage in or across rivers, harbors, or lakes, when performed by or on behalf of rail carriers.

Sec. 10. Any alteration, addition, or erasure in this bill of lading which shall be made without the special notation hereon of the agent of the carrier issuing this bill of lading, shall be without effect, and this bill of lading shall be enforceable according to its original tenor.

Appendix 6

Ocean Bill of Lading Contract Terms and Conditions

(These contract terms and conditions are typical; the various steamship lines do not necessarily use identical terms and conditions; some of them use short forms which adopt their long forms by reference, as do the railroads and others.)

Received from the Shipper hereinafter named, the goods or packages said to contain goods hereinafter mentioned, in apparent good order and condition, unless otherwise indicated in this bill of lading, to be transported subject to all terms of this bill of lading with liberty to proceed via any port or ports within the scope of the voyage described herein, to the port of discharge or so near thereunto as the ship can always safely get and leave, always afloat at all stages and conditions of water and weather, and there to be delivered or transshipped on payment of the charges thereon. If the goods in whole or in part are shut out from the ship named herein for any cause, the Carrier shall have liberty to forward them under the terms of this bill of lading on the next available ship of this line or substitute therefor, or at the carrier's option, of any line.

It is agreed that the custody and carriage of the goods are subject to the following terms on the face and back hereof which shall govern the relations, whatsoever they may be, between the shipper, consignee, and the Carrier, Master and ship in every contingency, wheresoever and whensoever occurring, and also in the event of deviation, or of unseaworthiness of the ship at the time of loading or inception of the voyage or subsequently, and none of the terms of this bill of lading shall be deemed to have been waived by the Carrier unless by express waiver signed by a duly authorized agent of the Carrier:

1. This bill of lading shall have effect subject to the provisions of the Carriage of Goods by Sea Act of the United States of America, effective July 15, 1936, which shall be deemed to be incorporated herein, and nothing herein contained shall be deemed a surrender by the Carrier of any of its rights or immunities or an increase of any of its responsibilities or liabilities

under said Act. The provisions stated in said Act (except as may be otherwise specifically provided herein) shall govern before the goods are loaded on and after they are discharged from the ship and throughout the entire time the goods are in the custody of the Carrier. The Carrier shall not be liable in any capacity whatsoever for any delay, non-delivery or misdelivery, or loss of or damage to the goods occurring while the goods are not in the actual custody of the carrier. If this bill of lading is issued in a locality where there is in force a Carriage of Goods by Sea Act or Ordinance or Statute of a similar nature to the International Convention for the Unification of Certain Rules Relating to Bills of Lading at Brussels of August 25, 1924, it is subject to the provisions stated in such Act, Ordinance and rules thereto annexed which may be in effect where this bill of lading is issued.

(a) The Carrier shall be entitled to the full benefit of, and right to, all limitations of, or exemptions from, liability authorized by any provisions of Sections 4281 to 4286 of the Revised Statutes of the United States and amendments thereto and of any other provisions of the laws of the United States or of any other country whose laws shall apply. If the ship is not owned by or chartered by demise to the Federal Maritime Commission or the Company designated herein (as may be the case notwithstanding anything that appears to the contrary) this bill of lading shall take effect only as a contract with the owner or demise charterer, as the case may be, as principal, made through the agency of the Federal Maritime Commission or the Company designated herein which acts as agent only and shall be under no personal liability whatsoever in respect thereof. If, however, it shall be adjudged that any other than the owner or demise charterer is carrier and/or bailee of the goods all limitations of and exonerations from liability provided by law or by the terms hereof shall be available to such other.

2. In this bill of lading the word "ship" shall include any substituted vessel, and any craft, lighter or other means of conveyance owned, chartered or operated by the Carrier used in the performance of this contract; the word "Carrier" shall include the ship, her owner, master, operator, demise charterer, and if bound hereby the time charterer, and any substituted carrier, whether the owner, operator, charterer, or master shall be acting as carrier or bailee; the word "shipper" shall include the person named as such in this bill of lading and the person for whose account the goods are shipped; the word "consignee" shall include the holder of the bill of lading, properly endorsed, and the receiver and the owner of the goods; the word "charges" shall include freight and all expenses and money obligations incurred and payable by the goods, shipper, consignee, or any of them.

3. The scope of voyage herein contracted for shall include usual or

OCEAN BILL OF LADING

customary or advertised ports of call whether named in this contract or not, also ports in or out of the advertised, geographical, usual or ordinary route or order, even though in proceeding thereto the ship may sail beyond the port of discharge or in a direction contrary thereto or return to the original port, or depart from the direct or customary route, and includes all canals, straits and other waters. The ship may call at any port for the purpose of the current voyage or of a prior or subsequent voyage. The ship may omit calling at any port or ports whether scheduled or not, and may call at the same port more than once; may for matters occurring before loading the goods, known or unknown at the time of such loading and matters occurring after such loading, either with or without the goods or passengers on board, and before or after proceeding toward the port of discharge, adjust compasses, dry dock, with or without cargo aboard go on ways or to repair yards, shift berths, make trial trips or tests, take fuel or stores, remain in port, sail with or without pilots, tow and be towed, and save or attempt to save life or property; and all of the foregoing are included in the contract voyage.

In view of the necessity for the expeditious employment of all the available Merchant Marine, the exercise by the carrier or master or any of the liberties granted herein with respect to loading, departure, scope of voyage, arrival routes, ports of call, stopping, discharge, destination, surrender, delivery, or otherwise, shall be presumed to be for the purposes of conserving and utilizing war time, sea mileage or shipping space, and therefore prima facie reasonable and necessary in the assembling, transportation or distribution of materials essential to the war effort.

4. In any situation whatsoever and wheresoever occurring and whether existing or anticipated before commencement of or during the voyage, which in the judgment of the Carrier or the Master is likely to give rise to risk of capture, seizure, detention, damage, delay or disadvantage to or loss of the ship or any part of her cargo, to make it unsafe, imprudent, or unlawful for any reason to commence or proceed on or continue the voyage or to enter or discharge the goods at the port of discharge, or to give rise to delay or difficulty in arriving, discharging at or leaving the port of discharge or the usual or agreed place of discharge in such port, the Carrier may before loading or before the commencement of the voyage, require the shipper or other person entitled thereto to take delivery of the goods at port of shipment and upon failure to do so, may warehouse the goods at the risk and expense of the goods; or the Carrier or the Master, whether or not proceeding toward or entering or attempting to enter the port of discharge or reaching or attempting to reach the usual place of discharge therein or attempting to discharge the goods there, may discharge the goods into depot, lazaretto, craft, or other place; or the ship may proceed or return, directly or indirectly, to or stop at any port or

place whatsoever as the Master or the Carrier may consider safe or advisable under the circumstances, and discharge the goods, or any part thereof, at any such port or place; or the Carrier or the Master may retain the cargo on board until the return trip or until such time as the Carrier or the Master thinks advisable and discharge the goods at any place whatsoever as herein provided; or the Carrier or the Master may discharge and forward the goods by any means, rail, water, land, or air at the risk and expense of the goods. The Carrier or the Master is not required to give notice of discharge of the goods or the forwarding thereof as herein provided. When the goods are discharged from the ship, as herein provided, they shall be at their own risk and expense; such discharge shall constitute complete delivery and performance under this contract and the Carrier shall be freed from any further responsibility. For any services rendered to the goods as hereinabove provided, the Carrier shall be freed from any further responsibility. For any services rendered to the goods as hereinabove provided, the Carrier shall be entitled to a reasonable extra compensation.

5. The Carrier, Master and ship shall have liberty to comply with any orders or directions as to loading, departing, arrival, routes, ports of call, stoppages, discharge, destination, delivery or otherwise howsoever given by the government of any nation or department thereof or any person acting or purporting to act with the authority of such government or of any department thereof, or by any committee or person having, under the terms of the war risk insurance on the ship, the right to give such orders or directions. Delivery or other disposition of the goods in accordance with such orders or directions shall be a fulfillment of the contract voyage. The ship may carry contraband, explosives, munitions, warlike stores, hazardous cargo, and may sail armed or unarmed and with or without convoy.

In addition to all other liberties herein the Carrier shall have the right to withhold delivery of, reship to, deposit or discharge the goods at any place whatsoever, surrender or dispose of the goods in accordance with any direction, condition or agreement imposed upon or exacted from the carrier by any government or department thereof or any person purporting to act with the authority of either of them. In any of the above circumstances the goods shall be solely at their risk and expense and all expenses and charges so incurred shall be payable by the owner or consignee thereof and shall be a lien on the goods.

6. Unless otherwise stated herein, the description of the goods and the particulars of the packages mentioned herein are those furnished in writing by the shipper and the Carrier shall not be concluded as to the correctness of marks, number, quantity, weight, gauge, measurement, contents, nature, quality of value. Sincle pieces or packages exceeding 4,480 lbs. in weight shall be liable to pay extra charges in accordance with tariff rates in effect at time of shipment for loading, handling, transshipping or

OCEAN BILL OF LADING

discharging and the weight of each such piece or package shall be declared in writing by the shipper on shipment and clearly and durably marked on the outside of the piece or package. The shipper and the goods shall also be liable for, and shall indemnify the Carrier in respect of, any injury, loss or damage arising from shipper's failure to declare and mark the weight of any such piece or package or from inadequate or improper description of the goods or from the incorrect weight of any such piece or package or from inadequate or improper description of the goods or from the incorrect weight of any such piece or package having been declared or marked thereon, or from failure fully to disclose the nature and character of the goods.

7. Goods may be stowed in poop, forecastle, deck house, shelter deck, passenger space or any other covered in space commonly used in the trade and suitable for the carriage of goods, and when so stowed shall be deemed for all purposes to be stowed under deck. In respect of goods carried on deck, all risks of loss or damage by perils inherent in such carriage shall be borne by the shipper or the consignee but in all other respects the custody and carriage of such goods shall be governed by the terms of this bill of lading and the provisions stated in said Carriage of Goods by Sea Act notwithstanding Sec. 1 (c) thereof, or the corresponding provision of any Carriage of Goods by Sea Act that may be applicable. Specially heated or specially cooled stowage is not to be furnished unless contracted for at increased freight rate. Goods or articles carried in such compartment are at the sole risk of the owner thereof and subject to all the conditions, exceptions and limitations as to the Carrier's liability and other provisions of this bill of lading; and further the Carrier shall not be liable for any loss or damage occasioned by the temperature, risks or refrigeration, defects or insufficiency in or accidents to or explosion, breakage, derangement or failure of any refrigerator plant or part thereof, or by or in any material or the supply or use thereof used in the process of refrigeration unless shown to have been caused by negligence of the Carrier from liability for which the Carrier is not by law entitled to exemption.

8. Live animals, birds, reptiles and fish are received and carried at shipper's risk of accident or mortality, and the Carrier shall not be liable for any loss or damage thereto arising or resulting from any matter mentioned in Section 4, Sub-section 2, a to p, inclusive of said Carriage of Goods by Sea Act or similar sections of any Carriage of Goods by Sea Act that may be applicable, or from any other cause whatsoever not due to the fault of the Carrier, any warranty of seaworthiness in the premises being waived by the shipper. Except as provided above such shipments shall be deemed goods, and shall be subject to all terms and provisions in this bill of lading relating to goods.

9. If the ship comes into collision with another ship as a result of the negligence of the other ship and any act, neglect or default of the Master,

INDUSTRIAL TRAFFIC MANAGEMENT

mariner, pilot or the servants of the Carrier in the navigation or in the management of the ship, the owners of the goods carried hereunder will indemnify the carrier against all loss or liability to the other or non-carrying ship or her owners in so far as such loss or liability represents loss of, or damage to, or any claim whatsoever of the owners of said goods, paid or payable by the other or non-carrying ship or her owners to the owners of said goods and set-off, recouped or recovered by the other or non-carrying ship or her owners as part of their claim against the carrying ship or Carrier.

The foregoing provision shall also apply where the owners, operators or those in charge of any ship or ships or objects other than, or in addition to, the colliding ships or objects are at fault in respect of a collision or contact.

10. General average shall be adjusted, stated and settled, according to York-Antwerp Rules 1950, except Rule XXII thereof, at such port or place in the United States as may be selected by the carrier, and as to matters not provided for by these Rules, according to the laws and usages at the port of New York. In such adjustment disbursements in foreign currencies shall be exchanged into United States money at the rate prevailing on the dates made and allowances for damage to cargo claimed in foreign currency shall be converted at the rate prevailing on the last day of discharge at the port or place of final discharge of such damaged cargo from the ship. Average agreement or bond and such additional security, as may be required by the Carrier, must be furnished before delivery of the goods. Such cash deposit as the Carrier or his agents may deem sufficient as additional security for the contribution of the goods and for any salvage and special charges thereon, shall, if required, be made by the goods, shippers, consignees or owners of the goods to the carrier before delivery. Such deposit shall, at the option of the Carrier, be payable in United States money and be remitted to the adjuster. When so remitted the deposit shall be held in a special account at the place of adjustment in the name of the adjuster pending settlement of the general average and refunds or credit balances, if any, shall be paid in United States money.

In the event of accident, danger, damage, or disaster, before or after commencement of the voyage resulting from any cause whatsoever, whether due to negligence or not, for which, or for the consequence of which, the Carrier is not responsible, by statute, contract, or otherwise, the goods, the shipper and the consignee, jointly and severally, shall contribute with the Carrier in general average to the payment of any sacrifices, losses, or expenses of a general average nature that may be made or incurred, and shall pay salvage and special charges incurred in respect of the goods. If a salving ship is owned or operated by the Carrier, salvage shall be paid for as fully and in the same manner as if such salving ship or ships belonged to strangers.

11. Whenever the Carrier or the Master may deem it advisable or in any

OCEAN BILL OF LADING

case where the goods are consigned to a point where the ship does not expect to discharge, the Carrier or Master may, without notice, forward the whole or any part of the goods before or after loading at the original port of shipment, or any other place or places even though outside the scope of the voyage or the route to or beyond the port of discharge or the destination of the goods, by any vessel, vessels or other means of transportation by water or by land or by air or by any such means, whether operated by the Carrier or by others and whether departing or arriving or scheduled to depart or arrive before or after the ship expected to be used for the transportation of the goods. This Carrier, in making arrangements for any transshipping or forwarding vessel or means of transportation not operated by this Carrier shall be considered solely the forwarding agent of the shipper and without any other responsibility whatsoever.

The carriage by any transshipping or forwarding carrier and all transshipment or forwarding shall be subject to all the terms whatsoever in the regular form or bill of lading, freight note, contract or other shipping document used at the time by such carrier, whether issued for the goods or not, and even though such terms may be less favorable to the shipper or consignee than the terms of this bill of lading and may contain more stringent requirements as to notice of claim or commencement of suit and may exempt the on-carrier from liability for negligence. The shipper expressly authorizes the Carrier to arrange with any such transshipping or forwarding carrier that the lowest valuation of the goods or limitation of liability contained in the bill of lading or shipping document of such carrier shall apply even though lower than the valuation or limitation herein, provided that the shipper shall not be compelled to pay a rate higher than that applicable to the valuation contained in such bill of lading. Pending or during transshipment the goods may be stored ashore or afloat at their risk and expense and the Carrier shall not be liable for detention.

12. The port authorities are hereby authorized to grant a general order for discharging immediately upon arrival of the ship and the Carrier without giving notice either of arrival or discharge, may discharge the goods directly they come to hand, at or onto any wharf, craft or place that the Carrier may select, and continuously, Sundays and holidays included, at all such hours by day or by night as the Carrier may determine no matter what the state of the weather or custom of the port may be. The Carrier shall not be liable in any respect whatsoever if heat or refrigeration or special cooling facilities shall not be furnished during loading or discharge or any part of the time that the goods are upon the wharf, craft, or other loading or discharging place. All lighterage and use of craft in discharging shall be at the risk and expense of the goods. Landing and delivery charges and pier dues shall be at the expense of the goods unless included in the freight herein provided for. If the goods are not taken away by the consignee by the expiration of the next working day after the goods are at his

disposal, the goods may at Carrier's option and subject to Carrier's lien, be sent to store or warehouse or be permitted to lie where landed, but always at the expense and risk of the goods. The responsibility of the Carrier in any capacity shall altogether cease and the goods shall be considered to be delivered and at their own risk and expense in every respect when taken into the custody of customs or other authorities. The Carrier shall not be required to give any notification of disposition of the goods.

13. The Carrier shall not be liable for failure to deliver in accordance with marks unless such marks shall have been clearly and durably stamped or marked by the shipper before shipment upon the goods or packages, in letters and numbers not less than two inches high, together with name of the port of discharge. Goods that cannot be identified as to marks or numbers, cargo sweepings, liquid residue and any unclaimed goods not otherwise accounted for shall be allocated for completing delivery to the various consignees of goods of like character, in proportion to any apparent shortage, loss of weight or damage. Loss or damage to goods in bulk stowed without separation from other goods in bulk of like quality, shipped by either the same or another shipper, shall be divided in proportion among the several shipments.

14. The goods shall be liable for all expense of mending, cooperage, baling or reconditioning of the goods or packages and gathering of loose contents of packages; also for any payment, expense, fine, dues, duty, tax, impost, loss, damage or detention sustained or incurred by or levied upon the Carrier or the ship in connection with the goods, howsoever caused, including any action or requirement of any government or governmental authority or person purporting to act under the authority thereof, seizure under legal process or attempted seizure, incorrect or insufficient marking, numbering or addressing of packages or description of the contents, failure of the shipper to procure consular, Board of Health or other certificates to accompany the goods or to comply with laws or regulations of any kind imposed with respect to the goods by the authorities at any port or place or any act or omission of the shipper or consignee.

15. Freight shall be payable on actual gross intake weight or measurement or, at Carrier's option, on actual gross discharge weight or measurement. Freight may be calculated on the basis of the particulars of the goods furnished by the shipper herein but the Carrier may at any time open the packages and examine, weigh, measure and value the goods. In case shipper's particulars are found to be erroneous and additional freight is payable, the goods shall be liable for any expense incurred for examining, weighing, measuring and valuing the goods. Full freight shall be paid on damaged or unsound goods. Full freight hereunder to port of discharge named herein shall be considered completely earned on shipment whether the freight be stated or intended to be prepaid or to be collected at destination; and the Carrier shall be entitled to all freight and charges due here-

under, whether actually paid or not, and to receive and retain them irrevocably under all circumstances whatsoever ship and/or cargo lost or not lost or the voyage broken up or abandoned. If there shall be a forced interruption or abandonment of the voyage at the port of shipment or elsewhere, any forwarding of the goods or any part thereof shall be at the risk and expense of the goods. All unpaid charges shall be paid in full and without any offset, counterclaim or deduction in the currency of the port of shipment, or, at Carrier's option, in the currency of the port of discharge at the demand rate of New York exchange as quoted on the day of the ship's entry at the Custom House of her port of discharge. The Carrier shall have a lien on the goods, which shall survive delivery, for all charges due hereunder and may enforce this lien by public or private sale and without notice. The shipper and the consignee shall be jointly and severally liable to the Carrier for the payment of all charges and for the performance of the obligation of each of them hereunder.

16. Neither the Carrier nor any corporation owned by, subsidiary to or associated or affiliated with the Carrier shall be liable to answer for or make good any loss or damage to the goods occurring at any time and even though before loading on or after discharge from the ship, by reason or by means of any fire whatsoever, unless such fire shall be caused by its design or neglect.

17. In case of any loss or damage to or in connection with goods exceeding in actual value $500 lawful money of the United States, per package, or, in case of goods not shipped in packages, per customary freight unit, the value of the goods shall be deemed to be $500 per package or per unit, on which basis the freight is adjusted and the Carrier's liability, if any, shall be determined on the basis of a value of $500 per package or per customary freight unit, or pro rata in case of partial loss or damage, unless the nature of the goods and a valuation higher than $500 shall have been declared in writing by the shipper upon delivery to the Carrier and inserted in this bill of lading and extra freight paid if required and in such case if the actual value of the goods per package or per customary freight unit shall exceed such declared value, the value shall nevertheless be deemed to be the declared value and the Carrier's liability, if any, shall not exceed the declared value and any partial loss or damage shall be adjusted pro rata on the basis of such declared value.

Whenever the value of the goods is less than $500 per package or other freight unit, their value in the calculation and adjustment of claim for which the Carrier may be liable shall for the purpose of avoiding uncertainties and difficulties in fixing value be deemed to be the invoice value, plus freight and insurance if paid, irrespective of whether any other value is greater or less.

18. Unless notice of loss or damage and the general nature of such loss or damage be given in writing to the Carrier or his agent at the port of

discharge before or at the time of the removal of the goods into the custody of the person entitled to delivery thereof under the contract of carriage, such removal shall be prima facie evidence of the delivery by the Carrier of the goods as described in the bill of lading. If the loss or damage is not apparent, the notice must be given within three days of the delivery. The Carrier shall not be liable upon any claim for loss or damage unless written particulars of such claim shall be received by the Carrier within thirty days after receipt of the notice herein provided for.

19. In any event the Carrier and the ship shall be discharged from all liability in respect of loss or damage unless suit is brought within one year after the delivery of the goods or the date when the goods should have been delivered. Suit shall not be deemed brought until jurisdiction shall have been obtained over the Carrier and/or the ship by service of process or by an agreement to appear.

20. To avoid or alleviate preventions or delays in prosecution or completion of the voyage incident to the existence of hostilities, the Carrier has liberty and is authorized by the shipper and the owner of the goods to agree with the representatives of any government to submit the goods to examination at any place or places whatsoever and to delay delivery of the same until any restriction asserted by any governmental authority shall have been removed. The Carrier may put the goods in store ashore or afloat at the risk and expense of the owner of the same pending examination; and thereupon the Carrier's responsibility shall end. Any damage or deterioration occasioned by such examination or by delay and other risks of whatsoever nature shall be solely for account of the owner of the goods. All expenses incurred by the Carrier in relation to such detention of the goods shall be paid by the shipper or consignee or owner of the goods.

21. This bill of lading shall be construed and the rights of the parties thereunder determined according to the law of the United States.

22. Cargo skids and labor on quay are to be provided by ship's agent for account of consignee at current rates, and any cargo which may be ordered for delivery into fiscal deposits, must be taken by an official cartman appointed by the agent of the ship, at current rates for account and risk of consignee.

23. If any bagged or baled goods are landed slack or torn, receiver and/or consignee shall accept its proportion of the sweepings. Ship not responsible for loss of weight in bags or bales torn, mended or with sample holes.

24. *Cotton.* Description of the condition of the cotton does not relate to the insufficiency of or torn condition of the covering, nor to any damage resulting therefrom, and Carrier shall not be responsible for damage of such nature.

25. *Specie.* Specie will not be shipped or landed by the Carrier; it must be put on board by the shipper, and will only be delivered on board on

OCEAN BILL OF LADING

presentation of the bills of lading properly endorsed; it may be carried on at consignee's risk if delivery is not taken during the ship's stay in port, and in every case the liability of the Carrier shall cease when the specie leaves the ship's deck.

26. *Specified Dock Discharge:* If the carrier makes a special agreement, whether by stamp hereon or otherwise, to deliver the goods hereby receipted for at a specified dock or wharf at the port of discharge, it is mutually agreed that such agreement shall be construed to mean that the Carrier is to make such delivery only if, in the sole judgment of the Master, the ship can safely under her own power, proceed to, lie at, and return from said dock or wharf, always afloat at any time of tide, and only if such dock or wharf is available to the ship immediately the ship is ready to discharge the goods and, that otherwise, the ship shall discharge the goods in accordance with Clause 12 of this bill of lading, whereupon Carrier's responsibility shall cease.

27. All agreements of freight engagements for the shipment of the goods are superseded by this bill of lading, and all its terms, whether written, typed, stamped, or printed, are accepted and agreed by the shipper to be binding as fully as if signed by the shipper, any local customs or privileges to the contrary notwithstanding. Nothing in this bill of lading shall operate to limit or deprive the Carrier of any statutory protection or exemption from or limitation of liability. If required by the Carrier, one signed bill of lading duly endorsed must be surrendered to the agent of the ship at the port of discharge in exchange for delivery order.

Appendix 7

A Selection of Traffic Publications

BASIC PUBLICATIONS ESSENTIAL TO ALL TRAFFIC FUNCTIONS

In drawing up a list of this kind, some worthwhile traffic publication are inevitably (though unintentionally) omitted. Subscription prices for those publications that offer them are often lower for two-or-more-year than for single-year subscriptions. Foreign subscriptions are usually available upon request to the publishers, at an increased cost. The subscription rates shown here were correct at the time of publishing, but are, of course, subject to change.

Many of the publication listed are widely advertised and information on subscriptions, etc., is readily available from their advertising departments.

A world of caution: the simple fact that something is published does not mean that it should be uncritically accepted, without further thought or attention to what is being said.

TRAFFIC PUBLICATIONS

Publication	Address	Subscription Price
Association of American Railroads' various publications	Secretary, Assoc. of Amer. RRs., 59 E. Van Buren St., Chicago, Ill. 60605	Write for price list of publications of AAR. It covers operations, research, management and economics of RRs.
Bullinger's Postal and Shippers Guide (issued annually)	Bullinger's Guides, Inc. 63 Woodland Ave., Westwood, N.J. 07675	$75.00 per annual rental copy
Coordinated Freight Classification (affects New England Territory only) issued approximately every 2 years	New England Motor Rate Bureau 14 New England Executive Park Burlington, Mass. 01803	$18.00 per subscription
National Classification Board Docket (Proposals for changes in rules, descriptions, ratings and minimum weights) issued five times per year	National Classification Board, 1616 P Street, N.W., Washington, D.C. 20036	Published periodically in the Traffic Bulletin
National Motor Freight Classification No. 100F or reissues (reissued occasionally, and supplemented regularly)	American Trucking Associations, Inc. 1616 P Street, N.W. Washington, D.C. 20036	$19.45 for each issue and supplements
Official Guide of the Railways (issued bi-monthly)	424 West 33rd St., New York, N.Y. 10001	$55.00 per annum
Official Intermodal Equipment Register	424 West 33rd St., New York, N.Y. 10001	(issued quarterly) $20.00 per annum
Official List of Open and Pre-Pay Stations (reissued annually on November 1st)	A.P. Leland, Tariff Publishing Officer Station List Publishing Co. 827 Syndicate Trust Bldg. 915 Olive St. St. Louis, Mo. 63101	$26.00 per annum (includes November 1 reissue and supplements on first and fifteenth of each month)
Official Railway Equipment Register	424 West 33rd St. New York, N.Y. 10001	(issued quarterly) $50.00 per annum
Pocket List of Railroad Officials, The (issued quarterly)	424 West 33rd St. New York, N.Y. 10001	$16.00 per annum

INDUSTRIAL TRAFFIC MANAGEMENT

Publication	Address	Subscription Price
Postal Services Manual Chapters 1-6	United States Government Printing Office Washington, D.C. 20402	$16.50 per copy
Uniform Classification Docket (proposals for changes in rules, descriptions, ratings and minimum weights)	Uniform Classification Committee G.F. Garl 222 So. Riverside Plaza Chicago, Ill. 60606	Published weekly in the Traffic Bulletin
Uniform Freight Classification (reissued occasionally and supplemented regularly)	Uniform Classification Committee G.F. Garl 222 So. Riverside Plaza Chicago, Ill. 60606	$18.50 for each issue and supplements

BASIC PUBLICATIONS REQUIRED FOR MANY TRAFFIC FUNCTIONS

Aviation Law Reports (loose leaf-kept currently up to date)	Commerce Clearing House 4025 W. Peterson Ave. Chicago, Ill. 60646	$505.00 per annum
Branham Automobiles Reference Book	Branham Publishing Company 1422 Beckwith St. Los Angeles, Calif. 90049	$6.35 per annum
Daily Traffic World	The Traffic Service Corp. 1435 G Street, N.W. Suite 815 Washington, D.C. 20005	$220.00 per annum
Decisions and Reports of the Interstate Commerce Commission	Superintendent of Documents Government Printing Office Washington, D.C. 20402	Vary from $9.60 to $13.75 per copy
Directory of Common Motor Carrier Agency Tariffs and Directory of Forwarding Company Tariffs (loose leaf—kept currently up-to-date)	Transportation Consulting & Service Corp. 4704 W. Irving Park Rd. Chicago, Ill. 60641	$12.00 and $7.50 respectively
Federal Carriers Reports (loose leaf—kept currently up-to-date)	Commerce Clearing House 4025 W. Peterson Ave. Chicago, Ill. 60646	$300.00 per annum
Federal Motor Carrier Safety Regulations	Private Carrier Conf., Inc. 1616 P St., N.W. Washington, D.C. 20036	$1.00 per copy (quantity discount)

TRAFFIC PUBLICATIONS

Publication	Address	Subscription Price
MVMA Motor Vehicle Facts & Figures (issued annually—July)	Motor Vehicle Manufacturers Association 300 New Center Bldg. Detroit, Mich. 48202	Single copies free on request
National Industrial Traffic League Circulars and the weekly newsletter, The Legislator (issued regularly)	Office of the Executive Vice President National Industrial Traffic League 1909 K St., N.W. Washington, D.C. 20006	Included in dues
Official Airline Guide No. American Quick Reference; International Quick Reference; Pocket Flight Guide; Travel Planner & Hotel/Motel Guide; and Air Cargo Guide	Reuben H. Donnelley Corp. 2000 Clearwater Bldg., Oak Brook, Ill. 60521	$30.00 to $92.00 per annum, dependent on edition
Official Directory of Industrial & Commercial Traffic Executives, The (issued annually)	The Traffic Service Corporation 1435 G St., N.W. Suite 815 Washington, D.C. 20005	$30.00 per copy
Railway Line Clearances	424 West 33rd St. New York, N.Y. 10001	(issued annually) $12.00 per copy
Rail Carrier, Motor/Forwarder, Civil Aeronautics Board and Federal Maritime Commission Services and Aviation Court Cases (loose leaf—kept currently up to date)	Hawkins Publishing Co. 933 N. Kenmore St. Arlington, Va. 22201 (703) 525-9090	Prices vary with services and range from $70.00 to $175.00 per year
Rand, McNally Commercial Atlas and Marketing Guide (issued annually)	Rand, McNally & Company Commercial Atlas Service P.O. Box 7600 Chicago, Ill. 60641	Write for price
Reciprocity Guide for Private Motor Carriers	Private Carrier Conf., Inc. 1616 P St., N.W. Washington, D.C. 20036	$15.00 - 1st copy $ 7.50 each additional copy
Shippers Advisory Board, Reports of Quarterly Meetings	Individual Shippers Advisory Boards	Free to members
State Motor Carrier Guide (loose leaf—kept currently up to date)	Commerce Clearing House 4025 W. Peterson Avenue Chicago, Ill. 60646	$340.00 per annum (see Federal Carriers Reports)

INDUSTRIAL TRAFFIC MANAGEMENT

Publication	Address	Subscription Price
Summary of Size and Weight Limits and Reciprocity Authority	Private Carrier Conf., Inc. 1616 P St., N.W. Washington, D.C. 20036	$1.00 each (quantity discount)
Traffic Bulletin (issued weekly)	The Traffic Service Corporation 1435 G Street, N.W. Suite 815 Washington, D.C. 20005	$140.00 per annum
What's Happening In Transportation and other bulletins	Transportation Association of America 1100 17th St., N.W. Washington, D.C. 20036	Free to members (write for new member subscription costs)

RATE OR ROUTING GUIDES

ATA American Motor Carrier Directory (official ATA publication)	Guide Services, Inc. P.O. Box No. 13446 Atlanta, Ga. 30324	$70.50 per annum (2 semi-annual issues) (incl two issues)
Illinois & Missouri Edition (issued annually—March)	Guide Services, Inc. P.O. Box No. 13446 Atlanta, Ga. 30324	$16.50 per annum
Leonard's Guides for Parcel Post, Express, Motor Carrier Rates, and Routing	G.R. Leonard & Co. 2121 Shermer Rd. Northbrook, Ill. 60062	Prices range from $20.00 to $140.00 for regular publications and are higher in special cases
Middle Atlantic Edition (issued annually—August)	Guide Services, Inc. P.O. Box No. 13446 Atlanta, Ga. 30324	$16.50 per annum
National Highway & Airway Carriers and Routes (issued semi-annually)	National Highway Carriers Directory, Inc. 936 So. Betty Dr. Buffalo Grove, Ill. 60090	$38.00 per year (2 semi-annual issues)
New England Edition (issued annually—September)	Guide Services, Inc. P.O. Box No. 13446 Atlanta, Ga. 30324	$16.50 per annum
Pacific States Directory (issued annually—February)	Guide Services, Inc. P.O. Box No. 13446 Atlanta, Ga. 30324	$16.50 per annum
Southeastern Edition (Issued annually—April)	Guide Services, Inc. P.O. Box 13446 Atlanta, Ga. 30324	$16.50 per annum

TRAFFIC PUBLICATIONS

Publication	Address	Subscription Price
Specialized Carrier Services (issued annually—January)	Guide Services, Inc. P.O. Box No. 13446 Atlanta, Ga. 30324	$16.50 per annum
Tariff Guide No. 11	The Traffic Service Corporation 1435 G Street, N.W. Suite 815 Washington, D.C. 20005	$6.95
Warehouse Directory (issued annually—July)	Guide Services, Inc. P.O. Box No. 13446 Atlanta, Ga. 30324	$16.50 per annum

PERIODICALS

Air Cargo (issued monthly)	887 7th Avenue New York, N.Y. 10019	Free of charge
Container News (issued monthly)	6285 Barfield Avenue Atlanta, Ga. 30328	Free of charge
Distribution Worldwide (issued monthly)	Chilton Way Radnor, Pa. 19089	$40.00 per annum
Federal Register (issued Tuesday thru Saturday)	Superintendent of Documents Government Printing Office Washington, D.C. 20402	$50.00 per year
Handling & Shipping (issued monthly)	1111 Chester Avenue Cleveland, Ohio 44114	Free of charge
ICC Practitioners' Journal (issued bi-monthly)	Association of ICC Practitioners 1112 ICC Building Washington, D.C. 20025	Free to members; $30.00 per annum to non-members
Railway Age (issued semi-monthly)	350 Broadway New York, N.Y. 10013	$10.00 per annum
Traffic Management (issued monthly)	221 Columbus Avenue Boston, Mass. 02116	Free of charge
Traffic World (issued weekly)	The Traffic Service Corporation 1435 G Street, N.W. Suite 815 Washington, D.C. 20005	$75.00 per annum

INDUSTRIAL TRAFFIC MANAGEMENT

Publication	Address	Subscription Price
Transport Topics (issued weekly)	American Trucking Associations, Inc. 1616 P Street, N.W. Washington, D.C. 20036	$30.00 per annum
Transportation 2000 (issued bi-monthly)	Transportation 2000 4 Maria Loretoct Novato, Calif. 94947	$11.00 per annum
Warehousing and Physical Distribution Productivity Report (issued monthly)	221 National Press Building Washington, D.C. 20004	$35.00 per annum

PACKING & SHIPPING

Publication	Address	Subscription Price
Material Handling Engineering (issued monthly)	1111 Chester Ave. Cleveland, Ohio 44114	Free of Charge
Modern Materials Handling (issued monthly)	221 Columbus Ave. Boston, Mass. 02116	$8.00 per annum
National Defense Transportation Journal (issued bi-monthly)	1612 K Street, N.W. Washington, D.C.	Free to members $10.00 per annum to non-members
Private Carrier, The (issued semi-monthly)	The Private Carrier Conference American Trucking Associations, Inc. 1616 P Street, N.W. Washington, D.C. 20036	Free to Conference members
Transportation Journal (issued quarterly)	American Society of Traffic and Transportation, Inc. 22 West Madison St. Chicago, Ill. 60602	Free to members

EXPORT AND IMPORT INFORMATION

Publication	Address	Subscription Price
American Import-Export Bulletin (includes list of licensed foreign freight forwarders and customs house brokers) (issued monthly)	North American Publishing Co. 401 N. Broad St. Philadelphia, Pa. 19108	$15.00 per annum (domestic) Free with Custom House Guide

TRAFFIC PUBLICATIONS

Publication	Address	Subscription Price
Brandon's Shipper & Forwarder (issued weekly)	Brandon Publications, Suite 1927, One World Trade Center New York, N.Y. 10048 (212) 423-0750	$7.50 per annum (domestic) $13.50 (abroad)
Custom House Guide (issued annually)	North American Publishing Co. 401 N. Broad St. Philadelphia, Pa. 19108	$95.00 per copy (plus postage and binder)
Maritime Reporter & Engineering News (issued bi-monthly)	107 East 31st Street New York, N.Y. 10016	controlled circulation to qualified recipients
Export Document and Shipment Preparation by Durward L. Brooks	The Traffic Service Corporation 1435 G Street, N.W. Suite 815 Washington, D.C. 20005	Write for price

MISCELLANEOUS TEXTS AND PUBLICATIONS

ICC Rules of Practice	Rules Service Co. 4341 Montgomery Avenue Bethesda, Md. 20014	Loose-leaf $14.00 per annum
Law of Freight Loss and Damage Claims (Miller's) by Richard R. Sigmon	William C. Brown Company Dubuque, Iowa	429 pages 4th Ed., 1974
Model Legal Forms for Shippers by Stanley Hoffman	The Traffic Service Corporation 1435 G Street, N.W. Suite 815 Washington, D.C. 20005	508 pages, 1970 $12.00
Practice and Procedure Before Rate-making Associations by G.E. Lowe	The Traffic Service Corporation 1435 G Street, N.W. Suite 815 Washington, D.C. 20005	62 pages 3rd Ed., 1967, $3.75
Red Book of Hazardous Materials by Lawrence W. Bierlein	Cahners Books International 221 Columbus Avenue Boston, Mass. 02116	$65.00
Revised Interstate Commerce Act (published November 1978)	West Publishing Co. 50 W. Kellogg Blvd. St. Paul, Minn. 55165	307 pages plus the Act Write for price

INDUSTRIAL TRAFFIC MANAGEMENT

Publication	Address	Subscription Price
Transportation Regulation by Marvin L. Fair & John Guandolo	William C. Brown Company Dubuque, Iowa	595 pages 7th Ed., 1972
Traffic World's Question & Answers Book	The Traffic Service Corporation 1435 G Street, N.W. Suite 815 Washington, D.C. 20005	issued periodically

Appendix 8

Public Law 93-633
93rd Congress, H.R. 15223
January 3, 1975

An Act

To regulate commerce by improving the protections afforded the public against risks connected with the transportation of hazardous materials, and for other purposes.

Be it enacted by the Senate and House of Representatives of the United States of America in Congress assembled, That this act may be cited as the "Transportation Safety Act of 1974."

Title I—Hazardous Materials

Short Title

Sec. 101. This title may be cited as the "Hazardous Materials Transportation Act."

Declaration of Policy

Sec. 102. It is declared to be the policy of Congress in this title to improve the regulatory and enforcement authority of the Secretary of Transportation to protect the Nation adequately against the risks to life and property which are inherent in the transportation of hazardous materials in commerce.

Definitions

Sec. 103. As used in this title, the term—
  (1) "commerce" means trade, traffic, commerce, or transportation, within the jurisdiction of the United States, (A) between a place in a State and any place outside of such State, or (B)

which affects trade, traffic, commerce, or transportation described in clause (A);

(2) "hazardous material" means a substance or material in a quantity and form which may pose an unreasonable risk to health and safety or property when transported in commerce;

(3) "Secretary" means the Secretary of Transportation, or his delegate;

(4) "serious harm" means death, serious illness, or severe personal injury;

(5) "State" means a State of the United States, the District of Columbia, the Commonwealth of Puerto Rico, the Virgin Islands, American Samoa, or Guam;

(6) "transports" or "transportation" means any movement of property by any mode, and any loading, unloading, or storage incidental thereto; and

(7) "United States" means all of the States.

Designation of Hazardous Materials

Sec. 104. Upon a finding by the Secretary, in his discretion, that the transportation of a particular quantity and form of material in commerce may pose an unreasonable risk to health and safety or property, he shall designate such quantity and form of material or group or class of such materials as a hazardous material. The materials so designated may include, but are not limited to, explosives, radioactive materials, etiologic agents, flammable liquids or solids, combustible liquids or solids, poisons, oxidizing or corrosive materials, and compressed gases.

Regulations Governing Transportation of Hazardous Materials

Sec. 105. (a) General—The Secretary may issue, in accordance with the provisions of section 553 of title 5, United States Code, including an opportunity for informal oral presentation, regulations for the safe transportation in commerce of hazardous materials. Such regulations shall be applicable to any person who transports, or causes to be transported or shipped, a hazardous material, or who manufactures, fabricates, marks, maintains, reconditions, repairs, or tests a package or container which is represented, marked, certified, or sold by such person for use in the transportation in commerce of certain hazardous materials. Such regulations may govern any safety aspect of the transportation of hazardous materials which the Secretary deems necessary or appropriate, including, but not limited to, the packing, repacking, handling, labeling, marking, pacarding, and routing (other than with respect to pipelines) of hazardous materials, and the manufacture, fabrication, marking, maintenance, reconditioning, repairing, or testing of a package or container which is represented, marked, certified, or sold by such person for use in the transportation of certain hazardous materials.

(b) Cooperation—In addition to other applicable requirements, the Secretary shall consult and cooperate with representatives of the Interstate Commerce Commission and shall consider any relevant suggestions made by such Commission, before issuing any regulation with respect to the routing of hazardous materials. Such Commission shall, to the extent of its lawful authority, take such action as is necessary or appropriate to implement any such regulation.

(c) Representation—No person shall, by marking or otherwise, represent that a container or package for the transportation of hazardous materials is safe, certified, or in compliance with the requirements of this Act, unless it meets the requirements of all applicable regulations issued under this Act.

Handling of Hazardous Materials

Sec. 106. (a) Criteria.—The Secretary is authorized to establish criteria for handling hazardous materials. Such criteria may include, but need not be limited to, a minimum number of personnel; a minimum level of training and qualification for such personnel; type and frequency of inspection; equipment to be used for detection, warning, and control of risks posed by such materials; specifications regarding the use of equipment and facilities used in the handling and transportation of such materials; and a system of monitoring safety assurance procedures for the transportation of such materials. The Secretary may revise such criteria as required.

(b) Registration—Each person who transports or causes to be transported or shipped in commerce hazardous materials or who manufactures, fabricates, marks, maintains, reconditions, repairs, or tests packages or containers which are represented, marked, certified, or sold by such person for use in the transportation in commerce of certain hazardous materials (designated by the Secretary) may be required by the Secretary to prepare and submit to the Secretary a registration statement not more often than once every 2 years. Such a registration statement shall include, but need not be limited to, such person's name; principal place of business; the location of each activity handling such hazardous materials; a complete list of all such hazardous materials handled; and an averment that such person is in compliance with all applicable criteria established under subsection (a) of this section. The Secretary shall by regulation prescribe the form of any such statement and the information required to be included. The Secretary shall make any registration statement filed pursuant to this subsection available for inspection by any person, without charge, except that nothing in this sentence shall be deemed to require the release of any information described by subsection (b) of section 552 of title 5,

INDUSTRIAL TRAFFIC MANAGEMENT

United States Code, or which is otherwise protected by law from disclosure to the public.

(c) Requirement—No person required to file a registration statement under subsection (b) of this section may transport or cause to be transported or shipped extremely hazardous materials, or manufacture, fabricate, mark, maintain, recondition, repair, or test packages or containers for use in the transportation of extremely hazardous materials, unless he has on file a registration statement.

Exemptions

Sec. 107. (a) General—The Secretary, in accordance with procedures prescribed by regulation, is authorized to issue or renew, to any person subject to the requirements of this title, an exemption from the provisions of this title, and from regulations issued under section 105 of this title, if such person transports or causes to be transported or shipped hazardous materials in a manner so as to achieve a level of safety (1) which is equal to or exceeds that level of safety which would be required in the absence of such exemption, or (2) which would be consistent with the public interest and the policy of this title in the event there is no existing level of safety established. The maximum period of an exemption issued or renewed under this section shall not exceed 2 years, but any such exemption may be renewed upon application to the Secretary. Each person applying for such an exemption or renewal shall, upon application, provide a safety analysis as prescribed by the Secretary to justify the grant of such exemption. A notice of an application for issuance or renewal of such exemption shall be published in the Federal Register. The Secretary shall afford access to any such safety analysis and an opportunity for public comment on any such application, except that nothing in this sentence shall be deemed to require the release of any information described by subsection (b) of section 552 of title 5, United States Code, or which is otherwise protected by law from disclosure to the public.

(b) Vessels—The Secretary shall exclude, in whole or in part, from any applicable provisions and regulations under this title, any vessel which is excepted from the application of section 201 of the Ports and Waterways Safety Act of 1972 by paragraph (2) of such section (46 U.S.C. 391a(2)), or any other vessel regulated under such Act, to the extent of such regulation.

(c) Firearms and Ammunition—Nothing in this title, or in any regulation issued under this title, shall be construed to prohibit or regulate the transportation by any individual, for personal use, of any firearm (as defined in paragraph (4) of section 232 of title 18, United States Code) or any ammunition therefor, or to prohibit any transportation of firearms or ammunition in commerce.

(d) Limitation on Authority—Except when the Secretary determines

HAZARDOUS MATERIALS ACT

that an emergency exists, exemptions or renewals granted pursuant to this section shall be the only means by which a person subject to the requirements of this title may be exempted from or relieved of the obligation to meet any requirements imposed under this title.

Transportation of Radioactive Materials on Passenger-carrying Aircraft

Sec. 108. (a) General—Within 120 days after the date of enactment of this section, the Secretary shall issue regulations, in accordance with this section and pursuant to section 105 of this title, with respect to the transportation of radioactive materials on any passenger-carrying aircraft in air commerce, as defined in section 101(4) of the Federal Aviation Act of 1958, as amended (49 U.S.C. 1301(4)). Such regulations shall prohibit any transportation of radioactive materials on any such aircraft unless the radioactive materials involved are intended for use in, or incident to, research, or medical diagnosis or treatment, so long as such materials as prepared for and during transportation do not pose an unreasonable hazard to health and safety. The Secretary shall further establish effective procedures for monitoring and enforcing the provisions of such regulations.

(b) Definition.—As used in this section, "radioactive materials" means any materials or combination of materials which spontaneously emit ionizing radiation. The term does not include materials in which (1) the estimated specific activity is not greater than 0.002 microcuries per gram of material; and (2) the radiation is distributed in an essentially uniform manner.

Powers and Duties of the Secretary

Sec. 109. (a) General—The Secretary is authorized, to the extent necessary to carry out his responsibilities under this title, to conduct investigations, make reports, issue subpoenas, conduct hearings, require the production of relevant documents, records, and property, take depositions, and conduct, directly or indirectly, research, development, demonstration, and training activities. The Secretary is further authorized, after notice and an opportunity for a hearing, to issue orders directing compliance with this title or regulations issued under this title; the district courts of the United States shall have jurisdiction, upon petition by the Attorney General, to enforce such orders by appropriate means.

(b) Records.—Each person subject to requirements under this title shall establish and maintain such records, make such reports, and provide such information as the Secretary shall by order or regulation prescribe, and shall submit such reports and shall make such records and information available as the Secretary may request.

(c) Inspection.—The Secretary may authorize any officer, employee, or agent to enter upon, inspect, and examine, at reasonable times and in a

INDUSTRIAL TRAFFIC MANAGEMENT

reasonable manner, the records and properties of persons to the extent such records and properties relate to—

(1) the manufacture, fabrication, marking, maintenance, reconditioning, repair, testing, or distribution of packages or containers for use by any person in the transportation of hazardous materials in commerce; or

(2) the transportation or shipment by any person of hazardous materials in commerce.

Any such officer, employee, or agent shall, upon request, display proper credentials.

(d) Facilities and Duties.—The Secretary shall—

(1) establish and maintain facilities and technical staff sufficient to provide, within the Federal government, the capability of evaluating risks connected with the transportation of hazardous materials and materials alleged to be hazardous;

(2) establish and maintain a central reporting system and data center so as to be able to provide the law-enforcement and firefighting personnel of communities, and other interested persons and government officers, with technical and other information and advice for meeting emergencies connected with the transportation of hazardous materials; and

(3) conduct a continuing review of all aspects of the transportation of hazardous materials in order to determine and to be able to recommend appropriate steps to assure the safe transportation of hazardous materials.

(e) Annual Report.—The Secretary shall prepare and submit to the President for transmittal to the Congress on or before May 1 of each year a comprehensive report on the transportation of hazardous materials during the preceding calendar year. Such report shall include, but need not be limited to—

(1) a thorough statistical compilation of any accidents and casualties involving the transportation of hazardous materials;

(2) a list and summary of applicable Federal regulations, criteria, orders, and exemptions in effect;

(3) a summary of the basis for any exemptions granted or maintained;

(4) an evaluation of the effectiveness of enforcement activities and the degree of voluntary compliance with applicable regulations;

(5) a summary of outstanding problems confronting the administration of this title, in order of priority; and

(6) such recommendations for additional legislation as are deemed necessary or appropriate.

HAZARDOUS MATERIALS ACT

Penalties

Sec. 110.(a) Civil.—(1) Any person (except an employee who acts without knowledge) who is determined by the Secretary, after notice and an opportunity for a hearing, to have knowingly committed an act which is a violation of a provision of this title or of a regulation issued under this title, shall be liable to the United States for a civil penalty. Whoever knowingly commits an act which is a violation of any regulation, applicable to any person who transports or causes to be transported or shipped hazardous materials, shall be subject to a civil penalty of not more than $10,000 for each violation, and if any such violation is a continuing one, each day of violation constitutes a separate offense. Whoever knowingly commits an act which is a violation of any regulation applicable to any person who manufactures, fabricates, marks, maintains, reconditions, repairs or tests a package or container which is represented, marked, certified, or sold by such person for use in the transportation in commerce of hazardous materials shall be subject to a civil penalty of not more than $10,000 for each violation. The amount of any such penalty shall be assessed by the Secretary by written notice. In determining the amount of such penalty, the Secretary shall take into account the nature, circumstances, extent, and gravity of the violation committed and, with respect to the person found to have committed such violation, the degree of culpability, any history of prior offenses, ability to pay, effect on ability to continue to do business, and such other matters as justice may require.

(2) Such civil penalty may be recovered in an action brought by the Attorney General on behalf of the United States in the appropriate district court of the United States or, prior to referral to the Attorney General, such civil penalty may be compromised by the Secretary. The amount of such penalty, when finally determined (or agreed upon in compromise), may be deducted from any sums owed by the United States to the person charged. All penalties collected under this subsection shall be deposited in the Treasury of the United States as miscellaneous receipts.

(b) Criminal.—A person is guilty of an offense if he willfully violates a provision of this title or a regulation issued under this title. Upon conviction, such person shall be subject, for each offense, to a fine of not more than $25,000, imprisonment for a term not to exceed 5 years, or both.

Specific Relief

Sec. 111. (a) General.—The Attorney General, at the request of the Secretary, may bring an action in an appropriate district court of the United States for equitable relief to redress a violation by any person of a provision of this title, or an order or regulation issued under this title. Such district courts shall have jurisdiction to determine such actions and may grant such relief as is necessary or appropriate, including mandatory or prohibitive injunctive relief, interim equitable relief, and punitive damages.

(b) Imminent Hazard.—If the Secretary has reason to believe that an imminent hazard exists, he may petition an appropriate district court of the United States, or upon his request the Attorney General shall so petition, for an order suspending or restricting the transportation of the hazardous material responsible for such imminent hazard, or for such other order as is necessary to eliminate or ameliorate such imminent hazard. As used in this subsection, an "imminent hazard" exists if there is substanital likelihood that serious harm will occur prior to the completion of an administrative hearing or other formal proceeding initiated to abate the risk of such harm.

Relationship to Other Laws

Sec. 112. (a) General.—Except as provided in subsection (b) of this section, any requirement, of a State or political subdivision thereof, which is inconsistent with any requirement set forth in this title, or in a regulation issued under this title, is preempted.

(b) State Laws.—Any requirement, of a State or political subdivision thereof, which is not consistent with any requirement set forth in this title, or in a regulation issued under this title, is not preempted if, upon the application of an appropriate State agency, the Secretary determines, in accordance with procedures to be prescribed by regulation, that such requirement (1) affords an equal or greater level of protection to the public than is afforded by the requirements of this title or of regulations issued under this title and (2) does not unreasonably burden commerce. Such requirement shall not be preempted to the extent specified in such determination by the Secretary for so long as such State or political subdivision thereof continues to administer and enforce effectively such requirement.

(c) Other Federal Laws.—The provisions of this title shall not apply to pipelines which are subject to regulation under the Natural Gas Pipeline Safety Act of 1968 (49 U.S.C. 1671 et seq.) or to pipelines which are subject to regulation under chapter 39 of title 18, United States Code.

Conforming Amendments

Sec. 113. (a) Section 4472 of title 52 of the Revised Statutes of the United States, as amended (46 U.S.C. 170) is amended—

(1) by inserting, in the first sentence of paragraph (14) thereof, "criminal" before the word "penalty" and "or imprisoned not more than 5 years, or both" before the phrase "for each violation"; and

(2) by adding at the end thereof the following new paragraph: "(17)(A) Any person (except an employee who acts without

knowledge) who is determined by the Secretary, after notice and an opportunity for a hearing, to have knowingly committed an act which is a violation of any provision of this section, or of any regulation issued under this section, shall be liable to the United States for a civil penalty of not more than $10,000 for each day of each violation. The amount of such civil penalty shall be assessed by the Secretary by written notice. In determining the amount of such penalty, the Secretary shall take into account the nature, circumstances, extent, and gravity of the violation committed and, with respect to the person found to have committed such violation, the degree of culpability, any history of prior offenses, ability to pay, effect on ability to continue to do business, and such other matters as justice may require.

"(B) Such civil penalty may be recovered in an action brought by the Attorney General on behalf of the United States, in the appropriate district court of the United States or, prior to referral to the Attorney General, such civil penalty may be compromised by the Secretary. The amount of such penalty, when finally determined (or agreed upon in compromise), may be deducted from any sums owed by the United States to the person charged. All penalties collected under this subsection shall be deposited in the Treasury of the United States as miscellaneous receipts."

(b) Section 901(a)(1) of the Federal Aviation Act of 1958 (49 U.S.C. 1471(a)(1)) is amended—

(1) by inserting immediately before the period at the end of the first sentence thereof and inserting in lieu thereof: ",except that the amount of such civil penalty shall not exceed $10,000 for each such violation which relates to the transportation of hazardous materials."; and

(2) by deleting in the second sentence thereof ":*Provided,* That this" and inserting in lieu thereof the following: ".The amount of any such civil penalty which relates to the transportation of hazardous materials shall be assessed by the Secretary, or his delegate, upon written notice upon a finding of violation by the Secretary, after notice and an opportunity for a hearing. In determining the amount of such penalty, the Secretary shall take into account the nature, circumstances, extent, and gravity of the violation committed and, with respect to the person found to have committed such violation, the degree of culpability, any history of prior offenses, ability to pay, effect on ability to continue to do business, and such other matters as justice may require. This".

(c) Section 902(h) of the Federal Aviation Act of 1958, as amended (49 U.S.C. 1472(h)) is amended to read as follows:

INDUSTRIAL TRAFFIC MANAGEMENT

"Hazardous Materials

"(h)(1) In carrying out his responsibilities under this Act, the Secretary of Transportation may exercise the authority vested in him by section 105 of the Hazardous Materials Transportation Act to provide by regulation for the safe transportation of hazardous materials by air.

"(2) A person is guilty of an offense if he willfully delivers or causes to be delivered to an air carrier or to the operator of a civil aircraft for transportation in air commerce, or if he recklessly causes the transportation in air commerce of, any shipment, baggage, or other property which contains a hazardous material, in violation of any rule, regulation, or requirement with respect to the transportation of hazardous materials issued by the Secretary of Transportation under this Act. Upon conviction, such person shall be subject, for each offense, to a fine of not more than $25,000, imprisonment for a term not to exceed 5 years, or both.

"(3) Nothing in this subsection shall be construed to prohibit or regulate the transportation by any individual, for personal use, of any firearm (as defined in paragraph (4) of section 232 of title 18, United States Code) or any ammunition therefor."

(d) Section 6(c)(1) of the Department of Transportation Act (49 U.S.C. 1655(c)(1)) is amended by inserting in the first sentence thereof after "aviation safety" and before "as set forth in" the following: (other than those relating to the transportation, packaging, marking, or description of hazardous materials)".

(e)(1) Section 6(f)(3)(A) of the Department of Transportation Act (49 U.S.C. 1655(f)(3)(A)) is amended by striking out the period at the end thereof and by inserting in lieu thereof "(other than subsection (e)(4)).".

(2) Section 6((f)(3)(B) of the Department of Transportation Act (49 U.S.C. 1655(f)(3)(B)) is amended by striking out the period at the end thereof and by inserting in lieu thereof "(other than subsection (e)(4).".

(f) Subsection (6) of section 4472 of the Revised Statutes, as amended (46 U.S.C. 170(6)), is amended—

(1) in paragraph (a) thereof, by striking out "inflammable" each place it appears and inserting in lieu thereof at each such place "flammable"; by inserting before "liquids" the following: "or combustible"; and by deleting the colon and the proviso in its entirety and by inserting in lieu thereof a period and the following two new sentences: "The provisions of this subsection shall apply to the transportation, carriage, conveyance, storage, stowing, or use on board any passenger vessel of any barrel, drum, or other package containing any flammable or combustible liquid which has a lower flash point than that which is defined as safe pursuant to regulations establishing the defining flash-point criteria for

flammable and combustible liquids. Such regulations shall be prescribed, and revised as necessary, by the Secretary of Transportation.".

(2) in paragraph (b) thereof, by striking out in clause (iv) thereof "inflammable liquids" and inserting in lieu thereof "flammable or combustible liquids".

(g) The Hazardous Materials Transportation Control Act of 1970 (Pub. L. 91-458, title III; 49 U.S.C. 1761-1762) is repealed.

Effective Date

Sec. 114. (a) Except as provided in this section, the provisions of this title shall take effect on the date of enactment.

(b)(1) Except as provided in section 108 of this title or paragraph (2) of this subsection, any order, determination, rule, regulation, permit, contract, certificate, license, or privilege issued, granted, or otherwise authorized or allowed, prior to the date of enactment of this title, pursuant to any provision of law amended or repealed by this title, shall continue in effect according to its terms or until repealed, terminated, withdrawn, amended, or modified by the Secretary or a court of competent jurisdiction.

(2) The Secretary shall take all steps necessary to bring orders, determinations, rules, and regulations into conformity with the purposes and provisions of this title as soon as practicable, but in any event no permits, contracts, certificates, licenses, or privileges granted prior to the date of enactment of this title, or renewed or extended thereafter, shall be of any effect more than 2 years after the date of enactment of this title, unless there is full compliance with the purposes and provisions of this Act and regulations thereunder.

(c) Proceedings pending upon the date of enactment of this title shall not be affected by the provisions of this title and shall be completed as if this title had not been enacted, unless the Secretary makes a determination that the public health and safety otherwise require.

Authorization for Appropriations

Sec. 115. There is authorized to be appropriated for the purposes of this title, not to exceed $7,000,000 for the fiscal year ending June 30, 1975.

Appendix 9

Hazardous Materials Transportation

Hazardous Materials Definitions

The following definitions have been abstracted from the Code of Federal Regulations, Title 49—Transportation, Parts 100 to 199. Refer to the referenced sections for complete detail. *Note:* Rule making proposals are outstanding or are contemplated concerning some of these definitions.

Hazardous Material means a substance or material which has been determined by the Secretary of Transportation to be capable of posing an unreasonable risk to health, safety, and property when transported in commerce, and which has been so designated. (Sec. 171.8)

Multiple Hazards—A material meeting the definitions of more than one hazard class is classed according to the sequence given in Sec. 173.2.

Hazard Class	Definitions
	An Explosive—Any chemical compound, mixture, or device, the primary or common purpose of which is to function by explosion, i.e., with substantially instantaneous release of gas and heat, unless such compound, mixture, or device is otherwise specifically classified in Parts 170-189 of this subchapter. (Sec. 173.50)
Class A Explosive	Detonating or otherwise of *maximum hazard.* The types of Class A explosives are defined in Sec. 173.53.
Class B Explosive	In general, function by rapid combustion rather than detonation and include some explosive devices such as special fireworks, flash powders, etc. *Flammable hazard.* (Sec. 173.88)
Class C Explosive	Certain types of manufactured articles containing Class A or Class B explosives, or both, as components but in restricted quantities, and certain types of fireworks. *Minimum hazard.* (Sec. 173.100)

INDUSTRIAL TRAFFIC MANAGEMENT

Combustible Liquid	Any liquid with a *flash point* from 100/F to 200/F as measured by the tests specified in Sec. 173.115, except any mixture having one component or more with a flash point at 200/F or higher, that makes up at least 99 percent of the total volume of the mixture. (Sec. 173.115(b))
Corrosive Material	Any liquid or solid that causes destruction of human skin tissue or a liquid that has a severe corrosion rate on steel. (See Sec. 173.240(a) and (b) for details.)
Flammable Liquid	Any liquid with a *flash point less than 100/F* as measured by the tests specified in Sec. 173.115, with the following exceptions: (i) A flammable liquid with a vapor pressure greater than 40 psia at 100/F, as defined in Sec. 173.300; (ii) Any mixture having one component or more with a flash point of 100/F or higher that makes up at least 99 percent of the total volume of the mixture; and (iii) A water-alcohol solution containing 24 percent or less alcohol by volume if the remainder of the solution does not meet the definition of a hazardous material contained in this subchapter.
	Pyroforic Liquid—Any liquid that ignites spontaneously in dry or moist air at or below 130/F. (Sec. 173.115)
	Compressed Gas—Any material or mixture having in the container a pressure *exceeding* 40 psia at 70/F, or a pressure exceeding 104 psia at 130/F; or any liquid flammable material having a vapor pressure exceeding 40 psia at 100/F. (Sec. 173.300(a))
Flammable Gas	Any compressed gas meeting the requirements for lower flammability limit, flammability limit range, flame projection, or flame propagation criteria as specified in Sec. 173.300(b).
Nonflammable Gas	Any compressed gas other than a flammable compressed gas.
Flammable Solid	Any solid material, other than an explosive, which is liable to cause fires through friction, absorption of moisture, spontaneous chemical changes, retained heat from manufacturing or processing, or which can be ignited readily and when ignited burns so vigorously and persistently as to create a serious transportation hazard. (Sec. 173.150)
Organic Peroxide	An organic compound containing the bivalent -0-0- structure and which may be considered a derivative of hydrogen peroxide where one or more of the hydrogen atoms have been replaced by organic radicals must be classed as an organic peroxide unless—(See Sec. 173.151(a) for details)
Oxidizer	A substance such as chlorate, permanganate, inorganic peroxide, nitro carbo nitrate, or a nitrate, that yields oxygen readily to stimulate the combustion of organic matter. (Sec. 173.151)

HAZARDOUS MATERIALS DEFINITIONS

Poison A	*Extremely Dangerous Poisons*—Poisonous gases or liquids of such nature that a very small amount of the gas, or vapor of the liquid, mixed with air is *dangerous to life.* (Sec. 173.326)
Poison B	*Less Dangerous Poisons*—Substances, liquids, or solids (including pastes and semi-solid), other than Class A or Irritating materials, which are known to be so toxic to man as to afford a hazard to health during transportation; or which in the absence of adequate data on human toxicity, are presumed to be *toxic to man.* (Sec. 173.343)
Irritating Material	A liquid or solid substance which upon contact with fire or when exposed to air gives off dangerous or intensely irritating fumes, but *not including any poisonous material, Class A.* (Sec. 173.381)
Etiologic Agent	An etiologic agent means a viable micro-organism, or its toxin which causes or may cause human disease. (Sec. 173.386) (Refer to the Department of Health, Education and Welfare Regulations, Title 42, CER, Sec. 72.75(c) for details.)
Radioactive Material	Any material, or combination of materials, that spontaneously emits ionizing radiation, and having a specific activity greater than 0.002 microcuries per gram. (Sec. 173.389) *Note:* See Sec. 173.389(a) through (1) for details.
	ORM-A, B or C (Other Regulated Materials)—Any material that does not meet the definition of a hazardous material, other than a combustible liquid in packagings having a capacity of 110 gallons or less, and is specified in Sec. 172.101 as an ORM material or that possesses one or more of the characteristics described in ORM-A through D below. (Sec. 173.500) *Note:* An ORM with a *flash point of 100/F to 200/F,* when transported with more than 110 gallons in one container shall be *classed as a combustible liquid.*
ORM-A	A material which has an anesthetic, irritating, noxious, toxic, or other similar property and which can cause extreme annoyance or discomfort to passengers and crew in the event of leakage during transportation (Sec. 173.500(a)(1)).
ORM-B	A material (including a solid when wet with water) capable of causing significant damage to a transport vehicle or vessel from leakage during transportation. Materials meeting one or both of the following criteria are ORM-B materials: (i) A liquid substance that has a corrosion rate exceeding 0.250 inch per year (IPY) on aluminum (nonclad 7075-T6) at a test temperature of 130/F. An acceptable test is described in NACE Standard TM-01-69, and (ii) Specifically designated by name in Sec. 172.101 of this subchapter. (Sec. 173.500(a)(2))

INDUSTRIAL TRAFFIC MANAGEMENT

ORM-C	A material which has other inherent characteristics not described as an ORM-A or ORM-B but which make it unsuitable for shipment, unless properly identified and prepared for transportation. Each ORM-C material is specifically named in Sec. 172.101 of this subchapter. (Sec. 173.500(a)(4))
ORM-D	A material such as a consumer commodity which, though otherwise subject to the regulations of this subchapter, presents a limited hazard during transportation due to its form, quantity and packaging. They must be materials for which exceptions are provided in Sec. 172.101 of this subchapter. A shipping description applicable to each ORM-D material or category of ORM-D materials is found in Sec. 172.101 of this subchapter. (Sec. 173.500(a)(4))

The following are offered to explain additional terms used in preparation of hazardous materials for shipment. (Sec. 171.8)

Consumer Commodity (See ORM-D)	Means a material that is packaged or distributed in a form intended and suitable for sale through retail sales agencies or instrumentalities for consumption by individuals for purposes of personal care or household use. This term also includes drugs and medicines.
Flash Point	Means the minimum temperature at which a substance gives off flammable vapors which in contact with spark or flame will ignite. (Sec. 173.115)
Forbidden	The hazardous material is one that may *not be offered or accepted* for transportation. (Sec. 172.100(d))
Limited Quantity	Means the maximum amount of a hazardous material; as specified in those sections applicable to the particular hazard class, for which there are *specific exceptions* from the requirements of this subchapter. See Sec. 173.118, 173.118a, 173.153, 173.244, 173.306, 173.345 and 173.364.
Spontaneously Combustible Material (Solid)	Means a solid substance (including sludges and pastes) which may undergo spontaneous heating or self-ignition under conditions normally incident to transportation or which may upon contact with the atmosphere undergo an increase in temperature and ignite.
Water Reactive Material (Solid)	Means any solid substance (including sludges and pastes) which, by interaction with water, is likely to become spontaneously flammable or to give off flammable or toxic gases in dangerous quantities.

INDEX

A

Am. Institute of Shipping Assocs., Inc., 152
Am. Soc. for Testing & Materials (ASTM), 168
Assembly & Distribution Rates, 96

B

Bill of Lading, form, 67
  function, 72
  kind
    air, 79
    government, 78
    livestock, 70
    ocean, 78
    R.R. order notify, 70
    R.R. straight, 69
  preparing, 74
  who issues, 70
Bill of Lading terms, 184
Bill of Lading act, 67
boxes, specifications, 156-157
  certificate, 157-158

C

Civil Aeronautics Act (1938), 9
Civil Aeronautics Board organization chart, 32
claims, background, 183
  concealed, 196
  forms, 189-194
  ICC responsibility, 187-198
  loss & damage, 190-197
  overcharge, 187
  preparing, 192
  records, 199
class rates, 84
classification, rules, 55-66
committees, Congressional, transportation, 16-17
commodity rates, 91
consolidated freight classification #1, 40
consolidating in routing, 150
containerization, 179
contract carriage, 238
coordinated motor freight classification, 43
Cummins Amendment, 191
custom house brokers, 397
  role, 394

D

data processing, use, 286-91
  freight bills, 294
  language, 287
  reviews, 294

SPLC, 288
STCC, 288
demurrage, average, 266
    calendar, 263
    charges, 262
    definition, 261
    records, 269
    tariff
Department of Transportation, 9-15
Detention, definition, 271
    Ex Parte MC 88, 271
    free time, 271
discrimination, 5
distribution, bill of lading, 77
distribution in routing, 150
distribution, definition, 203
diversion, 275
docket 28300, 42, 89, 91
docket 28310, 41

E

embargoes, 273
exception rates, 80
expediting, 245
    air, 253
    carload, 248
    freight forwarder, 253
    LTL, 252
    trailer load, 250
export insurance, claims, 366
    FPA, 362
    FPA AC, 363
    FPA, EC, 363
    general average, 365
    open policy, 359-60
    premiums, 364
    W.A., 363
export packaging, 397
export payments, authority to purchase, 386
drafts, 379
letter of credit, 381
open account, 377
sight draft, 380
export sales terms, 348

F

foreign freight forwarders, 387
foreign trade zones, 391
freight bills, 79
freight claim, freight loading & container, 176
freight classification, 37
    Illinois, 39
    Official, 39
    Southern, 39
    Western, 39

G

government traffic, Department of Defense, 339
    documentation, 343
    G.B.L., 344-45
    General Services Administration, 339
    overseas, 341
    Section 10721 (PL 95-473), 337

H

hazardous materials, importing & exporting, 413
    via air, 414
household carriers, 137

I

imports, financing, 392
indemnity agreement, 190
inland waterways, 142
intercorporate hauling, 238

INDEX

Interstate Commerce Commission (ICC), 21-26
item 22, NMFC, 161

L

landbridge, 126
letters of credit, irrevocable, 381
 revocable, 381
liability, carrier, 186
 released valuation, 186
 warehouse, 186
lighter aboard ship (LASH), 401
loading, 175

M

manuals, traffic, 119
Maritime Commission, Federal, 28
material handling, 172-173
Merchant Marine Act (1936), 9
minibridge, 126
motor carrier classification, 42
motor carrier rates, 97
Murray Co. of Texas v. Morroe, 54 MCC 442, 118

N

National Motor Freight Classification, 42
National Safe Transit Committee (NSTC), 155, 168
national transportation policy, 24
National Council of Physical Distribution Management (NCPDM), 203

O

ocean carriers, 349
ocean documents, bills of lading, 368-70
certificate of origin, 374
dock, 367-68
shippers export declaration, 372
ocean rates, conference, 352-53
 kinds, 354
 tariffs, 349

P

packaging, domestic, 153-59
 export, 170
pallet loading, 178
parcel post, 140
physical distribution management,
 definitions, 418-19
 organization, 420-21
 responsibilities, 423-27
piggyback, 127-129
pipelines, 149
placement, R.R., active, 262
 constructive, 262
private carriage, 219
protests, shippers right, 103
Public Law 95-473, 23, 134, 135
public warehouse functions, 207

R

rates, auditing, 107
rate checking, 107
rate-making procedure, 105
rates, 83, 84
 auditing, 107
 how published, 98
 NEMRB, 89
reasonable dispatch, 72
reconsignment, 275
Reed-Bulwinkle Act, 99
regulatory agencies, federal transportation, CAB, 18
 FMC, 19
 ICC, 18

527

INDUSTRIAL TRAFFIC MANAGEMENT

released valuation rates, 47
   liability, 186, 191
roll on/roll off (RO/RO), 404
routing, right to, 118
   air, 145
   motor, 135
   rail, 121
routing services, 125
Rule 41-UFC, 160

S

sales terms, 111
Seabee, 403
shipment terms, 185
shipper's liability in routing, 135
standard contract terms conditions, 208
state commissions, 27
switching, 272

T

Transportation Arbitration Board (TAB), 197
tariff agencies, 93
terms, bill of lading, 72
TOFC, 92, 127, 128, 129
tonnage, distribution, 118-119
   reports, 120
tracing, 247
track agreements, 276-277
traffic department, cooperative functions, 325-30
   education, 304

equipping, 321
filing, 304
functions, 300
operations, 313
organization, 303
public relations, 331
responsibilities, 308-11
tariff filing, 323
traffic department, industrial, 2
traffic management, principles, 5
transit privileges, 273-75
transportation obligations, 5
transportation responsibilities, government, 12-14
Transportation Safety Act 1974, 405
   compliance program, 410-13
   hazardous materials list, 408-09
   PL 93-633, 406
   Title 49, 407
truck leasing, 222-230

U

Uniform Commercial Code, 208
Uniform Freight Classification, 41
United Parcel Service, 148

W

warehouses, location, 212
   role, 206
   types, 204
waybills, 79
weight agreement, 279

About the Authors

The authors have spent their entire careers as working traffic men. This Handbook, now appearing in its sixth and completely revised edition, fills a need that can only be realized by such experience—a need for a down-to-earth, practical guide to the day-to-day operation of an industrial traffic department.

Leon Wm. Morse, who updated this new edition, is a partner in the firm of Morse, Stoner, Travis & Associates, which specializes in transportation, traffic and distribution management consulting. He began his career in 1935 with a manufacturer of burlap and cotton bags where he handled a traffic operation involving volume shipments to all parts of the world. In 1937 he joined Food Fair Stores, Inc. and established the company's first Traffic Department. As its GTM, he oversaw rail activity of over 5,000 cars per year, and was closely involved in scheduling, rate negotiations, routing, tariff interpretation and consolidation.

During World War II he served as Captain in the Army Transportation Corps and when discharged, stayed on as their civilian Chief of Movement Control. For the first time in the history of the Army, 'distribution' sense was put into the procurement of vehicles, resulting in annual savings of $18 million.

In 1956 he joined William H. Rorer, Inc. and rose rapidly through the ranks to General Traffic Manager. When he began, the traffic department employed two people. When he retired twenty years later, there were 125 people. In that same period, the shipping activity grew from two doors inbound and outbound to nine doors inbound and 15 outbound.

Morse holds a Doctorate in Business Administration and has found time during his busy career to teach and lecture widely on the subjects of transportation and distribution. Currently he is Adjunct Professor at Pennsylvania State University, teaching business logistics and transportation economics. He was named Delaware Valley Traffic Manager of the Year and holds memberships in the Association of Interstate Commerce Commission Practitioners, Canadian Association of Physical Distribution Management, National Council of Physical Distribution Management, American Society of Traffic & Transportation, International Society of Law & Sciences, Traffic Club of Philadelphia, Delta Nu Alpha Transportation Fraternity, and the Society of Logistics Engineers. At one time or another he has acted as president or board member for chapters of nearly all of these organizations.

Charles H. Wager, responsible for having revised the previous (5th) edition, portions of which remain in this 6th edition, began his industrial transportation career in 1938 in the Traffic Department of Shell Oil Company. He had previously been with the Company's General Counsel and in its Legal Department and President's Office. In the ensuing thirty-plus years he was made General Traffic Manager in New York. His assignments have involved an an exceptionally wide range of industry and government transportation and distribution functions.

Along with his company transportation responsibilities, he was president of the National Industrial Traffic League and held chairmanship posts for both the American Petroleum Institute Industrial Transportation Committee and the Manufacturing Chemists Association Traffic Committee. He is a Past President of the Traffic Club of New York. In 1968-69 he was president of the newly-formed Transportation Data Coordinating Committee, Washington, D.C.